T0327569

INTELLIGENT TRANSPORT SYSTEMS

INTELLIGENT TRANSPORT SYSTEMS

TECHNOLOGIES AND APPLICATIONS

Asier Perallos
Faculty of Engineering
University of Deusto
Bilbao
Spain

Unai Hernandez-Jayo
Faculty of Engineering
University of Deusto
Bilbao
Spain

Enrique Onieva
Faculty of Engineering
University of Deusto
Bilbao
Spain

Ignacio Julio García-Zuazola
University of Loughborough
Leicestershire
United Kingdom

WILEY

This edition first published 2016

Registered Office
John Wiley & Sons, Ltd, The Atrium, Southern Gate, Chichester, West Sussex, PO19 8SQ, United Kingdom

For details of our global editorial offices, for customer services and for information about how to apply for permission to reuse the copyright material in this book please see our website at www.wiley.com.

Library of Congress Cataloging-in-Publication Data

Intelligent transport systems (2016)
Intelligent transport systems : technologies and applications / [edited by] Asier Perallos, Unai Hernandez-Jayo, Enrique Onieva, Ignacio Julio García-Zuazola.
pages cm
Includes bibliographical references and index.
ISBN 978-1-118-89478-1 (cloth)
1. Intelligent transportation systems. I. Perallos, Asier. II. Hernandez-Jayo, Unai. III. Onieva, Enrique. IV. García-Zuazola, Ignacio Julio. V. Title.
TE228.3.I5757 2015
388.3′12–dc23

2015017662

A catalogue record for this book is available from the British Library.

Cover image: Willowpix/Getty

Set in 10/12pt Times by SPi Global, Pondicherry, India

1 2016

The editors would like to express their heartfelt condolences to the family and friends of Dr Rus Leelaratne, who passed away unexpectedly on 16 April 2015. He is survived by his amazing wife Sharon and their two beautiful daughters, Ruchi and Ishani. His experience and reputation in the field of automotive antennas makes his contribution to this book of great importance, and his untimely death is a significant loss to the scientific world. May his heart and soul find peace and comfort.

Contents

About the Editors

Dr Asier Perallos holds a PhD, MSc and BSc in Computer Engineering from the University of Deusto (Spain). Since 1999 he has been working as a lecturer in the Faculty of Engineering at the University of Deusto, being now accredited by Spanish Government as Associate Professor. His teaching focuses on software design and distributed systems, having taught several BSc, MSc and PhD courses. He is currently the Head of the Computer Engineering Department and in the past has been the director of several MSc in Software Engineering. He is also Head Researcher at the DeustoTech Mobility Unit at the Deusto Institute of Technology (DeustoTech), where he coordinates the research activities of around 25 researchers. This research unit promotes the application of ICT to address smarter transport and mobility. In particular, Perallos' research background is focused on telematic systems, vehicular communication middleware and intelligent transportation systems. He has over a decade of experience in R&D management, has led around 50 projects and technology transfer actions, has published more than 20 JCR indexed publications, made more than 50 other contributions in the area of intelligent transport systems, and has supervised two PhD dissertations.

Dr Unai Hernandez-Jayo holds a PhD in Telecommunications from the University of Deusto. Since 2004 he has been working as a lecturer in the Faculty of Engineering at the University of Deusto. He teaches both undergraduate and graduate degrees in the area of acoustics, electronics and analogue and digital communication systems. He also works as researcher and project manager at the Deusto Institute of Technology (DeustoTech), where he is responsible for the work area on the development of wireless communications solutions in the field of vehicular communications. Moreover, he is part of the WebLab-Deusto (www.weblab.deusto.es) research team, leading the design and development of remote laboratories focused on analogue electronics. He has an extensive experience in R&D management, working in projects and technology transfer actions. As part of them, he has more than 70 publications in relevant international conferences and journals on vehicular communications, Intelligent Transport Systems, VANETs, remote laboratories, advanced learning technologies, including indexed journal articles.

Dr Enrique Onieva received a BE degree in Computer Science Engineering, an ME degree in Soft Computing and Intelligent Systems and a PhD degree in Computer Science from the University of Granada, Spain, in 2006, 2008 and 2011, respectively. From 2007 to 2012, he has been with the AUTOPIA Program at the Centre of Automation and Robotics, Consejo Superior de Investigaciones Científicas, Madrid, Spain. During 2012 he has been with the Models of Decision and Optimization group, at the University of Granada. Since the beginning of 2013, he has been with the Mobility Unit at the Deusto Institute of Technology (DeustoTech), where he carries out cutting-edge research in the application of soft computing techniques to the field of intelligent transportation systems. He has participated in more than 15 research projects and authored more than 80 scientific articles. From them, more than 25 are published in journals of the highest level. His research has been recognized and awarded several times in international conferences. Currently, he is one of the most prolific researchers in his area. His research interest is based on the application of Soft Computing Techniques to Intelligent Transportation Systems, including fuzzy-logic based decision and control and evolutionary optimization.

Dr Ignacio Julio García-Zuazola completed a PhD in Electronics (Antennas) part-time program in 2008; viva in 2010, University of Kent, Canterbury, UK, and received a BEng (with honours) in Telecommunications Engineering, Queen Mary University of London in 2003, an HND degree in Telecommunications Engineering, College of North West London in 2000, and an FPII degree in Industrial Electronics, School of Chemistry & Electronics of Indautxu, Spain in 1995. He has been employed as a Research Associate, University of Kent (2004), Research Engineer, Grade 9/9, University of Wales, Swansea, UK (2006), Research Associate, University of Kent (2008), Senior Research Fellow, University of Deusto, Bilbao, Spain (2011) and Visiting Senior Research Fellow at the I3S, University of Leeds, UK (2011). He holds educational awards in Electrical Wiring, Pneumatic and Hydraulic Systems, and Robotics and possesses relevant industrial experience, having been hired for Babcok & Wilcox (1993), Iberdrola (1995), Telefonica (1997), Thyssen Elevators (1998), and Cell Communications (2000). He engaged in an SME in electrical wiring in 1996, is currently a Research Associate at Loughborough University, UK (2014) and combines his full-time job with part-time EMBA studies. He led the antennas research line, supervised and mentored students at Deusto and currently bids for EU grants and contributes to the scientific and technological development of Loughborough by promoting and increasing research and innovation. He was included in the *Marquis Who's Who in the World* 2010 edition and has published work in international journals such as *IEEE Transactions*, *IET Proceedings*, and *Electronics Letters*. Current research interests include single-band and multiband miniature antennas, and the use of Electromagnetic-Band Gap (EBG) structures and Frequency-Selective Surfaces (FSS). He is an active member of the IET and IEEE.

List of Contributors

Nerea Aguiriano
PhD in Electronics and Communication Engineering
University of Navarra
San Sebastián, Spain

Daniele Alessandrelli
Institute of Communication
Information and Perception Technologies
Pisa, Italy

António Amador
Technology Development Manager
Brisa Innovation
Porto Salvo, Portugal

Andrea Azzarà
Institute of Communication
Information and Perception Technologies
Pisa, Italy

Stefano Bocchino
Institute of Communication
Information and Perception Technologies
Pisa, Italy

Alfonso Brazalez
PhD in Physics
University of Navarra
San Sebastián, Spain

Sergio Campos
Senior ITS Researcher
TECNALIA. ICT-ESI Division
Derio, Spain

Tomé Canas
Research Manager
Brisa Innovation
Porto Salvo, Portugal

Rui Dias
Project Manager
Brisa Innovation
Porto Salvo, Portugal

Tiago Dias
Project Manager
Brisa Innovation
Porto Salvo, Portugal

Pekka Eloranta
MSc (Tech.)
Mobisoft Oy
Tampere, Finland

Joaquim Ferreira
Instituto de Telecomunicações
ESTGA-Universidade de Aveiro
Portugal

José Fonseca
Instituto de Telecomunicações
DETI-Universidade de Aveiro
Portugal

David Gonzalez
PhD, RITS Team
National Institute for Research in Computer Science and Control (INRIA)
Le Chesnay, France

Martin Gregurić
Intelligent Transport System Department
Faculty of Traffic and Transport Science
University of Zagreb
Zagreb, Croatia

Clifford D. Heise
Vice President
Iteris, Inc.
Sterling, VA, USA

Edouard Ivanjko
Intelligent Transport System Department
Faculty of Traffic and Transport Science
University of Zagreb
Zagreb, Croatia

Victor C.S. Lee
PhD
City University of Hong Kong
Hong Kong

Rus Leelaratne†

Kai Liu
PhD
Chongqing University
China

Diego López-de-Ipiña
PhD in Computer Engineering
University of Deusto
Bilbao, Spain

Lester Low
MIRA Ltd
Nuneaton, UK

Luca Maggiani
Institute of Communication
Information and Perception Technologies
Pisa, Italy

Massimo Magrini
Signals and Images Lab
Institute of Information Science and Technologies (ISTI)
National Research Council of Italy (CNR)
Pisa, Italy

† Deceased

Sadko Mandžuka
Intelligent Transport System Department
Faculty of Traffic and Transport Science
University of Zagreb
Zagreb, Croatia

Luis Matey
PhD in Industrial Engineering
University of Navarra
San Sebastián, Spain

Tiago Meireles
Instituto de Telecomunicações
Universidade da Madeira
Portugal

Vicente Milanés
PhD, RITS Team
National Institute for Research in Computer Science and Control (INRIA)
Le Chesnay, France

Begoña Molinete
Head of ITS Research Team
TECNALIA. ICT-ESI Division
Derio, Spain

Asier Moreno
BSc in Computer Engineering
University of Deusto
Bilbao, Spain

Davide Moroni
Signals and Images Lab
Institute of Information Science and Technologies (ISTI)
National Research Council of Italy (CNR)
Pisa, Italy

Ignacio (Iñaki) Olabarrieta
Senior ITS Researcher
TECNALIA. ICT-ESI Division
Derio, Spain

Enrique Onieva
PhD
University of Deusto
Bilbao, Spain

Paolo Pagano
National Laboratory of Photonic Networks
National Inter-University Consortium for Telecommunications
Pisa, Italy

Joshué Pérez
PhD, RITS Team
National Institute for Research in Computer Science and Control (INRIA)
Le Chesnay, France

Matteo Petracca
National Laboratory of Photonic Networks
National Inter-University Consortium for Telecommunications
Pisa, Italy

Gabriele Pieri
Signals and Images Lab
Institute of Information Science and Technologies (ISTI)
National Research Council of Italy (CNR)
Pisa, Italy

Alastair R. Ruddle
MIRA Ltd
Nuneaton, UK

Itziar Salaberria
PhD in Computer Engineering
University of Deusto
Bilbao, Spain

Claudio Salvadori
Institute of Communication
Information and Perception Technologies
Pisa, Italy

Ovidio Salvetti
Signals and Images Lab
Institute of Information Science and Technologies (ISTI)
National Research Council of Italy (CNR)
Pisa, Italy

Pero Škorput
Intelligent Transport System Department
Faculty of Traffic and Transport Science
University of Zagreb
Zagreb, Croatia

Timo Sukuvaara
Lic.Sc. (Tech.)
Finnish Meteorological Institute
Sodankylä, Finland

Ana Isabel Torre
ITS Researcher
TECNALIA. ICT-ESI Division
Derio, Spain

Miroslav Vujić
Intelligent Transport System Department
Faculty of Traffic and Transport Science
University of Zagreb
Zagreb, Croatia

David D. Ward
MIRA Ltd
Nuneaton, UK

Xiao Zhang
PhD student
City University of Hong Kong
Hong Kong

Foreword

Asier Perallos[1], Unai Hernandez-Jayo[1], Enrique Onieva[1]
and Ignacio Julio García Zuazola[2]
[1] *University of Deusto, Bilbao, Spain*
[2] *Loughborough University, Leicestershire, United Kingdom*

Introduction

Computers, electronics, satellites and sensors are playing an increasingly important role in our transport systems, as instruments used for different purposes under different conditions. Intelligent Transportation Systems (ITS) apply information and communication technologies to every transport mode (road, rail, air, water) and provide services which can be used by both passenger and freight transport. Nowadays, the main challenge lies in the integration of existing technologies with the aim of making transport more sustainable, which involves a compromise between efficiency, eco-friendship and safety.

This book presents a holistic approach to ITS, combining academic and industrial contributions. It attempts to merge some of the most effective contributions and technical approaches in ITS, which are currently under development in referenced research institutions and universities. Also, the present book reflects on how these works can be deployed in real scenarios thanks to the experiences of collaboration.

The book is divided into five parts and 16 chapters. Each part and chapter delimits its own field in order to provide a well-connected and correlated narrative thread, in which the reader observes a workflow from research to deployment of ITS. First, it includes an overview of reference architectures developed within the main European and American research projects. Then, it enquires into each of the layers presented in architectures, from physical to application layer, describing the technological challenges which are currently faced by some of the most important ITS research groups. Some of these technological issues are related to areas such as wireless communications in vehicular environments, sensors networks and surveillance, or data processing techniques. It contributes to provide the desired holistic vision in this area. The book concludes with some end applications and services for users and traffic managers deployed by industrial partners.

Around 50 highly qualified authors have contributed to this book. Their studies provide a widespread, heterogeneous and international vision of ITS. They are from three continents, representing nine countries: United Kingdom, Spain, Portugal, Italy, France, Finland, Croatia, USA and China. Authors are affiliated to some of the world's leading institutions and companies in the ITS area. In total, 13 different affiliations are represented, with a fair balance between industrial and academic contributions.

Intelligent Transportation Systems

The term ITS encompasses the set of applications that make use of information and communication technologies in the field of transport with the aim of obtaining economic, social and energy benefits. ITS can be applied to any means of transport and considers any of the involved agents: vehicle, infrastructure and user (driver or passenger).

As mentioned above, the main function of ITS is to improve the performance of transport systems, as well as to help, on the one hand, the management of the infrastructure by means of its exploitation and decision-making systems and, on the other hand, the users, in order to obtain an overall satisfaction with the transport system.

For this reason, ITS consist of systems responsible for collecting information concerning the state of the scenario, systems in charge of processing and integrating the information and, finally, systems that are responsible for providing results to end users, as shown in the below figure. In this way, the information collected in real time by ITS may be used to

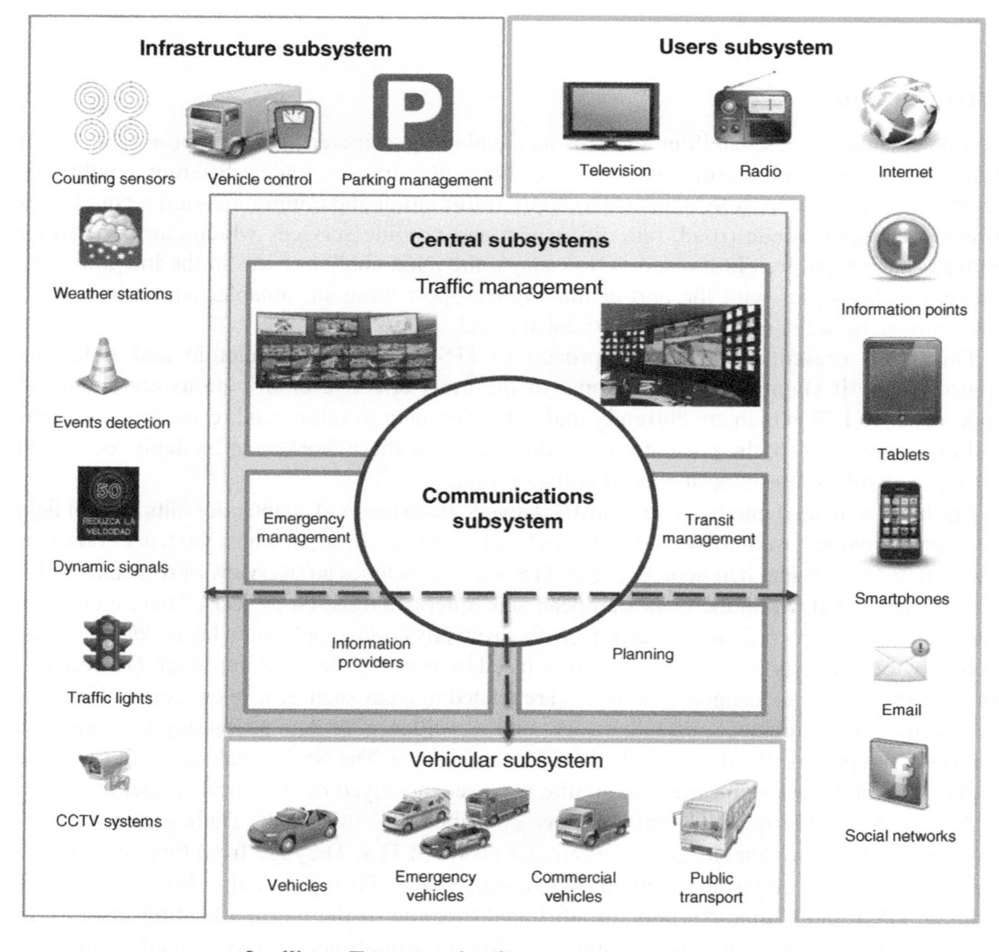

Intelligent Transportation Systems conceptual model.

determine the state of the communications network, to plan a trip, to manage dynamically the traffic in a certain area, to report data from a logistics operator to the customer or to show traffic events in Geographic Information Systems (GIS). In summary, all the actors have access to more information and more tools that help them to process it in order to carry out a more coordinated and intelligent operation of a transport system, whether it be on road, rail, water or air.

If a definition should be given, it can be said that ITS are a set of solutions based on the combination of telecommunications and computer technologies designed and developed to improve the management, maintenance, monitoring, control and safety of transport. At the same time, an intelligent system can be defined as optimizing processes or resources for obtaining a desired product, providing this information in real time, for its follow-up, evaluation and control, and allowing introduced variations during the course of their management.

For this reason, in addition to the main objectives listed above, ITS provide a number of benefits derived from the improvement in the operational efficiency and reliability of offered services, production improvement in the management of transport infrastructure, as well as increased safety, reduction of environmental impact and a variety of information services provided to the users of the transport. Therefore, ITS include different tools and services resulting from the application of telematics concepts in the area of transport; among many others, the following can be mentioned:

- automatic traffic management systems;
- public transport information services;
- traveller information systems;
- fleet management and location systems;
- emergency management;
- electronic payment systems;
- cooperative vehicular systems.

As can be seen, the types of applications and services are very varied, mainly because ITS continuously evolve, and increasingly more efforts and resources are being devoted to their development and implementation.

On the other hand, public (and private) institutions play a crucial role in fostering policies that help and encourage the development and deployment of applications that improve the current ITS. An example is found in the European Union; the current programme of aid (called research Horizon 2020) includes work oriented towards 'Smart, green and integrated transport'[1], which encourages projects and ideas related to 'Mobility for Growth' or 'Green Vehicles'.

In recent years, a large number of projects, research and innovations have been focused on the issues mentioned here, all of them grouped under the ITS topic. In the following table some research projects and initiatives in this area are listed.

[1] Further information can be found at http://ec.europa.eu/programmes/horizon2020/en/h2020-section/smart-green-and-integrated-transport

Some international ITS research projects.

Project	Description
PREVENT Preventive and active safety applications contribute to the road safety goals on European roads	It aims to promote the development, deployment and use of Intelligent Integrated Safety Systems in Europe, helping drivers to avoid accidents. The key concepts are that depending on the significance and timing of the danger, the systems alert the drivers as early as possible, warn them and, if they do not react, actively assist or ultimately intervene. It addresses services such as Safe Speed and Safe Following, Lateral Support and Driver Monitoring, Intersection Safety and Vulnerable Road Users and Collision Mitigation.
AIDE Adaptive Integrated Driver-vehicle Interface	The objective is to generate the knowledge and to develop methodologies and human-machine interface technologies required for safe and efficient integration of Advanced Driver Assistance Systems (ADAS), In-Vehicle Information Systems (IVIS) and nomad devices into the driving environment. It tests innovative concepts and technologies to maximize the efficiency of ADAS, minimize the level of workload and distraction imposed by IVIS and nomad devices and enable the potential benefits of new in-vehicle technologies and nomad devices in terms of mobility and comfort, without compromising safety.
SAFESPOT Cooperative systems for road safety 'Smart Vehicles on Smart Roads'	The project aims to prevent road accidents by developing a 'Safety Margin Assistant' that detects in advance potentially dangerous situations and that extends in space and time drivers´ awareness of the surrounding environment. This is done by using both the infrastructure and vehicles as sources (and destinations) of safety-related information, defining an open, flexible and modular communications architecture, developing the key enabling technologies and infrastructure-based sensing techniques and testing scenario-based applications to evaluate the impacts and end-user acceptability.
COOPERS Co-operative networks for intelligent road safety	This project is focused on the long-term development of innovative applications for traffic management, by coordinating between vehicle and infrastructure. It provides vehicles and drivers with real-time local situation-based information and safety-related status information, distributed via dedicated infrastructure-to-vehicle communication links. It follows a three-step approach for implementation of I2V communication: improvement of road sensor infrastructure and traffic control applications for more precise situation based traffic information and driver advice, development of a communication concept and applications able to cope with the I2V requirements (reliability, real time capability and robustness) and demonstration of results on important sections of European motorways.

Project	Description
CVIS Co-operative Vehicle-Infrastructure Systems	It develops and integrates the essential basic and enabling technologies such as a multichannel communications and network platform readily adaptable for both vehicle and roadside, a highly accurate positioning and local map module and an open software environment for applications. These components allow a vehicle to share urgent information with nearby vehicles, and to dialogue with both the immediate roadside infrastructure and infrastructure operators and service providers.
SAFETRIP Satellite application for emergency handling, traffic alerts, road safety and incident prevention	This project provides an integrated system platform that allows any third party company to develop applications for the road market, promotes innovative satellite technologies and communication features and integrates in vehicles a device called 'Greenbox' offering a universal two-way communication system. In the project the following applications were tested: provision of real-time traffic information and warnings generated by the collection of data coming from other vehicles, emergency call system and tracking in real-time of vulnerable passenger transports.
PRE-DRIVE	Based on the overall description of a common European architecture for an inter-vehicle and vehicle-to-infrastructure communication system, this projects aims to develop a detailed system specification and a functionally verified prototype to be used in future field operational tests. Furthermore, it develops an integrated simulation model for cooperative systems to estimate the expected benefits of Car-2-X communication in terms of safety, efficiency and environment. Finally, all tools and methods necessary for functional verification and testing of cooperative systems under real traffic conditions are realized.
DRIVE C2X DRIVing implementation and Evaluation of C2X communication technology in Europe	Its objective is to carry out comprehensive assessment of cooperative systems through field operational tests in various places in Europe in order to verify their benefits and to pave the way for market implementation. This proposal builds strongly on previous and on-going work on cooperative systems, which are now considered to be mature enough for large-scale field operational tests. Essential activities in this project are the testing methodology and evaluation of the impact of cooperative driving functions on users, environment and society.

(*Continued*)

Project	Description
INTERACTIVE Accident avoidance by active intervention for Intelligent Vehicles	The project addresses the development and evaluation of next-generation safety systems for Intelligent Vehicles, based on active intervention. It is based on the concept that by integrating in-vehicle components, large amounts of information about the driver, state of the vehicle and the environment can be provided to all interested applications. The project develops next-generation safety systems based on three pillar concepts: continuous driver support, collision avoidance and collision mitigation.
ADAPTIVE Automated Driving Applications and Technologies for Intelligent Vehicles	The project develops new and integrated automated driving functions in cars and trucks to improve traffic safety by minimizing the effects of human errors and to enhance traffic efficiency due to smoother flows and reduced congestion. The approach is based on a shared control concept, assuring proper collaboration between the driver and the automation system. This is realized using cooperative vehicle technologies, advanced obstacle sensors and adaptive schemes where the level of automation dynamically responds to the situation and driver status.
ECOMOVE Cooperative Mobility Systems and Services for Energy Efficiency	This project creates an integrated solution for road transport energy efficiency by developing systems and tools to help drivers to eliminate unnecessary fuel consumption (routes and driver behaviour optimization), and road operators to manage traffic in the most energy-efficient way (network management optimization). By applying this combination of cooperative systems using vehicle-infrastructure communication, the project aims to reduce fuel consumption by 20% overall.

Technologies and Applications

About Part 1: Intelligent Transportation Systems

Due to the rapid advances in the area of telematics, ITS are increasingly more complex and therefore more difficult to develop and deploy. Thus, when ITS are designed, it is necessary to ensure the operation of each of their components separately and the full operation of the system when all are combined.

Today, ITS are multisystem structures that combine tasks of management, control, information collection and actuation systems, and which have to be perfectly related and synchronized to meet the objectives of the whole system. Therefore, when ITS solutions are designed, all the synergies among the subsystems and the interests of all the stakeholders (users, companies, governments, etc.) must be defined. Thus, the architecture of the system provides a common framework, designed on the basis of the requirements of the users and the scope for planning, definition and integration in a transport smart system.

One way to try to lessen the complexity of ITS is to use a reference architecture that provides a framework that contains a set of common assumptions for any ITS. In Part 1, the architectures of reference in Europe (Chapter 1) and USA (Chapter 2) are introduced. Both have a common objective: help local stakeholders involved in the design of ITS to plan and implement operational systems which will contribute to improve mobility experience, involving vehicles, drivers, passengers, road operators and governments, all interacting with each other and the environment.

About Part 2: Wireless Vehicular Communications

Cooperative systems allow vehicles to communicate with each other (Vehicle-to-Vehicle, V2V) or with the infrastructure (Vehicle-to-Infrastructure, V2I, or Infrastructure-to-Vehicle, I2V), and the goal is to improve road safety and/or traffic management.

Currently, examples of vehicles can be found that communicate with the infrastructure thanks to a wireless link; public buses are provided with devices that allow them to communicate with traffic lights in order to assign passing priority in an intersection. This is a cooperative system because both elements exchange information about their states, position and other facts, and according to this data a decision is taken. These kinds of systems are known as independent or autonomous systems, because they are based on an individual platform that has been developed only for this single application and which most likely would not be easily adapted to new ones. The novelty in the technology deployed in cooperative systems is that two-way communication on an open platform is possible while allowing multiple services and applications, distributed among several mobile nodes, to be added easily. Part 2, by means of Chapters 3, 4 and 5, introduces several wireless technologies and scenarios that enable ITS to deal with problems, specifically in vehicular environments, where there can be many mobile nodes (vehicles) that want to communicate with each other (using V2V links) and with the infrastructure (V2I or I2V links).

Antennas play a key role in optimizing the performance of wireless vehicular communication systems. In Chapter 6, the vehicle interaction with electromagnetic (EM) fields and associated implications for ITS development are presented, while Chapter 7 describes novel in-car integrated and roof-mounted antennas for the transportation industry. Both sections provide up-to-date and in-depth industrially focused advances in the subject from MIRA Limited and Harada Industry Co. Limited, respectively. In particular, Chapter 6 provides an overview of issues relating to channel characterization and investigations of human exposure to in-vehicle EM fields using software tools and bespoke techniques, as well as reporting a novel low disturbance scanner for in-vehicle EM field measurements. Complementary to this, Chapter 7 discusses new small-footprint roof-mounted automotive antennas and innovative hidden and integrated prototypes that offer the potential to increase diversity for superior performance, as well as predicting future telematics antennas that are both physically small and make use of intelligent solutions (i.e. optimize the antenna electronically).

About Part 3: Sensors Networks and Surveillance at ITS

Traditional methods used for traffic management, surveillance and control become inefficient in terms of performance, cost, maintenance and support with increased traffic. Sensor Networks are an emergent technology with an effective potential to overcome these difficulties, and which will have a great added value to ITS. Each year, more industrial and academic researchers are involved in its development due to its potential applications in both urban and highway scenarios. Sensor networks offer the potential to improve significantly the efficiency of existing transport systems, by collecting data with the purpose of planning and managing actuations over the road, as well as providing information for the users.

Information exchange among nodes in the network can be performed either through ad-hoc communication, or using infrastructure, or hybrid. In addition, two kinds of sensors are distinguished: on-road sensors and on-vehicle sensors. The combination of sensor types and communication paradigms gives birth to various sensor network architectures for ITS. In Part 3 some successful sensors networks with direct application to ITS are presented. In particular, they are focused in the wireless version of these networks, due to the advantages provided, with respect to the wired networks. Concretely, these are covered in Chapters 8 and 9.

About Part 4: Data Processing Techniques at ITS

The deployment of future Internet and communication technologies will provide ITS with huge volumes of (at some point real-time) data that need to be managed, communicated, interpreted, aggregated and analysed. Therefore, new data processing and mining as well as optimization techniques need to be developed and applied to support decision-making and control capabilities.

In general, all the participants of ITS act both as data generators and sources, leading to a huge amount of data with short update rates. This growth in data production is being driven by: individuals and novel types of sensors and communication capabilities in vehicle. Part 4, by means of Chapters 10 and 11, introduces applications that take advantage of the new technologies of data acquisition and processing, with the purpose of designing new methods and applications in the field of ITS. In Chapter 12 some related legal aspects are addressed.

About Part 5: Applications and Services for Users and Traffic Managers

Once several technologies from underlying architectonic layers have been covered, these are combined to provide software applications and services for users and traffic managers. Their aim is usually to improve the management of mobility and the quality of life in the cities.

From the traffic manager's point of view, one of the most important components of modern ITS is traffic management system. Its main goal is the efficient management of a road network, optimizing the traffic flow while maximizing road users' safety and comfort. Specifically, to improve traffic safety, a particularly potent approach was recognized in the possible application of cooperative systems in traffic. Part 5, in Chapter 13, introduces a conceptual framework of traffic management systems, in which data collection, data processing and analysis and information dissemination and actuation are established as pillars. Subsequently, some examples of the cooperative approach application in urban traffic are described (Chapter 14), such as those related to cooperative ramp metering. Traffic management discussion is completed by a description of different urban and highway traffic environments in which traffic management systems operate.

From the driver's point of view, several services can be provided inside the vehicle, while it is being driven. These services can be designed with different purposes: infotainment, safety and trip information, among others. In fact, along the road, drivers experience unexpected situations. For this reason, static and dynamic road traffic signalization, besides real-time traffic and travel information systems, attempt to warn drivers about all the events happening on the road. Such knowledge of the surrounding situation is necessary for creating the basis of decision-making in the driving task.

However, if all detected information is displayed to the driver on the instrument panel without any filter or prioritization process, the workload of the driver could be increased considerably. It is necessary to prioritize all the information given to the driver. Being aware of the importance of this prioritization, Chapter 15 presents a new methodology for an intelligent in-car traffic information management system.

Finally, from the user's point of view, new approaches in services for multimodal trip planning have arisen. During the last years there has been a remarkable change in transport habits and urban mobility, due to, at institutional level, promotion of the use of public transport as a priority. People are adopting gradually as a viable alternative the use of public transport versus private transport, that is, multimodal transport. In Chapter 16 a review of software solutions and services designed to promote the use of multimodal transport is presented. It describes an alternative solution to the management of transit data by applying the linked open data principles to the field of multimodal transport. Such data, made available to users and developers, are intended to enable the development of advanced software services related to multimodal mobility and oriented to the use of sustainable transport.

Conclusions

Intelligent Transportation Systems are becoming a key player to improve the mobility of people and freights in terms of safety, efficiency, sustainability and comfort. At this moment, there is a good combination between sci-fi and reality in ITS. On the one hand, a set of very promising technologies have arisen, which are undergoing preliminary exploration by research groups. On the other hand, some relevant ITS are already deployed and exploited by industrial companies.

This book presents a well-balanced combination of academic contributions and industrial applications in the field of ITS. The most representative technologies and research results have been achieved by some of the most relevant research groups working on ITS, in order to show the possibility of generating industrial solutions to deploy in real transport environments.

The book is aimed at researchers and PhD students looking for a baseline to start their research in ITS topics. It will provide them with a good overview of new trends in ITS (facing several layers, from physical to application layer) and then they can focus on the areas in which they are interested using the bibliographical references or other resources. It also provides a very useful introduction to hot topics in several areas related to Information and Communications Technology (ICT) and transport, as a first step to future research activities. Moreover, it completes this holistic approach by including real cases, described by industrial partners, in which previous research works have been undertaken on a real ITS product and/or service.

Hope you find this book interesting!

Acknowledgements

The editors would like to thank the European Commission for its support. This work was partially funded by the Seventh Framework Programme under grant agreement no. 317671 (ICSI – Intelligent Cooperative Sensing for Improved traffic efficiency). Several of the authors of this book and its editors participated in the ICSI project and some of the works described here were the result of this action. Thanks also to Elena Cordiviola from Intecs S.p.A for the coordination of the ICSI project.

Part 1

Intelligent Transportation Systems

1

Reference ITS Architectures in Europe

Begoña Molinete, Sergio Campos, Ignacio (Iñaki) Olabarrieta
and Ana Isabel Torre
TECNALIA. ICT-ESI Division, Derio, Spain

1.1 Introduction

Intelligent Transportation Systems (ITS) are complex systems which require a systematic basis for their planning and deployment processes. In view of this need, the European ITS Framework Architecture (FRAME) was set up to support ITS development and to foster their roll-out at the Member States by facilitating system integration, fostering interoperability, avoiding vendor lock-in situations and promoting the standardization of functionalities, interfaces and data models. The fast technological evolution and the growing interest in Cooperative Systems based on V2X communications unveiled new requirements uncovered by the existing architectures, leading to alternative approaches by certain research projects, which ended in an extended version of the FRAME architecture and complementary standardization processes to fulfil the needs posed by the connected vehicles and road infrastructures.

This chapter will outline the reference architectures conceived for ITS planning and deployment in Europe, their evolution through the inputs provided by major ITS projects and initiatives and some experiences from the authors when facing the definition of ITS architectures through our involvement in research projects, by following different approaches.

1.2 FRAME: The European ITS Framework Architecture

An ITS Architecture sets a framework to plan, analyse, define, deploy and integrate Intelligent Transportation Systems, allowing at the same time understanding of their business, organizational and technical implications. It is commonly depicted as a high-level design showing the

Intelligent Transport Systems: Technologies and Applications, First Edition. Asier Perallos, Unai Hernandez-Jayo, Enrique Onieva and Ignacio Julio García-Zuazola.

structure and operation of a certain system in a given context, which can be used as a basis for further low-level design phases.

An ITS architecture integrates three main elements:

- the functions that are required for ITS;
- the system partitioning into logical or functional entities such as subsystems, modules or components, where these functions reside;
- the information and data flows that connect these functions and physical subsystems together into an integrated system.

When talking about ITS architectures in Europe, the key reference is the European ITS Framework Architecture (often referred to as FRAME). The FRAME Architecture was created as an attempt to provide a common approach across the European Union so that the implementation of integrated and interoperable ITS can be planned. Its main objective is to foster the deployment of ITS in Europe and for that purpose, it defines a framework providing a systematic basis for ITS planning, easing integration between multiple systems and ensuring interoperability and consistency of information.

1.2.1 Background

The FRAME Architecture was created by the project KAREN (Keystone Architecture Required for European Networks), funded by the European Commission under the Fourth Framework Programme in the area of Transport Telematics and first published in October 2000. The need to keep the architecture up to date was soon identified, and this entailed a huge maintenance effort. During the following years (2001–2004), projects FRAME-NET and FRAME-S, funded by the European Commission under the Fifth Framework Programme, carried out this task successfully. As a result, not only did the architecture evolve but also user needs were updated, a methodology supported by computer-based tools was defined, a centre of knowledge created and the FRAME Forum was established so that users and stakeholders could exchange advice and experiences. Since 2006, due to the growing expectations created by the so-called Cooperative Systems based on V2X communications and strongly supported by the European Commission, the FRAME Architecture entered an adaptation process to provide support to the new set of user needs and requirements posed by connected vehicles and road infrastructures, which were not covered by the original version. The project E-FRAME (Extend FRAMEwork architecture for cooperative systems), funded by the European Commission under the Seventh Framework Programme, addressed this need between 2008 and 2011. As a result, the FRAME Architecture version 4 supporting Cooperative Systems was released in 2010.

Since the time of the KAREN project, the FRAME Architecture has been adopted and successfully used in different ways by many nations, regions, cities and projects throughout Europe. Reference best practices can be mentioned such as the French national ITS Architecture (ACTIF), the Italian national ITS Architecture (ARTIST) and other adopting nations such as Austria (TTS-A), the Czech Republic (TEAM), Hungary (HITS) and Romania (NARITS). In addition, specific ITS Architectures have been created in the UK including one for Transport for Scotland and another for the County of Kent, while Transport for London intended to use FRAME Architecture to plan its future ITS deployments. In some cases it has also been used by R&D projects such as VIKING, EASYWAY, COOPERS and MoveUs among others, and even in pre-commercial procurement programmes such as CHARM-PCP, participated by the Highways Agency (UK) and Rijkswaterstaat (the Netherlands).

1.2.2 Scope

Strictly speaking, FRAME is not always considered as an architecture, but more a framework targeted to help European countries and regions planning their own ITS architecture tailored to their particular needs. In this sense, the experience is quite similar to the one in the USA, where the main objective of the National ITS Architecture has been to guide the ITS development and deployment throughout the country at the federal, state and local levels.

The most recent release of the FRAME Architecture (version 4), published in 2010 with the updates from the E-FRAME project, covers the following ITS areas:

- Electronic Fee Collection;
- Emergency Notification and Response – Roadside and In-Vehicle Notification;
- Traffic Management – Urban, Inter-Urban, Simulation, Parking, Tunnels and Bridges, Maintenance, together with the Management of Incidents, Road Vehicle Based Pollution and the Demand for Road Use;
- Public Transport Management – Schedules, Fares, On-Demand Services, Fleet and Driver Management;
- In-Vehicle Systems – including Cooperative Systems;
- Traveller Assistance – Pre-Journey and On-Trip Planning, Travel Information;
- Support for Law Enforcement;
- Freight and Fleet Management;
- Support for Cooperative Systems – specific services not included elsewhere such as bus lane use, freight vehicle parking;
- Multimodal interfaces – links to other modes when required, e.g. travel information, multimodal crossing management.

When dealing with an ITS architecture, and also with FRAME, a number of different views can be considered:

- User Needs, which are always the starting point and collect the expectations to be covered by an ITS deployment and its associated set of services. The identification of these needs may involve different stakeholders such as public transport or freight operators, system integrators, national/regional governments and every kind of traveller.
- Functional Viewpoint, which defines the functionality to be provided by the ITS in order to meet the User Needs, usually structured into functional areas and further divided into specific functions. This is represented as Data Flow Diagrams containing the functions and

showing how they relate to each other, to data stores and terminators, as well as the data that flows between them.

- Physical Viewpoint, which describes how functions can be grouped into physical components allocated to modules or subsystems. Hence, detailed specifications for each component can be produced.
- Communications Viewpoint, which describes the communications links needed to support physical data flows. Once the functionalities have been allocated to physical modules, the location of the Functional Data Flows can be inferred and also the information flow between modules, thus representing communication channels. At this level, communication specifications can be produced and the use of an existing standard, or even the need to define a new one, may be agreed.

In addition to the main architectural views, use of the FRAME Architecture enables other kind of activities such as a Deployment Study, showing how to deploy the systems and communications derived from the ITS Architecture and the way to migrate existing systems to be compliant to FRAME; a Cost/Benefit Study, helping to predict the likely costs and the expected benefits from the ITS deployment; and a Standardization Study, identifying existing applicable standards related to the European ITS Framework Architecture and future standardization needs.

1.2.3 Methodology and Content

A general high-level diagram showing the methodology and the different views to be considered when creating an ITS Architecture is illustrated in Figure 1.1, where the specific scope of the FRAME Architecture has been highlighted.

The FRAME Architecture is intended to be used in the European Union, it does not impose any kind of structure on a Member State and comprises only a set of User Needs describing what ITS can provide and a Functional Viewpoint showing how it can be done. The methodology is supported by computer-based tools (a Browsing tool and a Selection tool), enabling the definition of logically consistent subsets of the FRAME Architecture Functional Viewpoint and the creation of subsequent Physical Viewpoints. It is worth noting that the FRAME Architecture is technology independent and does not entail the use of any specific technology or product in order to ensure that the ITS architectures and high-level

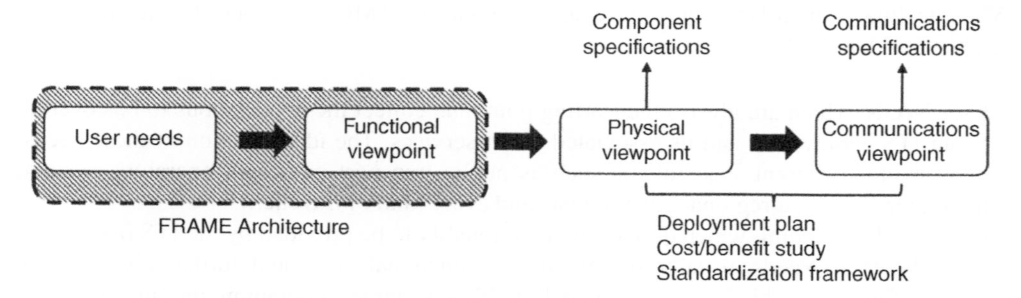

Figure 1.1 General methodology to create an ITS architecture and scope of FRAME architecture. © 2009 PJCL.

requirements planned using the methodology will not become obsolete despite the evolution of technology and product development.

The KAREN project generated about 550 User Needs to cover the ITS applications and services being considered for implementation at the end of the 1990s. Since then, the FRAME-related projects have continuously updated this set of User Needs with the most recent E-FRAME project, adding about 230 User Needs related to the use of Cooperative Systems.

The FRAME Architecture can be used in a number of scenarios, being one of the most ambitious to plan large-scale integrated ITS deployment in a country or region over a number of years. By collecting the vision of the different stakeholders involved, a suitable subset of the FRAME Architecture can then be used to provide a high-level model on the way to achieve it. When creating a subset ITS architecture, the most appropriate set of functionality to deliver the required services must be selected. The system structure obtained as a result should provide enough information to develop the services and deploy the equipment needed, all of them compliant with the overall architecture concept.

It can be stated that the FRAME architecture in its current version covers most ITS applications in place or considered for implementation in Europe without imposing any technical requirement on the development phase and therefore remaining technology independent. Comprehensive documentation and the tools to support its use, which are strongly recommended, are freely available [1].

1.3 Cooperative Systems and Their Impact on the European ITS Architecture Definition

Cooperative Systems based on V2X communications are still regarded as one of most promising solutions to address the current and future needs for increased safety and road traffic efficiency. Research on Cooperative Systems in Europe started at the end of the 1980s with the PROMETHEUS project. However, the technology required was not mature enough at that time and it was not until the last decade that real investment was made in this field through different initiatives and R&D projects, always with strong support from the European Commission.

Although initial research and awareness activity was carried out in a noncoordinated manner, a significant effort was made to align and harmonize the results from the different projects in order to provide an integrated view of the architectural needs to support Cooperative Systems. In addition, the European Commission defined a Policy Framework for ITS deployment and promoted an intense activity on standardization, not only limited to the EU but also looking outside to establish a fruitful cooperation with the USA and Japan. In the last years, the projects focused on the design and development of Cooperative System prototypes have given way to the so-called Field Operational Tests (FOTs) and Pilots, targeted to large-scale deployment, validation and impact assessment.

1.3.1 Research Projects and Initiatives

The preliminary framework for the support of research and technological development in European Cooperative ITS was set by the Intelligent Car Initiative [2] in 2006, targeted to coordinate research on smarter, cleaner and safer vehicles and to create awareness of

ICT-based solutions for safer and more efficient transport. One of its pillars was the eSafety Forum, now renamed as iMobility Forum [3], which aimed to support ITS deployment specifically targeted to overcome safety needs. In particular, the eSafety Forum Communications Working Group was entrusted to provide all the information required to advise the European Commission on the deployment of a harmonized EU wide communication system for Vehicle-to-Vehicle (V2V) and Vehicle-to-Infrastructure (V2I), paying special attention to spectrum issues, standardization and international cooperation.

On the industry side, key initiatives can also be mentioned such as the Car2Car Communication Consortium (C2C-CC) [4], an industrial nonprofit organization driven by European vehicle manufacturers and focused on V2V communication whose primary goal was to establish an open European industry standard for V2V communication systems based on wireless network components ensuring European-wide inter-vehicle compatibility. The consortium promoted the allocation of the frequency band at 5.9 GHz range for safety-critical automotive use. The C2C-CC is currently integrated in the Amsterdam Group [5], a broader stakeholder association targeted to Cooperative ITS deployment in corridors and cities, sharing knowledge and experiences with leading organizations such as CEDR (Conference of European Directors of Roads), ASECAP (European Association of Tolled Motorways) and POLIS (European Cities and Regions Networking for Innovative Transport Solutions).

Concerning research and development activity, it is worth mentioning projects such as CVIS, SAFESPOT, COOPERS, GEONET and PRE-DRIVE C2X, all funded by the European Commission under the Sixth and Seventh Framework Programmes. All of them dealt with Cooperative Systems development, prototyping and demonstration but putting the focus on different perspectives. CVIS [6] worked on a global view, developing and integrating the essential applications and enabling technologies for communications, planning and positioning in a common platform unit implementing the ISO CALM (Continuous Air-interface Long and Medium range) architecture; while SAFESPOT [7] focused on the development of a 'Safety Margin Assistant', a safety bubble around the vehicle in which full driver awareness to anticipate potentially dangerous situations could be achieved by means of cooperative communications; and COOPERS [8] looked for a better use of the available infrastructure capacity by using V2I communications for safety-relevant data exchange. Shortly after, GEONET [9] produced reference specifications for Cooperative Systems networking covering GeoNetworking and IPv6, which were later adopted for standardization, and PRE-DRIVE C2X [10] prototyped a common European V2X communication system and designed the necessary tools for operating a Field Operational Test with Cooperative Systems on a European level performing comprehensive assessment of the impacts.

1.3.2 Pilots and Field Operational Tests

As an evolution of the first R&D projects, which provided prototype implementations of Cooperative Systems allowing early testing of draft standards, Pilots and Field Operational Tests are intended to bridge the gap from demonstrations to system roll-out. For that purpose, it was deemed important to collect sound and valid data on the performance and cost–benefit ratio of these systems. Large-scale pilots and Field Operational Tests play an important role by providing comprehensive data on the performance and user acceptance of Cooperative Systems.

Some representative Field Operational Tests at national and regional level can be mentioned such as SIM-TD in Germany, SCORE@F in France, SPITS in the Netherlands, Easy Rider in Italy, TSN in Norway and SISCOGA in Spain, among others. As for the pilots, the most representatives at the moment are COMPASS4D [11], aimed to deploy three Cooperative ITS services (Red Light Violation Warning, Road Hazard Warning and Energy Efficiency Intersection Services) in seven European cities and CO-GISTICS [12], intended to deliver five Cooperative ITS services targeted to sustainable mobility of goods in seven European logistic hubs.

1.3.3 European Policy and Standardization Framework

As research activity evolved and the results from the different projects were released, the European Commission identified the need to set a Policy Framework for the development of ITS in Europe. In this sense, two major actions were taken:

- Action Plan for the Deployment of Intelligent Transport Systems (ITS) in Europe, COM(2008)88 (Dec. 2008) [13]. It suggests a number of targeted measures structured in six priority areas to speed up market penetration of rather mature ITS applications and services in Europe.
- Directive 2010/40/EU: Framework for the Coordinated and Effective Deployment and Use of Intelligent Transport Systems [14]. It is a key instrument for the coordinated implementation of ITS in Europe, aimed to establish interoperable and seamless ITS services while leaving Member States the freedom to decide which systems to invest in.

Regarding standardization, the European Commission issued the Mandate M/453 [15] under the 2010/40/EU Directive, inviting the European Standardization Organizations to prepare a coherent set of standards, specifications and guidelines to support European Community wide implementation and deployment of Cooperative ITS. The results of the European research and development projects as well as the Field Operational tests were included in the standardization process. Coordinated by the COMeSafety initiative [16], these projects developed a harmonized ITS communications architecture [17], which was provided to ETSI and ISO and resulted in published standards. The first standards package, the so-called 'Release 1 specifications' [18], produced by ETSI and CEN, has been adopted and issued in 2014.

In addition, international cooperation on standardization activities has been addressed since 2010 with the USA and Japan, searching for global harmonization of standards for Cooperative Systems. A joint ITS technical task force has been established between ETSI and the US Department of Transportation, while a Memorandum of Cooperation has been signed with the Japanese Ministry of Land, Infrastructure, Transport and Tourism.

1.3.4 Impact on FRAME Architecture

After an initial phase of research, development and prototyping in isolation, Cooperative Systems are reaching a fair maturity level to make steps towards integration with any other ITS application or service, since they need to interact with other elements to exchange information, e.g. a traffic management system.

Thanks to the E-FRAME project, the current FRAME Architecture (version 4) contains all the applications and services that were considered by CVIS, SAFESPOT and COOPERS projects and can therefore show how this integration may be achieved. Analogously, when a Physical Viewpoint has been created and the corresponding communications requirements are identified, the work of COMeSafety and the projects contributing to the standards produced by CEN and ETSI as a result of Mandate M/453, can be used to define the communications links in detail.

The measures and priority areas defined in the ITS Action Plan demand the delivery of specific ITS services and applications throughout Europe, which need to be supported by one (or more) architectures. These architectures can be now defined by using a subset of the current FRAME Architecture. By following the methodology, applicable existing standards or new standardization needs can be identified. As a result, a technology-independent view of each service or application can be generated, ensuring interoperability between products from different manufacturers, avoiding vendor lock-in situations and facilitating the further merging of several ITS architectures, even coming from different Member States, since they will be compliant with FRAME elements and terminology.

1.4 Experiences in ITS Architecture Design

In this section, part of our experience in the definition of ITS architectures is explained through our involvement in two R&D projects funded by the European Commission: Cybercars-2 and MoveUs. These projects are radically different, since they were conceived and carried out at different moments in the ITS evolution timeline. The resources available were not the same and therefore, different approaches for architecture definition were required and put into practice.

On the one hand, Cybercars-2 belongs to the wave of projects related to Cooperative Systems heavily invested on since 2006. At that time, the FRAME architecture was not yet ready to cover the requirements posed by this kind of systems, so an alternative methodology was used taking as a reference the ongoing initiatives and the flagship projects in this field. On the other hand, MoveUs is a still ongoing project dealing with the provision of personalized mobility services in the context of Smart Cities aiming to trigger behavioural changes towards sustainable mobility habits. In this case, the FRAME Architecture has been used as main reference in combination with other approaches and tools available, thus enabling all the functionalities to be covered and providing richer and more complete results at the design phase.

1.4.1 Cybercars-2: Architecture Design for a Cooperative Cybernetics Transport System

Cybercars-2 (Close Communications for Cooperation between Cybercars 2) was a project funded by the European Commission under the Sixth Framework Programme between 2006 and 2008. It was conceived as an evolution from its predecessors, CyberCars and CyberMove, which dealt with the development and evaluation of Cybernetic Transport Systems (CTS) comprising a number of individual driverless vehicles with the capability to travel on existing road infrastructure without the need for dedicated ways. However, the prototype vehicles developed in those projects were designed for low-demand road traffic environments and were neither able to communicate nor cooperate with each other. Hence, the main challenge for

Cybercars-2 was to empower these vehicles with the ability to cooperate through Vehicle-to-Vehicle and Vehicle-to-Infrastructure communication links in order for the CTS to enable higher traffic flows and improved network efficiency.

One of the main project tasks was to define, develop, deploy and test a Cooperative Cybernetic Transport System Architecture able to provide: interconnectivity and interoperability between different types of driverless vehicles, compatibility with ADASE (Advanced Driver Assistance Systems in Europe) Architecture and increased road traffic efficiency and safety. Since consortium partners brought their own driverless vehicles to the project, featuring different system architectures, great effort was invested on adapting the existing vehicle control architectures to the Cooperative Vehicles' Communication Architecture paradigm.

With these objectives in mind, specific cooperative communication needs were identified in the context of interoperability, operational safety, reliability and compatibility with ADASE Architecture. In particular, the following steps were made:

- The system architecture for each component of the CyberCars-2 fleet of vehicles (fully driverless Cybercars and Dual-Mode vehicles, manually or automatically driven) was analysed in depth.
- The Cybercars-2 Architecture was defined: use cases, Vehicle-to-Vehicle and Vehicle-to-Infrastructure communication requirements were identified and a proposal for a communications architecture based on protocol layers was developed.
- The main requirements for compatibility between Cybercars and vehicles compliant with ADASE Architecture in terms of functionality, security and safety were identified, articulated and examined.
- Enhanced Safety Certification procedures applicable to the scope of Cooperative Cybernetic Transport Systems were defined.
- A small-scale Cooperative Cybernetic Transport System was demonstrated in La Rochelle (France) in September 2008; it comprised a simplified Control Centre and a fleet of Cybercars and Dual-Mode vehicles driving in an 8-shaped test track.

At the same time frame, numerous R&D activities focusing on vehicular communications were undertaken under the umbrella of the European Commission and EU member countries. Many of them explored the use of V2V/V2I communication technologies for different application fields and purposes, e.g. improved safety or enhanced traffic efficiency. Representative examples are European projects PReVENT, CVIS, SAFESPOT and COOPERS, already introduced.

According to the project objectives and to the status of research activity on Cooperative Systems, it was soon agreed that the architecture to be proposed by CyberCars-2 project should be:

- focused on the safety application field, because the aim was to perform cooperative manoeuvres safely;
- mainly based on robust, low-latency and reliable V2V communications, which is the most efficient way to exchange information between the vehicles, since cooperative manoeuvres are time-critical actions;
- compliant with the architecture supported by any of the key reference projects previously mentioned, in order to achieve the highest possible degree of compatibility.

As a result, the SAFESPOT architectural approach was chosen as a reference for Cybercars-2 architecture design for the following reasons:

- It enabled advanced detection of potentially dangerous road traffic situations and made drivers aware of the surrounding environment.
- It was implicitly aligned to Car2Car-Communications Consortium Reference Architecture, which was focused on the creation and establishment of an open European industry standard for Car-to-Car communication systems based on wireless networking technologies.
- Being compatible with the ISO CALM architecture, it offered a standardized set of air interface protocols and parameters for medium and long range, high-speed ITS communication.

The methodology adopted to define the project communication requirements was based on the available reference from Vehicle Safety Communications Project (VSC) in the USA. As a result, spreadsheet tables were produced in which the columns represented the driving manoeuvres to be performed cooperatively by the vehicles, while the rows listed the set of communication parameters identified (e.g. message size, allowable latency, communication range), and suitable target indicators were given for each parameter. Other references from European projects were also taken into account, such as the communication requirements for an intersection scenario provided by INTERSAFE, a PReVENT subproject.

The main building blocks involved in the CyberCars-2 Architecture were: (a) the vehicles (both, driverless and dual-mode vehicles), which exchanged messages containing useful data to perform cooperative manoeuvres safely; (b) the control centres located at dedicated environments but not necessarily close to the roads, from which traffic was controlled and the vehicle fleet efficiently monitored; and (c) the infrastructure elements, which occasionally helped to improve system efficiency, e.g. transmitting a Differential GPS correction to achieve higher positioning accuracy, or monitoring and supervising the traffic flow from a control centre.

Built in line with SAFESPOT Reference Architecture, the Cybercars-2 Architecture consisted of three subarchitectures or viewpoints, namely:

- Functional Architecture, derived from system requirements and aimed at fulfilling user needs; it comprised functional modules, data structures and interfaces and is illustrated in Figure 1.2;
- Physical Architecture, describing the way in which the required functionality and system requirements are fulfilled;
- Communication Architecture, defining links between the Physical Architecture components and identifying suitable communication protocols to enable the data flows from one component to another.

Concerning communications, as one of the examples of safety-critical applications, the Cybercars-2 Architecture supported ongoing efforts from the European industry (C2C-CC) and standardization bodies (ETSI ERM TG37) for specifically protected frequency band allocations at 5.9 GHz range to guarantee European-wide inter-vehicle compatibility. More specifically, the Cybercars-2 Architecture dealt with three types of communication channels:

- Short-Medium range V2V/V2I communication channels delivering the data required by the vehicles to perform cooperative manoeuvres safely;

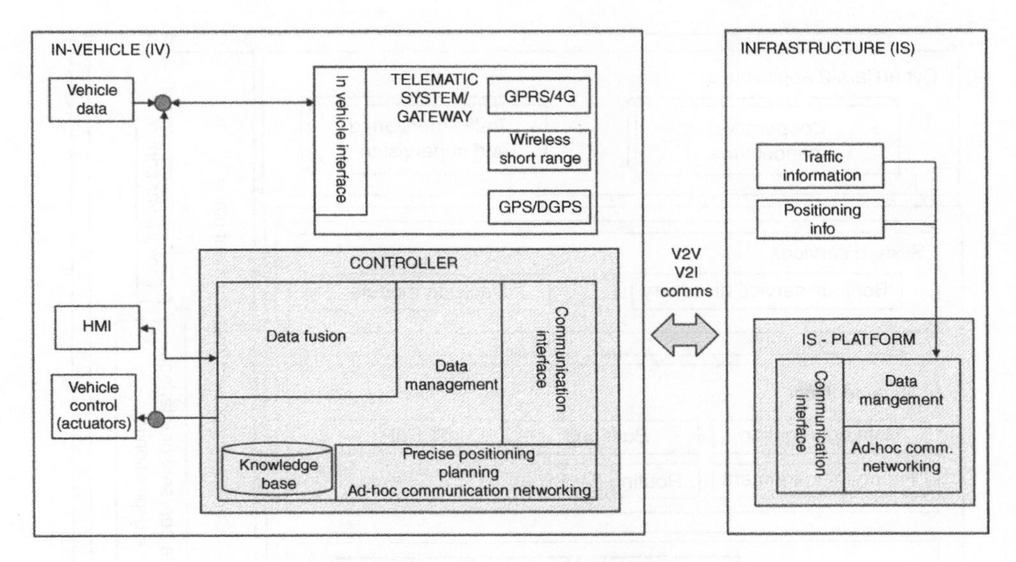

Figure 1.2 Cybercars-2 functional architecture [19].

- Short-Medium range V2I/I2V communication channels delivering either support information to the vehicles (e.g. Differential GPS correction) or information on the network status and traffic flow to/from an infrastructure unit nearby;
- Long range V2I/I2V communication channels delivering information on the network status and traffic flow to a remote infrastructure unit (V2I) and control commands back to the vehicles in order to improve traffic efficiency.

The layered communications protocol architecture proposed, also based on SAFESPOT approach, is illustrated in Figure 1.3.

In conclusion, it can be stated that the Cybercars-2 Cooperative Communication Architecture enabled communication between driverless vehicles of different kinds, keeping the alignment with some of the most relevant architectural approaches existing at that time, fostering interoperability and allowing architecturally different vehicles to perform driving manoeuvres safely in cooperation with each other. These results would contribute to improve driving safety as well as traffic efficiency in Cybernetic Transport Systems deployed in dedicated environments such as city centres, airports or theme parks.

Shortly after the end of Cybercars-2, the E-FRAME project started, which contributed to extending the FRAME architecture to include Cooperative Systems by collating system requirements from key projects in the field into a set of requirements in the format of FRAME User Needs. SAFESPOT, the main architectural reference for Cybercars-2, was one of the selected projects.

1.4.2 MoveUs Cloud-Based Platform Architecture

MoveUs (ICT Cloud-Based Platform and Mobility Services: Available, Universal and Safe for All Users) is a project funded by the European Commission under the Seventh

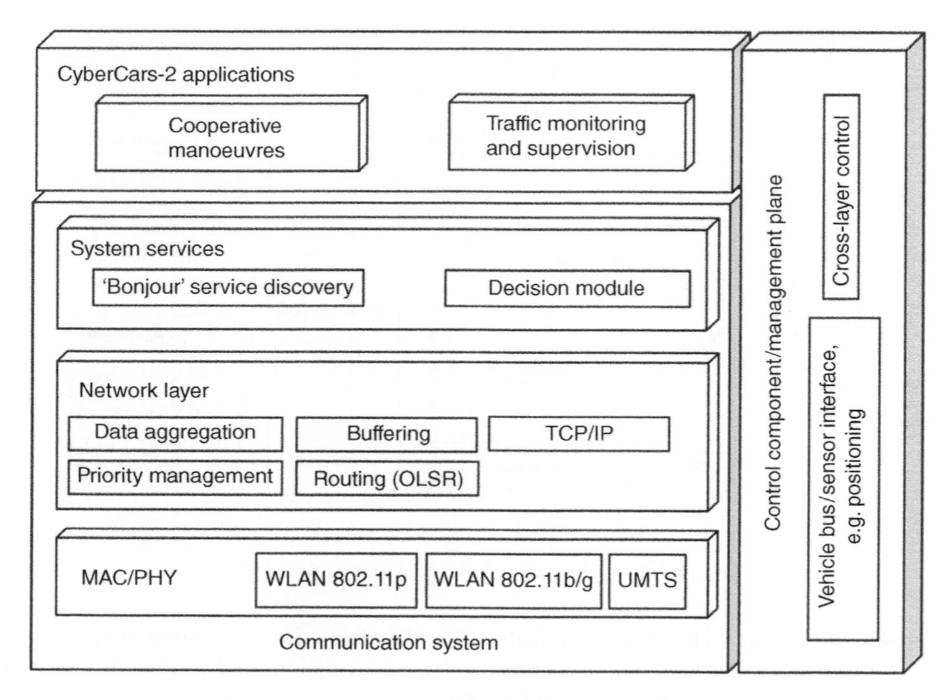

Figure 1.3 Cybercars-2 communications architecture [19].

Framework Programme between 2013 and 2016 in the context of Smart Cities. It aims to benefit from the huge potential of combined ITS and ICT to radically change the European users' mobility habits by offering intelligent and personalized travel information services, helping people to decide the best transport choice and providing meaningful feedback on the energy efficiency savings obtained as a result. Recommendations supported by incentives will be provided to foster 'soft' mobility modes and the use of shared and public transport modes (buses).

Information from a wide variety of transport modes and mobility systems such as public buses, car/bike sharing systems, traffic management systems, equipped vehicles to measure traffic density, and users' smartphones will be integrated and processed in a cloud-based, high-capacity computing platform that will allow to measure 'the pulse of urban mobility' from a global perspective; to obtain valuable information on how traffic density evolves and how public transport is used; and to learn how individual users can move along the city in a more eco-friendly way, thus improving energy efficiency.

The project has a pan-European approach, engaging three different smart-city pilots placed in Madrid, Tampere and Genoa, where the platform and the personalized mobility services will be deployed and tested in 2016.

In view of the project's ambitious goals, the definition of an architecture supporting the cloud platform operation and the delivery of mobility services envisaged, with a predominant role of ITS, entailed a huge effort. The main objective was to define a comprehensive architecture for the MoveUs cloud-based platform, the specifications of its core facilities and the high- level interfaces between the different platform components. The methodology adopted to define the specifications of this platform integrated insights and approaches from different engineering fields: traffic engineering, data analytics, software architecting and cloud-based computation deployment. Each view focused or emphasized specific aspects: functionality, performance and openness, allowing third parties access to exploit the information and extend the platform. As key challenges, the demanding real-time requirements, the vast amount of data to be handled, the scalability and availability needs and finally, the compliance of security/privacy normative are worth mentioning.

The concept of ITS reference architecture was present during this work by analysing existing solutions and by adopting and adapting selected best-practices and knowledge. The advantage of using a standard or de-facto reference architecture as a basis relies on interoperability, avoidance of vendor lock-in situations and in general, on the consistency of information delivered to end users while manufacturers and designers focus on added value (design optimization) aspects. For that reason, FRAME was initially chosen as ITS Reference Architecture in the context of MoveUs.

FRAME defines a methodology which starts from a well-defined set of functions, users, datasets and ends in a set of data stores and data flows ensuring consistent links. The FRAME Browsing and Selection tools were used by the three city pilot teams and worked fine at some stages of the process. However, some divergences identified between the common needs and requirements considered by FRAME and some of the peculiarities offered in MoveUs (e.g. incentive management, energy efficiency assessment and service customization), not explicitly addressed before, forced us to slightly adapt the methodology workflow and introduce alternative tools such as Enterprise Architect to produce the Data Flow Diagrams (DFD). Excel spreadsheets were also produced to handle a complete definition of requirements and functions in the three city pilots. Finally, the project extended this basis with the introduction of a service view, a view on the mobility-domain in terms of services linked to the functions that those services perform.

Hence, the final methodology adopted for the MoveUs platform architecture design covered the following steps:

1. Analysis of key references in the domain. ITS reference architectures and R&D projects with similar challenges were studied and their commonalities and differences with MoveUs identified.
2. Analysis of MoveUs requirements and use cases (producing Sequence Diagrams and Forms).
3. MoveUs Functional Architecture definition (Data Flow Diagrams). A sample is illustrated in Figure 1.4.
   a. elicitation of user needs per city pilot (according to FRAME methodology);
   b. elicitation of terminals (actors and external systems) per city pilot (according to FRAME methodology);
   c. selection of candidate functionalities (supported by FRAME tools);
   d. decision on what can be re-used; alignment and identification of remaining gaps. Identification of functions, data stores, data flows and terminators.

   At this stage, in order to ensure a common functional view of the whole system between different project tasks running at the same time, the resulting functionalities were mapped into MoveUs high-level functions.
4. Definition of the MoveUs platform Service Viewpoint.

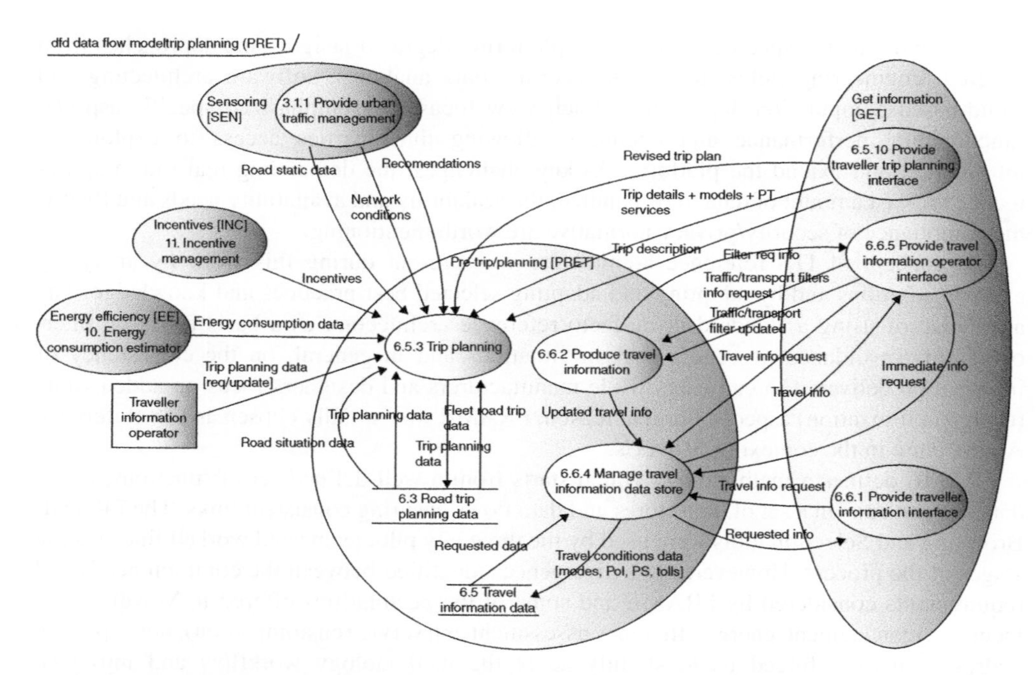

Figure 1.4 Sample of data flow diagram produced for MoveUs trip planning functionality [20].

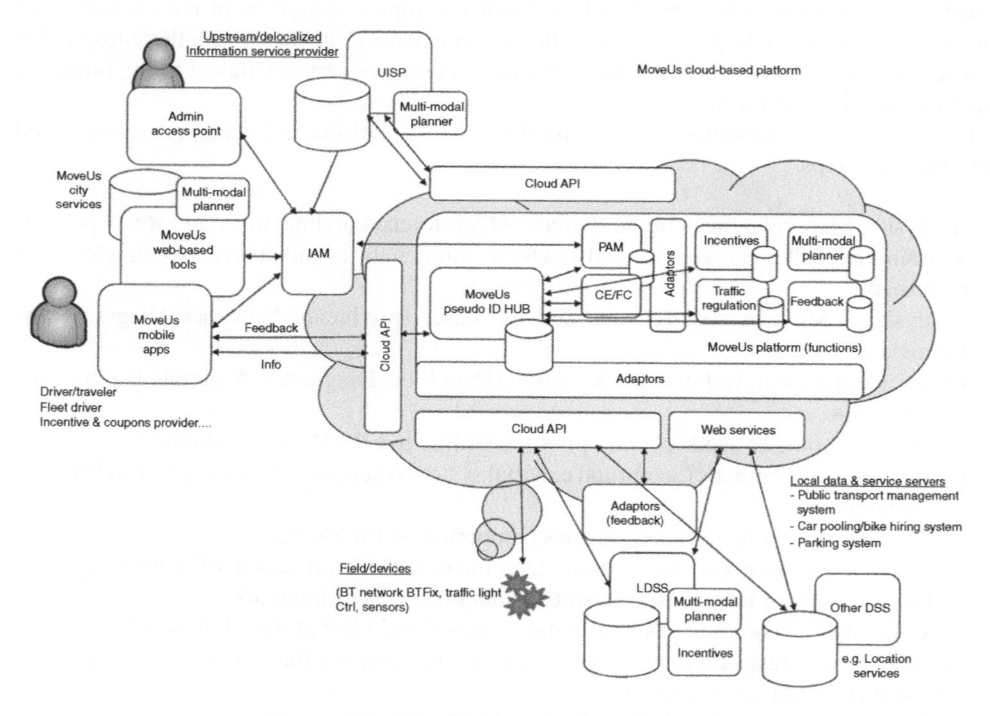

Figure 1.5 MoveUs operational scenario [20].

5. Technical/Application architecture definition (comprising subsystems and modules).
   a. subsystems containing elements of the functional architecture (functions, data stores) and communicating with another subsystem or terminators;
   b. modules; aggregation of related functions.
6. Mapping functionalities/services vs. modules.
7. Detailed functional description for each module.
8. Physical/deployment viewpoint definition.
9. Cloud-based platform definition.

At the time of writing, the architecture design for the MoveUs platform has not been fully completed yet, though the final steps are being undertaken. The process has remained technologically neutral so far and will produce its final results at the beginning of 2015, paving the way for the implementation phase. The current MoveUs operational scenario (a high-level architectural view), showing the main actors and roles, architecture elements and information flows is illustrated in Figure 1.5.

References

1. FRAME Architecture Resource Centre: http://www.frame-online.net (last accessed 23 April 2015).
2. i2010 Intelligent Car Initiative: http://europa.eu/legislation_summaries/information_society/other_policies/l31103_en.htm (last accessed 23 April 2015).
3. iMobility Forum: http://www.imobilitysupport.eu/imobility-forum (last accessed 23 April 2015).
4. Car2Car Communication Consortium (C2C-CC): https://www.car-2-car.org (last accessed 23 April 2015).
5. The Amsterdam Group: www.amsterdamgroup.eu (last accessed 23 April 2015).
6. FP6 Integrated Project CVIS: http://www.cvisproject.org (last accessed 23 April 2015).
7. FP6 Integrated Project SAFESPOT: http://www.safespot-eu.org (last accessed 23 April 2015).
8. FP6 Integrated Project COOPERS: http://www.coopers-ip.eu (last accessed 23 April 2015).
9. FP7 Project GeoNet: https://team.inria.fr/rits/projet/geonet (last accessed 23 April 2015).
10. FP7 Project PreDrive C2X: http://www.transport-research.info/web/projects/project_details.cfm?id=44542 (last accessed 23 April 2015).
11. CIP-PSP Pilot Project COMPASS4D: http://www.compass4d.eu (last accessed 23 April 2015).
12. CIP-PSP Pilot Project CO-GISTICS: http://cogistics.eu (last accessed 23 April 2015).
13. COM(2008) 886: Action Plan for the Deployment of Intelligent Transport Systems in Europe. Available at: http://eur-lex.europa.eu/legal-content/EN/TXT/?uri=CELEX:52008DC0886 (last accessed 23 April 2015).
14. Directive 2010/40/EU of the European Parliament and of the Council of 7 July 2010 on the framework for the deployment of Intelligent Transport Systems in the field of road transport and for interfaces with other modes of transport. Available at: http://eur-lex.europa.eu/legal-content/EN/TXT/?uri=CELEX:32010L0040 (last accessed 23 April 2015).
15. Mandate M/453 Standardisation Mandate addressed to CEN, CENELEC and ETSI in the field of Information and Communication Technologies to support the interoperability of Cooperative Systems for Intelligent Transport in the European Community. Available at: http://ec.europa.eu/enterprise/sectors/ict/files/standardisation_mandate_en.pdf (last accessed 23 April 2015).
16. COMeSafety initiative: http://www.comesafety.org (last accessed 23 April 2015).
17. European ITS Communication Architecture – Overall Framework – Proof of Concept Implementation, version 3.0.
18. ETSI/CEN Standards for Cooperative Intelligent Transport Systems. Available at: http://www.etsi.org/technologies-clusters/technologies/intelligent-transport (last accessed 23 April 2015).
19. FP6 Project Cybercars-2 (Grant Agreement No. 028062) – Deliverable D1.1: Architecture of the Cooperative Cybernetics Transport System.
20. FP7 Project MoveUs (Grant Agreement No. 608885) – Deliverable D3.2.1: MoveUs cloud-based platform: specification and architecture.

2

Architecture Reference of ITS in the USA

Clifford D. Heise
Iteris, Inc., Sterling, VA, USA

2.1 Introduction

Intelligent Transportation Systems (ITS) involve the integration of numerous transportation-related systems or subsystems. These systems must work together in an integrated and cooperative manner in order for the benefits of ITS to be realized. In 1993, the United States embarked on the development of a National ITS Architecture to guide the United States federal government's investment in research, interface standards development, and state/local ITS deployment. This National ITS Architecture was defined following a systems engineering process and based on stakeholder views of ITS services and capabilities, called ITS user services. The National ITS Architecture in the United States has evolved as ITS has matured and it is used to plan and guide ITS development and deployment throughout the country. It has gained increasing acceptance as ITS benefits have been experienced. The United States Department of Transportation has established regulations regarding its use for federally funded ITS projects. The Department of Transportation is expanding the National ITS Architecture to include support for the connected vehicle environment and the evolving research being done within that community. This chapter will review the origins of the National ITS Architecture in the United States, the basis of its definition, the impact it has had on ITS development in the United States at the federal, state and local levels as well as on major ITS initiatives, and the environment it creates to afford evolution as ITS and Connected Vehicle technologies mature.

Intelligent Transport Systems: Technologies and Applications, First Edition. Asier Perallos, Unai Hernandez-Jayo, Enrique Onieva and Ignacio Julio García-Zuazola.

2.2 National ITS Architecture in the USA

The National ITS Architecture in the United States is a functional representation of ITS. It provides a framework that is available to any stakeholder who is interested in planning, developing and deploying ITS to address the transportation needs of their state, region, or locality. The National ITS Architecture has two primary components that form the basis for supporting features. The first is a set of functions and data exchanges or data flows that represent "what" must be done to provide ITS services. There are 512 specific functions or process specifications defined in the architecture and they are connected together by 5857 data flows. The functions ingest data from other functions or from terminators (architecture components that are on the boundary of the ITS domain), process that data based on the function's purpose, and output the processed data to other functions or terminators.

With the functional definition, also referred to as the Logical Architecture, in place, the functions are grouped into physical collections reflecting what is common in the transportation world. Functions related to traffic management are grouped together, for instance. The physical groupings are called subsystems. When the functions are assigned to a subsystem, their data flows remain connected as they are in the Logical Architecture. The data flows that cross subsystem boundaries are called architecture flows which are interface considerations for standards development. The grouping or partitioning of functions into subsystems along with the architecture flows that connect the subsystems constitute the Physical Architecture.

Upon this foundation of Logical and Physical Architectures, several supporting components of the National ITS Architecture were developed. These "tools" of the National ITS Architecture provide the information that guides standards development and ITS project planning and development. The United States Department of Transportation funds and manages the National ITS Architecture Program that maintains the Architecture definition and provides stakeholder support for its application. The remainder of this chapter will explore the details of the National ITS Architecture, its use, and the stakeholder tools that are available.

2.3 Origins of ITS Architecture in the USA

In the early 1990s as ITS, then called Intelligent Vehicle-Highway Systems or IVHS, was initially being established as an emerging concept for improving transportation management and services, it was understood that system interfaces would be critical to its successful implementation. The benefits of ITS were to be realized only if information was shared among systems. That required integration to a level of information exchange.

In the United States, the federal government funds transportation projects through periodic legislation, specifically a highway bill. The Intermodal Surface Transportation Efficiency Act of 1991 (ISTEA) passed by the United States Congress in 1991, directed the United States Secretary of Transportation to establish standards that would support the widespread use of ITS.

ITS is the use of computer and communications technologies to collect transportation data, to process that data into useful information for transportation decisions concerning travel, transportation system management, emergency response, and operations, and to disseminate that information to travellers, system operators, emergency responders, and operations personnel. To efficiently exchange information among the various systems, interfaces would be required and the interfaces would need some level of standardization. Standards are very important, not only for information exchange among the systems involved, but to establish common interface requirements so that the limited funding available can be invested efficiently with lower risk.

To identify the interfaces for standardization, a functional-level system architecture was required to inform the Department of Transportation about the interfaces and the requirements that each interface would fulfil to support ITS. This was the point at which the National ITS Architecture development was initiated.

The Department of Transportation procured four separate teams made up of private industry consultants, universities, and public agencies to competitively develop four architecture definitions. The four architectures were different in many ways. Each team developed their architectures separately but was required to show their work in periodic stakeholder workshops in which their approaches were revealed and comments gathered from the stakeholders for incorporation. At the end of the procurement, the four architectures were evaluated by the Department of Transportation and two architectures were selected to be further refined into one National ITS Architecture. The initial version of the National ITS Architecture was released in 1996.

It was very important to the United States Department of Transportation to have stakeholder involvement in the development of the National ITS Architecture. The competitive development approach that was followed provided every opportunity possible for stakeholder engagement and input. The same emphasis on stakeholder feedback and consensus exists today as the National ITS Architecture evolves with ITS deployment, standards development, and new transportation research initiatives. Since its initial release in 1996, the National ITS Architecture definition has evolved through six major upgrades and several minor updates to Version 7.0 in 2013. The National ITS Architecture has been evolved and maintained by Iteris, Inc. under contract to the United States Department of Transportation since 1996.

2.4 US National ITS Architecture Definition

ITS is broad in scope and the National ITS Architecture development required a structured process to properly define the functionality and information exchanges needed to support it.

2.4.1 The Development Process

The development of the United States National ITS Architecture followed a systems engineering process. The general process steps followed are illustrated in Figure 2.1. The systems engineering process starts on a conceptual foundation referred to as a concept of operations which narratively describes the vision for what the system will ultimately do and how it will operate. This allows the owner or stakeholders to describe what they expect of the system without designing how it will be accomplished. From the concept of operations, system requirements are derived. In the case of the National ITS Architecture, these requirements

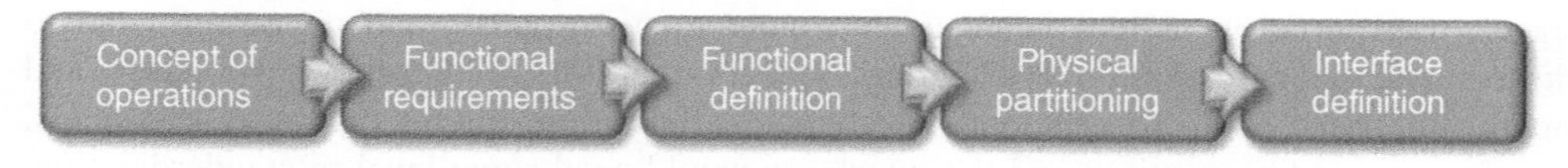

Figure 2.1 Systems engineering process steps for National ITS architecture development.

were functional in nature. They described “what” the system would do and did not include performance or environmental requirements. The functional requirements were used to define functions that would satisfy those requirements. The functions describe actions that are taken within the ITS environment. Data collected for the action and distributed as a result of the action described by the function were defined as data flows. The functions and data flows collectively formed a functional architecture.

In the transportation environment, the functions do not exist separately. Several functions will be performed within components of a deployed system. The interfaces between the system components provide the basis for ITS standards. System interface definitions drive standards development and need a physical reference point to which the eventual standard will apply. Interface standards are applied to interfaces between physical systems such as between a field device and a transportation center that controls and accesses information to and from that device. Therefore, the next step in the development of the National ITS Architecture was to partition the functions and data flows into physical entities called subsystems that would be commonly understood in the transportation environment. By grouping or partitioning the functions, the data flows that connected one function to another would either be inside a physical partition or cross the physical partition to a function in another physical entity/subsystem. The data flows that crossed over physical partitions became the interface definitions that would inform the standards development activities as illustrated in Figure 2.2 by the data flows grouped in the dashed circles between the physical entities. The interfaces

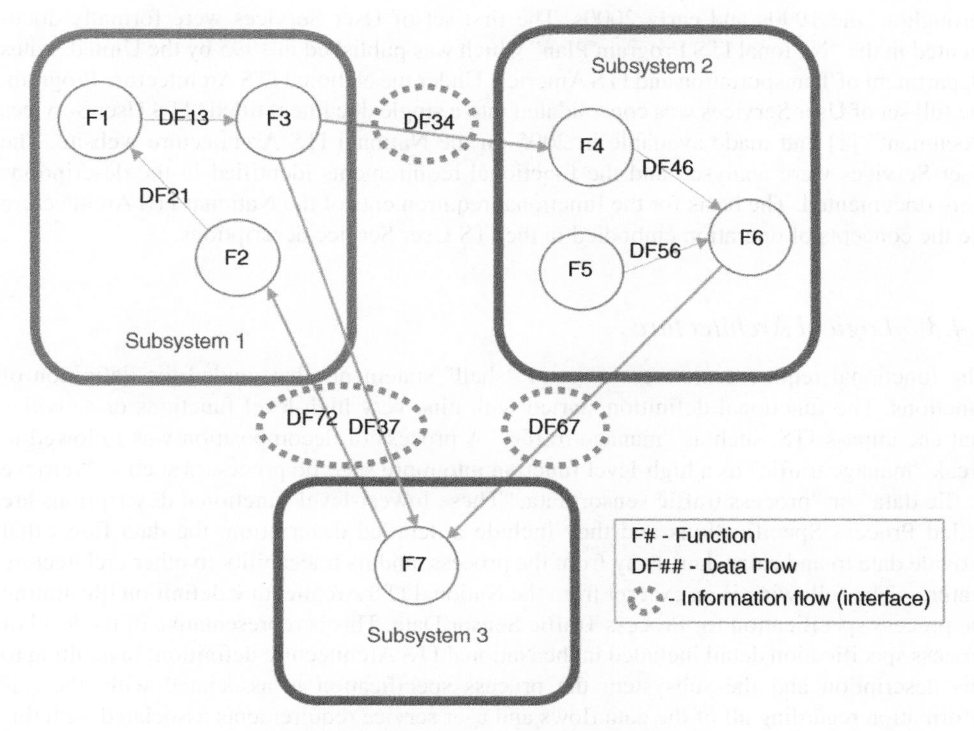

Figure 2.2 Physical partitioning of functions.

between the subsystems resulting from the partitioning became known as architecture flows or information flows. Essentially information flows are collections of data flows that cross subsystem boundaries.

The general process described above provides an overview of the process steps taken to develop the National ITS Architecture. It is important to also understand some of the process details and to become familiar with some of the other architecture features that were developed and ultimately supported the standards development. In addition, these same features supported the use of the National ITS Architecture in the ITS planning and development activities of state and local stakeholders in the United States.

2.4.2 User Services

The concept of operations, the first process step in Figure 2.1, took the form of a set of ITS User Services that were developed outside of the National ITS Architecture Program. The User Services are stakeholder views of ITS. They describe what systems (vehicles, information services, traffic management centers, cameras, etc.) would be available and what those systems would do to deliver services to a traveller, transportation manager, or transportation system user. The original National ITS Architecture development process was based on 29 User Services addressing views such as En-Route Driver Information, Traffic Control, Electronic Payment Services, and Public Transportation Management. The original 29 User Services were eventually expanded to 33 User Services as ITS was defined in more detail throughout the 1990s and early 2000s. The first set of User Services were formally documented in the "National ITS Program Plan" which was published in 1995 by the United States Department of Transportation and ITS America. Under the National ITS Architecture Program, the full set of User Services was consolidated into a single document titled "ITS User Services Document" [1] and made available in 2005 on the National ITS Architecture website. The User Services were analysed and the functional requirements identified in the descriptions were documented. The basis for the functional requirements of the National ITS Architecture are the concepts of operation embodied in the ITS User Service descriptions.

2.4.3 Logical Architecture

The functional requirements were a set of "shall" statements that guided the definition of functions. The functional definition started with nine very high-level functions or activities that encompass ITS, such as "manage traffic." A process of decomposition was followed to break "manage traffic" as a high-level function into more specific processes such as "retrieve traffic data" or "process traffic sensor data." These lowest-level functional descriptions are called Process Specifications and they include a detailed description, the data flows that provide data to and carry data away from the process, and its traceability to other architecture features. The following is an excerpt from the National ITS Architecture definition illustrating the process specification for Process Traffic Sensor Data. This is representative of the level of process specification detail included in the National ITS Architecture definition. In addition to this description and the subsystem the process specification is associated with, there is information regarding all of the data flows and user service requirements associated with this specification.

Process Traffic Sensor Data (Pspec) [2]

This process shall be responsible for collecting traffic sensor data. This data shall include traffic parameters such as speed, volume, and occupancy, as well as video images of the traffic. The process shall collect pedestrian images and pedestrian sensor data. The process shall collect reversible lane, multimodal crossing and high occupancy vehicle (HOV)/high occupancy toll (HOT) lane sensor data. Where any of the data is provided in analog form, the process shall be responsible for converting it into digital form and calibrating. The converted data shall be sent to other processes for distribution, further analysis and storage. The process shall accept inputs to control the sensors and return operational status (state of the sensor device, configuration, and fault data) to the controlling process.

This Pspec is associated with the Roadway Subsystem.

The functional definition of the National ITS Architecture is referred to as the Logical Architecture. It also includes a set of boundary entities that provide a connection with the non-ITS environment. These boundary entities are called "terminators." The National ITS Architecture does not describe functionality within the terminators. It only provides data flows to and from the terminators, indicating data exchange with systems such as financial institutions whose functionality is not in the ITS environment but interacts with the ITS environment for functions that involve financial transactions like tolling.

2.4.4 Physical Architecture

ITS is a collection of systems, not one very large system. The Logical Architecture represents the functionality required to deliver the User Services. In the transportation environment, ITS is implemented as numerous systems that are owned and operated by different stakeholders depending on their responsibility (transit, traffic, commercial vehicle, emergency management, etc.). To properly represent ITS, the functions of the Logical Architecture are partitioned into groups of functions called subsystems that are relatable to a particular domain such as traffic management. The interfaces that exist between these partitioned groupings consist of the data flows between the functions that are separated by the partitions. The interfaces between these physical subsystems define where standards are required.

Figure 2.2 illustrates the concept of the functional partitioning. The data flows between functions are retained but where those data flows cross a partition, the data flows contribute to interface definitions. The partitions are known as subsystems and groups of data flows between the subsystems are called architecture flows or information flows. The same terminators that exist in the Logical Architecture exist in the physical partitioning as well. The subsystems, information flows and terminators together represent the Physical Architecture.

The following is an excerpt from the National ITS Architecture as an example of two architecture flows related to the Roadway subsystem to illustrate the level of detail provided within the Physical Architecture and the relationship between the architecture flows and the data flows each of them envelop.

Roadway (Subsystem) [3]

Inputs/outputs: Architecture Flows and Data Flows

- Traffic flow (Architecture Flow) – Raw and/or processed traffic detector data which allows derivation of traffic flow variables (e.g., speed, volume, and density measures) and associated information (e.g., congestion, potential incidents). This flow includes the traffic data and the operational status of the traffic detectors.
  Data Flows:
  - incident_analysis_data
  - traffic_image_data
  - traffic_sensor_data
  - traffic_sensor_status
- Traffic images (Architecture Flow) – High fidelity, real-time traffic images suitable for surveillance monitoring by the operator or for use in machine vision applications.
  Data Flows:
  - dynamic_lane_video_image
  - incident_video_image
  - traffic_video_image
  - traffic_video_image_for_display
  - video_device_status

The information flows, supported by the details of the data flows that they envelop, inform the development of ITS standards and are kept in alignment with each other as standards are defined and new information is discovered in the standards development process. The partitioning is necessary to inform the standards development process and does not restrict particular functionality allocation when translated to real deployment. It was understood by the development team that in some cases functions allocated to one subsystem domain such as emergency management might also be implemented in other subsystems in a real deployment such as traffic management. No matter the subsystem in which a function is deployed, the same ITS standard to which it is related can be applied to the interface of the deployed subsystem. The standards convey the flexibility needed. The subsystems of the National ITS Architecture provide a guide to how deployment might be planned and implemented.

Not all functions in a subsystem are required for the subsystem to provide capabilities that a user or stakeholder may be seeking. For instance, if a user is only interested in collecting data from roadway sensors for use in traffic management, there are functions allocated to the Roadway subsystem that are not required for that data collection to occur. The Physical Architecture was further refined to group related functionality within a subsystem to reflect commonly deployed capabilities. These groupings are called Equipment Packages. An example of an Equipment Package is Roadway Basic Surveillance which includes the Process Traffic Sensor Data and the Process Traffic Images process specifications. There are 63 process specifications allocated to the Roadway subsystem but these two are all that is needed to monitor traffic conditions using fixed equipment such as loop detectors and CCTV cameras [4]. Equipment packages make it easier for a user to select the functionality in a subsystem that meets their needs.

The Physical Architecture is the most frequently referenced component of the National ITS Architecture because its definition is at a level equivalent to systems with which stakeholders are familiar. The subsystems described in the Physical Architecture represent systems such as Traffic Management, Transit Vehicle, and Roadway equipment which stakeholders deploy and manage within their transportation environments.

The Logical and Physical Architectures provide the fundamental definition of the National ITS Architecture in the United States and they are contained in Microsoft Access databases. From this foundation, other valuable components of the National ITS Architecture were developed to support standards, planning and development of ITS across the United States.

It should be noted that there is traceability among all of the components within the National ITS Architecture. User Services, User Service Requirements, Process Specifications, Data Flows, Subsystems, Architecture Flows, and Services (discussed in the next section) are all linked with references to the source or a component in the related artefact. For instance, each process specification is linked to one or more user service requirements that it addresses. The process specification is then linked to the subsystem that it is assigned to and to the data flows to which it is related. This traceability ultimately provides the connectivity that drives the National ITS Architecture website providing a hyperlinked site for a user to explore all features of the architecture related to their area of interest.

2.4.5 Services

Each subsystem by itself only offers the functionality that is contained within it. Transportation services are provided by linking functionality from multiple subsystems to provide benefit to a stakeholder, traveller, or user of ITS. For instance, Figure 2.3 illustrates the Network Surveillance Service from the National ITS Architecture. The larger purple boxes are subsystems (Traffic Management and Roadway). The Traffic Management subsystem cannot provide surveillance of the network by itself. It requires information to be collected from the roadway in the form of camera images and traffic sensors in order to process a status of the transportation network for use in other services such as traffic control or to pass along to a traveller information service provider. The service packages described in the National ITS Architecture are collections of subsystems and information flows that together provide service benefits. The specific functionality required to provide a service is defined by the equipment packages in each subsystem. In Figure 2.3, the equipment packages are the smaller white boxes inside of each subsystem. In the case of the Roadway subsystem, the Roadway Basic Surveillance equipment package collects the data needed from devices at the roadway for the service.

To stakeholders and users of the National ITS Architecture, the service packages provide access to the pieces of the architecture that meet their needs. The Physical Architecture has 22 subsystems and 535 architecture flows. Service packages allow a stakeholder or user to look at the architecture detail from a higher point and quickly identify the components that are most applicable to their needs.

2.4.6 Standards Mapping

The original intent of the National ITS Architecture was to guide ITS standards development and maintain a tight linkage between the architecture content and the related standards. The linkage between the National ITS Architecture and ITS Standards allows stakeholders to

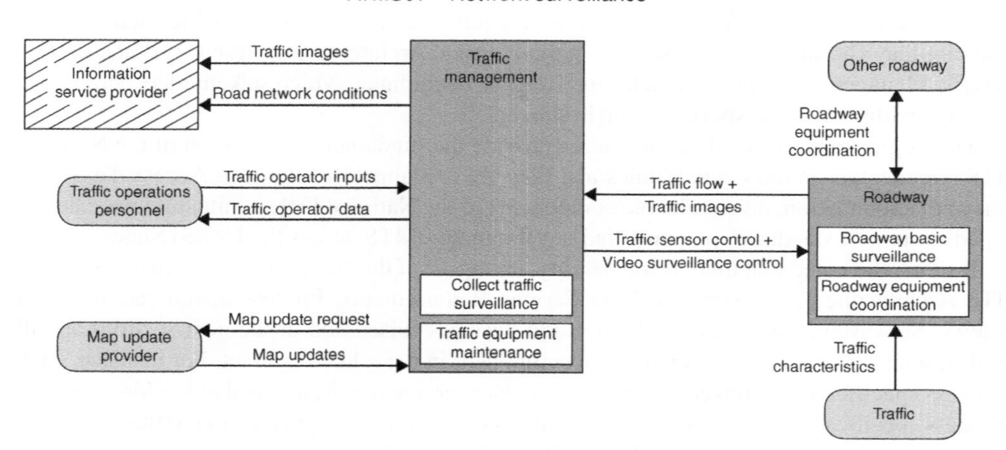

Figure 2.3 ITS service: network surveillance [5].

relate what is of interest to them in the architecture with their plans for ITS projects and to inform their project specifications with the details of the ITS standards. Standards promote an open environment for technology application and system development while supporting interoperability. The information flows in the National ITS Architecture have been analysed and are monitored against the current ITS standards and those in development to maintain a mapping of the information flows to the ITS standards that address those flows. Not every information flow is being standardized. For example, interfaces were analysed based on their criticality to interoperability of the subsystems they support and the commonality of the interface among multiple subsystems. In version 7.0 of the National ITS Architecture, 178 of the 535 information flows are mapped and tracked to ITS Standards.

As the architecture content is explored by a stakeholder and information flows that are linked to standards are identified, more information about that interface can be found within the referenced standard. The standards provide the interested stakeholder more detailed interface definition to support ITS implementation.

2.5 Impact on ITS Development in USA

The starting point of the National ITS Architecture Program that is in place today at the United States Department of Transportation was the ISTEA legislation that directed the development of ITS standards. The United States Department of Transportation took action to develop the National ITS Architecture to support standards planning and development. The program produced the initial version of the National ITS Architecture in 1996 and has maintained the architecture content since that time, resulting in multiple updates of the content to version 7.0 in 2013. The program has involved more than maintaining the architecture content. It supports the stakeholder community in their application of the National ITS Architecture with technical support, outreach, training, facilitated workshops, and guidance. The National ITS Architecture Program has been supported by Iteris, Inc. under contract to the United States Department of Transportation since 1996.

2.5.1 Architecture and Standards Regulation

The United States Department of Transportation has encouraged the use of the National ITS Architecture beyond its use as an ITS standards development guide. The architecture presents a functional representation of ITS providing stakeholders with a reference from which to plan and develop ITS in their state, region, or local environment. In the United States, federal funding is provided for transportation system development and maintenance. To manage the funding distributions, Metropolitan Planning Organizations (MPO) were established for areas with populations greater than 50 000. MPOs are typically not deployment or operating agencies. They manage federal funding for their state, metropolitan, and local members. They work with these members to plan transportation programs to address the needs of their region. ITS presents a different transportation solution than traditional infrastructure implementations such as roads and bridges.

ITS involves technology which evolves quickly. In order to incorporate ITS into the traditional transportation planning process that uses a 3-year horizon for the near-term plan and a 20-year horizon for the long-term plan, technology presented an obsolescence issue. However, the National ITS Architecture is a functional representation of ITS. By basing its definition in functions and their corresponding data exchanges, the architecture can be used as a reference that spans any time horizon. For instance, as illustrated in Figure 2.3, the functionality of Roadway Basic Surveillance that exists in the Roadway Subsystem in the Network Surveillance Service Package provides information to the Traffic Management Subsystem regarding Traffic Flow and Traffic Images. To provide the needed information, the Roadway Basic Surveillance functionality must Process Traffic Sensor Data and Process Traffic Images. The sensor data being collected can come from many different types of technology such as inductive loops, video detection, microwave detectors or acoustic sensors. If specific technology is used to define the implementation and the project is not deployed immediately, the technology may be obsolete if too much time passes.

If a stakeholder in a region defines a need for gathering information from the roadway and they define their solution in functional terms, such as provided by the National ITS Architecture, they can define a project in functional terms, place it in the planning process in the near- or long-term planning horizon and it will remain relevant in the future when the funding and priority of the project has elevated it to a deployment status in the plan.

By tying the project definition to a functional reference, such as the National ITS Architecture, the project is valid in any timeframe. If the stakeholder ties their project to a specific technology and that technology becomes obsolete, it is likely that the project is no longer valid at that time and its impact as a solution has been compromised.

In 2005, the United States Department of Transportation put in place a regulation requiring the development of an ITS architecture for any region planning the deployment of ITS projects using federal funding. The purpose of this regulation was to encourage MPOs, state and local agencies to plan their ITS projects so that they took into account the integration opportunities available with ITS and to properly coordinate those integration opportunities among the involved stakeholders. This would result in more efficient use of federal funding and facilitate more successful ITS project implementation. The so-called "regional ITS architectures" were developed from the National ITS Architecture as a reference which supported conformance between the region's ITS architecture and the National ITS Architecture. The regulation also required conformance with ITS standards. With all ITS architectures using the same reference point, ITS projects will have similar interface definitions and standards references broadening the available solutions and creating a broader supplier base for those implementations.

The impact of the regulation resulted in approximately 300 ITS architectures being developed across the United States to be used to guide ITS planning and project development.

2.5.2 *ITS Planning*

Transportation projects generally require large funding allocations due to the infrastructure and capital investments of roadway, bridges, and rolling stock that constitute transportation facilities and assets. In the United States, transportation planning is accomplished in long-term (10- or 20-year time horizons) and short-term (3-year time horizon). The ITS architecture for a particular region is useful in both planning timeframes.

In the planning process, projects are defined to address a transportation need. In the long-term timeframe, the ITS project definitions are generally high-level and conceptual in nature but it is important to understand how one project is related to other planned projects or to existing ITS systems within the planning region. An ITS architecture for the planning region allows for project definition at a functional level that shows how a future project relates/interfaces with current and other future projects. When the ITS architecture is defined, the stakeholders include the systems they operate today as well as the systems or services they envision in the future to address existing or anticipated transportation needs.

ITS architectures distinguish what exists currently from what is planned or future. As ITS projects are deployed, the status of the project elements are changed from planned to existing status. This allows for projects in the planning process to be properly scoped and to improve cost estimates for those projects based on the elements needed to address its scope. By defining ITS projects in compliance with their ITS architecture, the stakeholders have an understanding of the dependencies among their projects today as well as in the future. Technology choices are delayed until project development is ready to get underway.

Near-term planning at the state, regional (MPO), and local levels is often documented in the Transportation Improvement Plan (TIP). The TIP has a 3-year time horizon so it is more specific and financially constrained. The scope of each ITS project in the TIP needs to be detailed to support an understanding of the integration opportunities with other systems or projects in the region and to support an accurate cost estimate for the project. A project defined in conformance with the ITS architecture in the region will address these needs. For instance, in a metropolitan environment, the transit authority may express the need for better on-time performance of their bus service. After some research they determine that transit signal priority is a solution that they wish to implement. Unfortunately, they may be unable to implement the solution due to differing project priorities at the City's department of transportation which controls and maintains the traffic signal system. The project is put into the Long-Range Plan and eventually is moved to the TIP where it is allocated funding and priority in the installation schedules for the transit authority and the City department of transportation. The project architecture in Figure 2.4 is developed directly from the regional ITS architecture in which the City department of transportation and the transit authority are stakeholders.

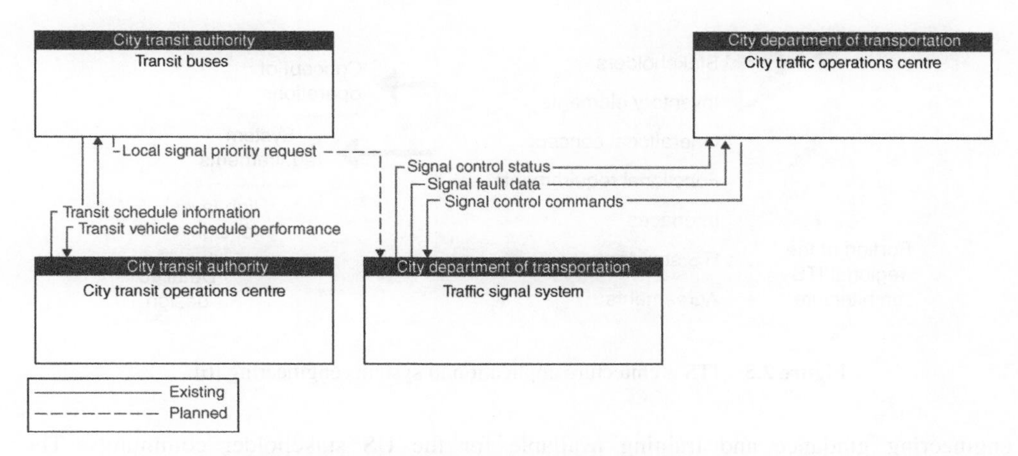

Figure 2.4 Transit Signal Priority project architecture example.

The project architecture illustrates what exists (solid line flows) and what is planned (dashed line flow). This informs stakeholders about what interface and functionality the project will be implementing and what other interfaces may be impacted in the event the new interface imposes further requirements beyond the initial interface. This provides the planner with an understanding of the scope of the project, other interfaces or systems that may be impacted, and the stakeholders that are involved.

In this example, no technology has been declared; only the information exchanges and functionality needed to drive those interfaces. When the project emerges from the planning process as a funded project, the development process including technology selection will be made based on a systems engineering analysis. If a technology solution had been determined when the project was initially planned, possibly 5 or more years before design took place, the technology may have been so obsolete that budget may not be available for newer technology or more effective solutions may be available that this project would not be able to take advantage of. By delaying the technology choice until the project emerges from the planning process, the best, most affordable and advantageous technology solution can be selected when the project is ready to move forward to implementation.

The ITS architecture is a tool for stakeholders to functionally characterize and document the projects they need in the planning process to the level of detail needed to effectively scope, estimate cost, and determine interdependencies that need to be resolved or taken into account. The architecture lowers the risk of project unknowns in these early stages of project evolution.

2.5.3 *ITS Project Development*

When an ITS project emerges from the planning process and enters development and implementation, the ITS architecture is again a valuable reference tool. From the project architecture used in the planning process, further information can be made available to support project development in the systems engineering process. The United States Department of Transportation supports the use of systems engineering for all ITS projects and makes systems

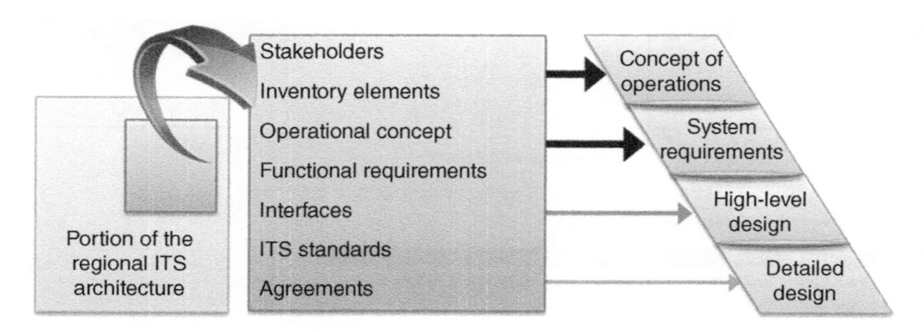

Figure 2.5 ITS architecture application in systems engineering [6].

engineering guidance and training available for the US stakeholder community. The Department has developed Systems Engineering Guidance for ITS project development which is available on the National ITS Architecture website which can be accessed through http://www.its.dot.gov/arch/index.htm. Figure 2.5 illustrates the portions of the systems engineering process which benefit most from the use of ITS architecture in project development.

The ITS architecture can be used in the development of concepts of operation, system requirements, high-level design, and detailed design.

- Concepts of Operation: The project architecture provides a basis for a description of how the system implemented with the project will operate. A narrative can be developed which describes the functions of each subsystem and the resulting movement of information from one subsystem to the next. Stakeholders assigned as owners of each subsystem have roles to play in the operation of their subsystems. This provides a vision for the system operation highlighting the dependencies among the subsystems and, therefore, the stakeholders involved.
- System Requirements: The functional requirements of each subsystem in the project architecture provide an initial set of requirements for the systems engineering process step. Requirements derivation from the concept of operation is a key source of requirements but functional requirements are readily available in the architecture to augment this activity. In the example of the transit signal priority project, the functional requirements associated with the transit signal priority functions in the transit and signal system elements can be used as the starting point for requirements analysis. Table 2.1 illustrates the functional requirements available from the National ITS Architecture for each subsystem in the example project.
- High-Level Design: The subsystems of the ITS Architecture provide a starting point for system block diagrams that can be further refined along with interface relationships among the subsystems. The functionality that underlies each subsystem offers high-level functionality that can be allocated among the high-level design components.
- Detailed Design: At this level the ITS architecture contributes standards mapping from the information flows in the architecture to applicable standards. The standards provide specification details that will support detailed design with interface specific information. Table 2.2 lists the standards related to the transit signal priority service that would be considered during detailed design.

Table 2.1 Functional requirements from ITS architecture.

Element name	Subsystem	Functional requirement
City Transit Operations Center	Transit Management	The center shall analyze transit vehicle schedule performance to determine the need for priority along certain routes or at certain intersections.
City Transit Operations Center	Transit Management	The center shall define business rules that govern use of transit vehicle signal priority, communicate these rules to the transit vehicle, and monitor transit vehicle requests for priority at signalized intersections.
City Transit Operations Center	Transit Management	The center shall provide transit operations personnel with the capability to control and monitor transit signal priority operations.
Traffic Signal System	Roadway	The field element shall respond to signal priority requests from transit vehicles.
Transit Buses	Transit Vehicle	The transit vehicle shall determine the schedule deviation and estimated times of arrival (ETA) at transit stops.
Transit Buses	Transit Vehicle	The transit vehicle shall send priority requests to traffic signal controllers at intersections, pedestrian crossings, and multimodal crossings on the roads (surface streets) and freeway (ramp controls) network that enable a transit vehicle schedule deviation to be corrected.
Transit Buses	Transit Vehicle	The transit vehicle shall send the schedule deviation data and status of priority requests to the transit vehicle operator and provide the capability for the transit vehicle operator to control the priority system.
Transit Buses	Transit Vehicle	The transit vehicle shall prevent a priority request from being sent when the transit vehicle cannot use the priority (e.g. when the transit vehicle makes a passenger stop on the approach to an intersection).

The systems engineering analysis that is done during project development is also the subject of the Architecture and Standards regulation. A systems engineering analysis is required by the regulation for any project using federal funding. In the early stages of this analysis, the ITS architecture is very useful.

From planning to project development the National ITS Architecture can be used as a reference to inform stakeholders of the considerations they should be making about their project's scope, understanding the costs of a project at a high level to support planning, and the integration opportunities that might be available to the stakeholders when considering how their systems will fit with others in their transportation environment. Technology choices are not made until project development and, more formally, until high-level and detailed design is completed. A single project architecture based on a regional ITS architecture in conformance with the National ITS Architecture can be used as a guide from planning to design. This allows for flexibility in the technology selection based on the latest capabilities the market has to offer.

Table 2.2 Standards mapped to project information flows.

Source element	Flow name	Destination element	Standard title
City Traffic Operations Centre	Signal control commands	Traffic Signal System	Field Management Stations (FMS) – Part 1: Object Definitions for Signal System Masters
City Traffic Operations Centre	Signal control commands	Traffic Signal System	Global Object Definitions
City Traffic Operations Centre	Signal control commands	Traffic Signal System	NTCIP Centre-to-Field Standards Group
City Traffic Operations Centre	Signal control commands	Traffic Signal System	Object Definitions for Actuated Traffic Signal Controller (ASC) Units
City Traffic Operations Centre	Signal control commands	Traffic Signal System	Object Definitions for Conflict Monitor Units (CMU)
City Traffic Operations Centre	Signal control commands	Traffic Signal System	Object Definitions for Signal Control and Prioritization (SCP)
City Transit Operations Centre	Transit schedule information	Transit Buses	Standard for Transit Communications Interface Profiles
Transit Buses	Local signal priority request	Traffic Signal System	Dedicated Short Range Communication at 5.9 GHz Standards Group
Transit Buses	Local signal priority request	Traffic Signal System	Dedicated Short Range Communication at 915 MHz Standards Group
Transit Buses	Local signal priority request	Traffic Signal System	Global Object Definitions
Transit Buses	Local signal priority request	Traffic Signal System	Object Definitions for Signal Control and Prioritization (SCP)
Transit Buses	Local signal priority request	Traffic Signal System	Standard for Transit Communications Interface Profiles
Transit Buses	Transit vehicle schedule performance	City Transit Operations Centre	Standard for Transit Communications Interface Profiles

2.5.4 Tools

The National ITS Architecture content is made available to stakeholders through various tools.

2.5.4.1 National ITS Architecture Website

The National ITS Architecture Website is available at www.its.dot.gov/arch/index.htm. This is the United States Department of Transportation's page for the National ITS Architecture and from this page multiple National ITS Architecture related links are available. The link to the National ITS Architecture content provides access to the Logical and Physical Architectures, Service Packages, and Standards linkages. It also provides information about the use of the National ITS Architecture specifically in planning and project development. All Architecture Products such as documents, databases, and website archives are available for download. A list of the National ITS Architecture training and workshops are available for review. All levels of the National ITS Architecture definition are available via the website.

2.5.4.2 Turbo Architecture™ Software Tool

Turbo Architecture™ is a software tool developed by the United States Department of Transportation to support the use of the National ITS Architecture as a reference in developing regional ITS architectures. Turbo Architecture™ allows a user to tailor the National ITS Architecture content to address the needs of the state, region, locality or project of interest. It provides a set of tools that guide the user through an architecture development process that will yield an ITS architecture upon which ITS project planning and development can be based.

Turbo Architecture™ is based on the same Microsoft Access databases as the National ITS Architecture. It affords references to state, regional, and local transportation planning objectives providing the resulting ITS architecture with linkages to the planning process and specific references to the needs of the region.

The user maps the current and planned ITS systems in their environment to subsystems within the National ITS Architecture. This mapping relates the general functionality of the user's existing or planned ITS to the National ITS Architecture subsystems that best match what is needed. Turbo Architecture™ allows the user to label each subsystem mapping with the name of the ITS system and the stakeholder who owns and/or operates that system. This creates an architecture representation that is readily identifiable and specific to the region. Figure 2.6 illustrates the subsystem mapping for the project illustrated in Figure 2.4 for Transit Signal Priority. Note that the "Element" is the stakeholder's name for the system, in this case the City Traffic Operations Center, and the stakeholder is identified as well. The subsystem that the Traffic Operations Center is mapped to is the Traffic Management Subsystem. These mappings are made for each element or system in the region. More than one subsystem can be mapped to an element or system since the system may perform functions that are provided by more than one subsystem. This idea of aggregation of subsystem functionality provides

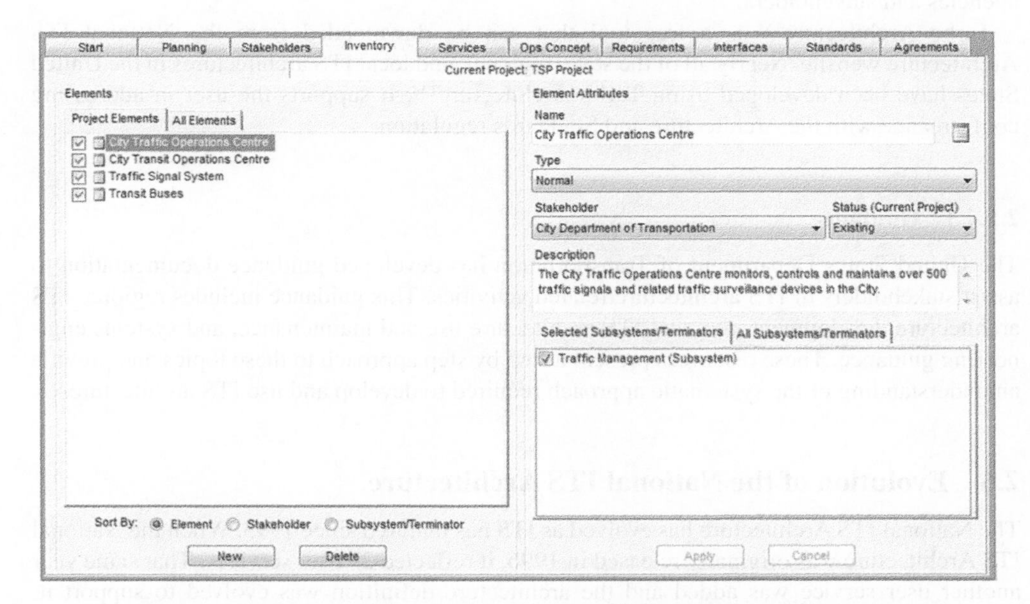

Figure 2.6 Transit Signal Priority project subsystem mapping in Turbo Architecture™.

flexibility in the architecture application that accommodates numerous ITS implementation approaches in different regions.

The user also identifies and maps ITS services in the National ITS Architecture to the services that exist or are planned in the state, region or locality, and address the needs identified. By selecting services, specific subsystems are identified. These subsystems may already be selected by the user when the subsystem mapping was completed. In some cases, the services may include subsystems that were not mapped by the user to existing or planned ITS in their region. In this case, the services provide the user with additional considerations regarding what is needed in their transportation system to provide the services they need. The user may need to identify an additional subsystem and the related functionality to realize the service in the future. In some cases, the subsystems may exist but are owned or operated by a stakeholder in the region that was not previously considered. In this case, additional outreach is required to that stakeholder to explore integration opportunities.

After the subsystem and service selections and mappings are complete, Turbo Architecture™ will build the interfaces for the ITS Architecture based on that information. The tool provides a list subsystem pairs and the information flows that connect them to be tailored by the user. In some cases, the user may not require a particular interface or subsystem functionality because it is taken care of by another pair of subsystems. The user simply selects what is needed and deselects what is not needed.

The breadth of the National ITS Architecture provides options and guidance on what to include in terms of ITS functionality to meet the needs identified. Standards are also mapped to the resulting interfaces in the ITS architecture providing the user with guidance on what standards to consider. Numerous reports and diagrams are available from Turbo Architecture™ to convey the ITS architecture definition to stakeholders and to allow them to comment and relate their operations to the ITS architecture furthering the consensus process among the agencies and stakeholders.

Turbo Architecture™ is a free tool that can be downloaded from the National ITS Architecture website. Nearly all of the state, regional, and local ITS architectures in the United States have been developed using Turbo Architecture™. It supports the user in addressing conformance with the Architecture and Standards regulation.

2.5.4.3 Guidance

The United States Department of Transportation has developed guidance documentation to assist stakeholders in ITS architecture related activities. This guidance includes regional ITS architecture development, regional ITS architecture use and maintenance, and systems engineering guidance. These resources present a step by step approach to these topics and provide an understanding of the systematic approach required to develop and use ITS architectures.

2.6 Evolution of the National ITS Architecture

The National ITS Architecture has evolved as ITS has matured since 1993. When the National ITS Architecture was originally released in 1996, it reflected 29 user services. That same year another user service was added and the architecture definition was evolved to support it. Throughout the next six years the architecture was updated to accommodate three more user services bringing the total to 33. The architecture was also evolved as a result of standards developments to keep it aligned with those results. As ITS research and deployment progressed, the architecture definition was modified to reflect the real-world examples allowing users the latest information for their ITS planning and project development activities. This stakeholder feedback is critical to maintaining a realistic, effective, and accepted National ITS Architecture.

Most recently, new initiatives such as Connected Vehicle in the United States, or Cooperative ITS as it is known internationally, have emerged and are changing the National ITS Architecture. Under the National ITS Architecture Program, the United States Department of Transportation has developed the Connected Vehicle Reference Implementation Architecture (CVRIA). The National ITS Architecture provided reference inputs to the CVRIA as a starting point. As the CVRIA definition matures, it will be merged with the National ITS Architecture fully integrating the expanding world of ITS. This integration will support Connected Vehicle planning and project development in the same way ITS planning and project development are supported in the transportation planning process today.

The Connected Vehicle/Cooperative ITS environment will require interoperability on a wider scale than previously needed for ITS. The CVRIA and National ITS Architecture will be primary reference points for this interoperability facilitating consideration of all integration and interoperability opportunities. This is the future of the National ITS Architecture in the United States.

References

1. Department of Transportation, US (2013, May 9). *National ITS Architecture Documents*. Retrieved from National ITS Architecture Website: http://www.iteris.com/itsarch/html/menu/documents.htm (last accessed April 23, 2015).
2. Department of Transportation, US (2013, May 9). *1.1.1.1-Process Traffic Sensor Data (PSPEC)*. Retrieved from National ITS Architecture Website: http://www.iteris.com/itsarch/html/pspec/5476.htm (last accessed April 23, 2015).
3. Department of Transportation, US (2013, May 9). *Roadway (Subsystem)*. Retrieved from National ITS Architecture Website: http://www.iteris.com/itsarch/html/entity/rs.htm#tab-5 (last accessed April 23, 2015).
4. Department of Transportation, US (2013, May 9). *Roadway Basic Surveillance (Equipment Package)*. Retrieved from National ITS Architecture Website: http://www.iteris.com/itsarch/html/ep/ep65.htm (last accessed April 23, 2015).
5. Department of Transportation, US (2013, May 9). *ATMS01 Network Surveillance Service Package*. Retrieved from National ITS Architecture: http://www.iteris.com/itsarch/html/mp/mpatms01.htm (last accessed April 23, 2015).
6. Department of Transportation, US (2013, May 9). *National ITS Architecture Use in Project Development*. Retrieved from National ITS Architecture Website: http://www.iteris.com/itsarch/html/archuse/projdev.htm (last accessed April 23, 2015).

Part 2

Wireless Vehicular Communications

3

Wireless Communications in Vehicular Environments

Pekka Eloranta[1] and Timo Sukuvaara[2]
[1] *Mobisoft Oy, Tampere, Finland*
[2] *Finnish Meteorological Institute, Sodankylä, Finland*

3.1 Background and History of Vehicular Networking

When we look back to the 1950s, automobiles were basically mechanical systems with no electronics. In the past few decades, however, electronics has become one of the major elements of a vehicle's value, reaching a mean share of around one-third of the total value of a modern car. The first generation of vehicle electronics was stand-alone in-vehicle systems, basically automating or supporting certain driving tasks. A typical example of such an achievement is the anti-lock braking system. The number of such ECU (Electronic Control Unit) systems on each car has increased from only a few in the 1990s to around 50 and more by 2010. ECUs control almost every activity in a modern vehicle, aiming to improve travel safety and comfort, as well as reducing fuel consumption [1]. Automobiles today are no longer vehicles with electronics, they could be characterized as "computers on wheels."

The next major step just emerging and happening is the wider scale adaptation and exploitation of wireless telecommunications. The main motivation for the applications of wireless networking to road traffic scenarios is to optimize driving with respect to safety, fluency and efficiency. While passive safety systems have proven to be effective in protecting passengers, they typically do not help in avoiding accidents in the first place. That is the key motivation for the development of active safety systems, very often relying on wireless communications. Safety, fluency and efficiency in traffic, as well as travel convenience can be enhanced with wireless networking advantages [1].

The term "wireless communication" in this chapter refers to the different concepts developed in the area of wireless (local area) networking, cellular networking dominated by mobile

Intelligent Transport Systems: Technologies and Applications, First Edition. Asier Perallos, Unai Hernandez-Jayo, Enrique Onieva and Ignacio Julio García-Zuazola.

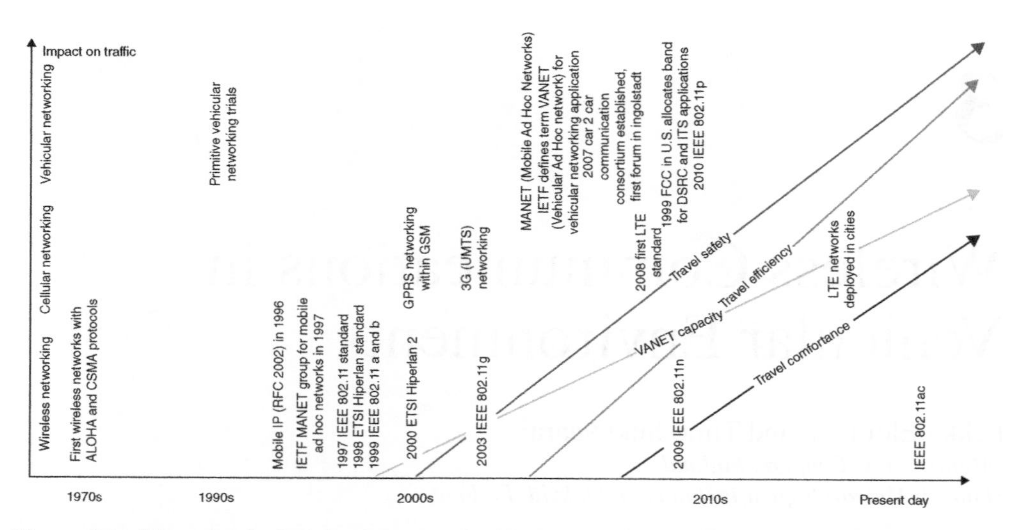

Figure 3.1 The technological development of vehicular networking related wireless communication.

phone systems and vehicular networking, respectively. The technological development of wireless communication is overviewed in Figure 3.1, with particular emphasis on car communications.

Wireless networking started to gain more popularity during the late 1990's as the communication devices on the markets become more affordable and more attractive also for general use and a the technologies needed in wireless communications became available. The concept of wireless networking was originally developed for the office environment, typically between computers and communication devices with static or nearly-static characteristics (negligible mobility). Wireless communication range between devices was generally assumed, typically requiring also LOS (Line-of-Sight) between counterparts. However, LOS was not always expected, and a more appropriate definition is short-range communication, referring to a range of few hundred meters.

The idea of ad-hoc networking connecting together independent computers located in the same office area, providing media for data exchange and communication with seemingly little effort was fulfilled. One step further was the concept of multihop networking, where the devices forwarded data packets from one partner to another.

The range of wireless network was thus significantly enhanced, with the cost of an additional network load leading to a decreased performance. However, with enhanced ad-hoc routing methods, the performance of multihop access networks made it possible to optimize into the appropriate level for seamless use of network resources, as long as network complexity, in terms of maximum hop-count and ad-hoc network members, remained low enough. When relatively stable and smooth operation in offices was reached, the interest in using wireless communication in more challenging environments grew. One of the ultimate targeted wireless communication environments was the ad-hoc network between moving vehicles.

The vehicular ad-hoc networking concept introduces a completely different and much more challenging wireless network. The dynamics in the communication environment and supporting infrastructure availability is relatively high, depending on how essential the road stretch is.

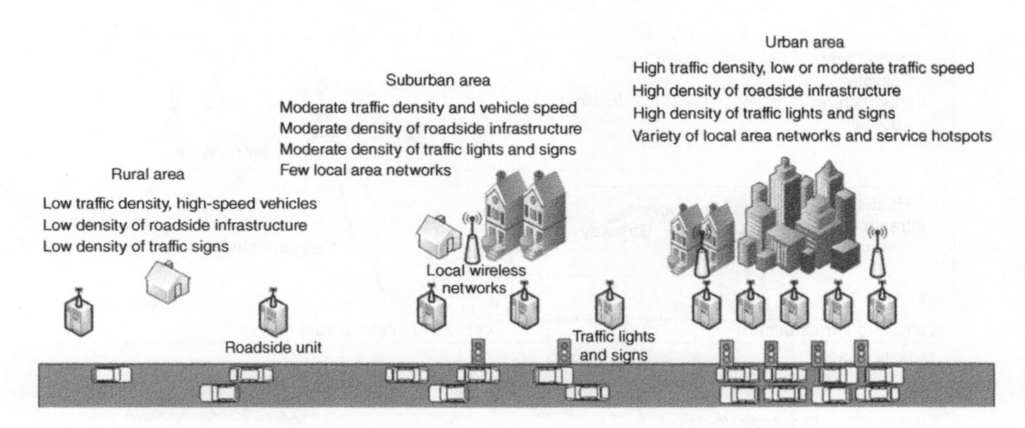

Figure 3.2 Different environments in vehicular networking.

The vehicular networking is typically divided into three different types, namely rural, suburban and urban area networking. The main properties of these entities are overviewed in the Figure 3.2.

In general, moving from rural areas towards urban areas decreases the traffic speed and increases the availability of a roadside infrastructure (roadside units, traffic lights etc.) and local communication entities. Outside urban areas vehicles may be moving at extremely high velocities, in either the same or opposite directions, providing extreme challenges in terms of delay requirements (with the short time the nodes are exposed). In the case of communication with a roadside unit, and especially with an oncoming vehicle, the time window for communication is very short. In general, the vehicular access network availability varies a lot compared to a traditional wireless network. Also the line-of-sight link between the counterparts is often blocked by other vehicles and roadside installations like bridges and buildings, making the signal dynamic. The Doppler effect appearing in communicating modules moving towards each other also may noticeably change the signal quality. Multihop communication can only be attempted between vehicles moving into the same direction, and even in that case traditional congestion avoidance methods are hard, or even impossible to implement due to the high variation in the network structure.

However, establishing an ad-hoc network between moving vehicles in close proximity has a number of important advantages. Vehicles can exchange their (environmental) observations and information about the traffic or weather conditions, and anomalies in the road, depending on the sensors implemented in the vehicle. Ultimately, a vehicle in a traffic accident can broadcast a warning to other vehicles approaching, and contributing to avoiding of further accidents. With exploitation of roadside instalments with a link to a fixed network, this data can be further forwarded, allowing vehicles to avoid road stretches occupied with incidents, accidents, queues or road construction works. In addition to this, an unlimited amount of commercial services like advertisements, guidance and general information can be delivered to and from the vehicles. Furthermore, it is important also to classify different types of vehicular area networking, in terms of communication range.

The taxonomy of vehicular networking is viewed in Figure 3.3, based on the automotive network domain presented in [1]. The shortest range of communication emerges in in-vehicle

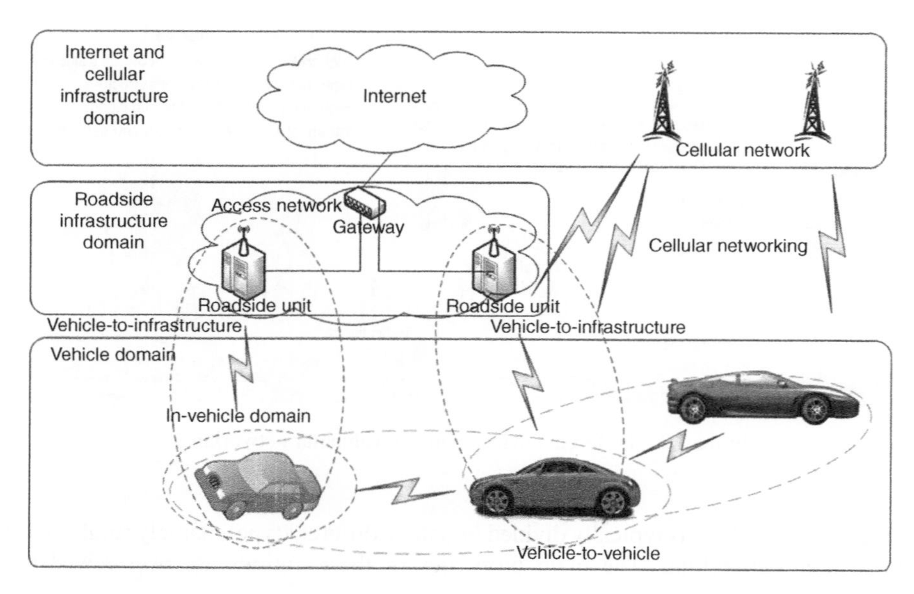

Figure 3.3 The taxonomy of vehicular networking.

communication, where wireless devices inside the vehicle form the network. A typical application for this is a wireless link between a mobile smart phone and vehicle systems, allowing the use of the microphone and the speakers of the car for a mobile phone conversation. Vehicle-to-vehicle (V2V) communication consists of data exchange with by-passing vehicle, networking between vehicles traveling to the same direction and emergency data broadcasting to the other vehicles nearby. Emergency data broadcasting is the most important application, from which the whole idea of vehicle networking is derived. Additionally, vehicle-to-infrastructure (V2I) communications employs the roadside infrastructure for data exchange and networking with the car. The roadside infrastructure usually has a permanent link to the fixed network, hypothetically allowing the Internet connectivity at least on a temporary basis. Both vehicle-to-vehicle and vehicle-to-infrastructure communications can be implemented with either radio or optical communications. However, wireless optical communication is an emerging technology, and has not been studied very much. A vehicle can also have a direct connection to the fixed network infrastructure through cellular network systems (typically mobile phone networks) allowing continuous connectivity not depending on the location. Combining these different networking types (excluding in-car communication) into a single architecture is one of the main objectives in the near future.

It is obvious that the kind of stable operating wireless network expected in office environments is not yet fully possible in a vehicle network's environment and numerous compromises need to be made. Truly continuous connectivity is one issue that is very hard to pull out, especially with low vehicle and roadside unit densities, and usually it is not expected to take place in vehicular networking. However, with the support of simultaneous cellular mobile communication, the connectivity can be significantly enhanced and sometimes even continuously supplemented. The price of this is the complexity of managing multiple communication systems simultaneously. The issues to be considered are smooth or even seamless handovers between the systems, adaptations to the high variations in the data throughput rates, and the quality of service (QoS), as well as the parallel maintenance of different systems. The main challenge in vehicular networking is to find a good engineering balance between conflicting requirements. The services developed for this environment must be tailored to cope with anything but a stable data channel. Highly probable communication blackouts must not significantly decrease the general performance. For vehicular communication, there are generally two fundamental approaches, short-range wireless local area networking and wide-area cellular based communication. Wireless Local Area Networking (WLAN) is independent of any network operator, and is also more suitable for instantaneous data exchange between parties relatively near to each other. Cellular communication offers wide area coverage with a relatively small data rate, and requires a network operator to host the communication.

The most advanced cellular networking systems, like LTE, provide the best data rates, comparable to WLAN data rates, but the coverage is lower than in "older" cellular networks due to the more limited cell size and lower density of service access points. However, as the LTE system is downward compatible to the GPRS communication system and ultimately to the 3G system, it provides in practice complete coverage with different quality of services, depending of the location. The global trend in vehicular networking approaches has been to focus on WLAN type of solutions, but with advanced data rates, cellular systems are gaining more and more interest. Naturally this also depends on the infrastructure and the operational environment. Nowadays, even some advanced vehicular applications relying on a cellular networking system do exist. For example, the co-operative WAZE application [2] allows vehicles to either drive with the application open on their phone to passively contribute to traffic and other road data, or take a more active (co-operative) role by sharing road reports on accidents or any other hazards along the way.

Together with accident warnings, road weather services are commonly recognized as key advantages available through vehicular networking, especially in communication between roadside infrastructure and vehicles. On the other hand, the road weather service justifies the use of bidirectional communication, as weather-related data gathered from moving vehicles can clearly enhance the accuracy of local weather forecasting and related services. Finally, with a functional local road weather service partially based on vehicle data the whole vehicular networking architecture can be justified and its operability in real-life usage verified. The research work of the authors draws an overview of the vehicular networking development during the past years. The work started within the Carlink project (Wireless Traffic Service Platform for Linking Cars) [3], established in 2006. The architecture development basis combined both vehicular ad-hoc network and infrastructure-based networking with roadside fixed network stations inherited from the self-configurable heterogeneous radio network concept [4]. The conceptual idea of multiprotocol access networking was used for combining Wi-Fi (Wireless Fidelity) and GPRS networking. As a result, the Carlink project designed and piloted one of the first operating V2V and V2I communication architectures.

The general state-of-the-art in the field of vehicular networking was composed of a number of somewhat separated component technologies. The first rudimentary vehicular services had already been launched, exploiting the mobile phone SMS-messaging (Short Message Service) system as the communication media. An example of such a service is the VARO-service designed by the Finnish Meteorological Institute (FMI), providing SMS weather warnings and route guidance to the end-user devices embedded into mobile phone in a car [5]. A variety of more general SMS services contained vehicle identification (based on registration plates) information request and primitive navigation services. On the other hand, few roadside weather stations were already installed to gather up-to-date local weather information to be used to enhance weather forecasts and warnings in the road areas. Obviously, communication with passing vehicles was not an issue in the first five years of the 21st century, but those road weather stations were equipped with a power supply and some means of collect and deliver station data to the network host supervising the stations. The concept of wireless networking was already a hot topic in telecommunications research, and especially the ad hoc networking in self-configurable networks was gaining considerable interest. In the field of ad hoc networking a great deal of different routing methods were proposed and studied, but the communication media was usually assumed to be the same, the so called Wi-Fi based wireless networking based on the IEEE 802.11 family of standards. The concept of hybrid vehicular access network architecture were successfully studied, developed and evaluated in the Carlink project. The general idea of the continuation project WiSafeCar (Wireless traffic Safety network between Cars) [6] was to overcome the limitations of communications by upgrading communication methodology, Wi-Fi with the special vehicular WAVE (Wireless Access in Vehicular Environments) system based on IEEE 802.11p standard amendment [7] and GPRS with 3G communication, respectively.

The architecture was employed with a set of more sophisticated services, tailored for traffic safety and convenience. The set of example services was also adjusted to be compliant with services proposed by the Car-2-Car Communication Consortium (C2C-CC) [8] and ETSI standardization for the "day one set of services" [9]. Especially the newly-found IEEE 802.11p based vehicular access network system underwent an extensive set of test measurements, both with V2V and V2I communications, respectively. The platform capacity and range were estimated and analyzed in the evaluation and field-testing of the system, presented in [5]. The project pilot platform was deployed with the example services in operation under realistic conditions. Based on the experience gained from both field measurements and pilot deployment, a realistic architecture deployment strategy for simple scenarios was presented, too. The measurements demonstrated that the IEEE 802.11p has clearly better general performance and behavior in the vehicular networking environment, compared to the traditional Wi-Fi solutions used for this purpose.

The peak performance in terms of data throughput was lower when using IEEE 802.11p, but still more than appropriate for the needs of vehicular access network. The pilot platform deployment proved that the new system operates also in practice, it was possible to provide defined pilot services properly. In the deployment, the overlay cellular network (3G) played an important role, and this hybrid method would be an attractive solution for the ultimate commercial architecture. One clear benefit was that exploiting 3G, the communication system would be available in a limited form already on day 1 of the deployment process, and with low implementation costs. It was concluded that the solution had clear potential for the comprehensive heterogeneous vehicular communication architecture, aiming at decreasing the amount of accidents and lives lost on the roads. The system deployment could be initiated in a cost-effective manner, relying purely on the existing 3G overlay network in the early deployment phase. As a result, the WiSafeCar project drew an outline for the commercially operating intelligent vehicular access network architecture, with a general deployment proposal.

Even if the commercial deployment did not yet take place, the developed system served as the basis for a more advanced project, CoMoSeF (Co-operative Mobility Services of the Future) project [10], along with other intelligent traffic related research. The focus in the CoMoSeF project was on near-the-market services and multistandard communication. The aim was not only to service vehicles, but also to exploit vehicle-originating data to ultimately enhance the very same services. Similarly, roadside units are not just serving the vehicles as connectivity points, but also host Road Weather Station (RWS) capabilities to provide additional data for the services.

Both of these goals are combined in the Finnish Meteorological Institute's approach to employing vehicular networking architecture to provide route weather information for vehicles passing by the combined RWS/RSU. The station is equipped with up-to-date road weather measurement instrumentation, compatible with (but not limited to) the equipment expected to be available also in the other demonstration sites' own permanent and locally owned RWSs. The procedure is to design, develop and test both the local road weather service generation, and the service data delivery between RWS and vehicles. The vehicle passing the combined RWS/RSU is supplemented wirelessly and automatically with up-to-date road weather related data and services, and at the same time possible vehicle-oriented measurement data is delivered upwards. IEEE 802.11p is the primary communication protocol, but also traditional Wi-Fi communication is supported. The station, together with the research vehicles, forms the pilot system in Sodankylä, Finland, acting as a real-life testbed for the future demonstration systems yet to come. Based on the common architecture, challenges and underlying goals, and considering defined assumptions, the approach of the hybrid wireless traffic service architecture between cars was presented in CoMoSeF.

The enabling technologies are an IEEE 802.11p vehicular networking approach and a 3G/GPRS mobile communications system. The system capacity and range was estimated and evaluated in the field tests, presented in detail later. Partially based on the capacity estimate, the successful project pilot platform deployment was constructed, with the specially designed example services in operation under realistic conditions. In Figure 3.4, the user interface of the WiSafeCar pilot system is presented. Furthermore, in Figure 3.5 an up-to-date user interface of the FMIs operative combined RWS/ RSU (identical to both Internet and vehicular networking users) is presented.

Figure 3.4 WiSafeCar pilot system in operation.

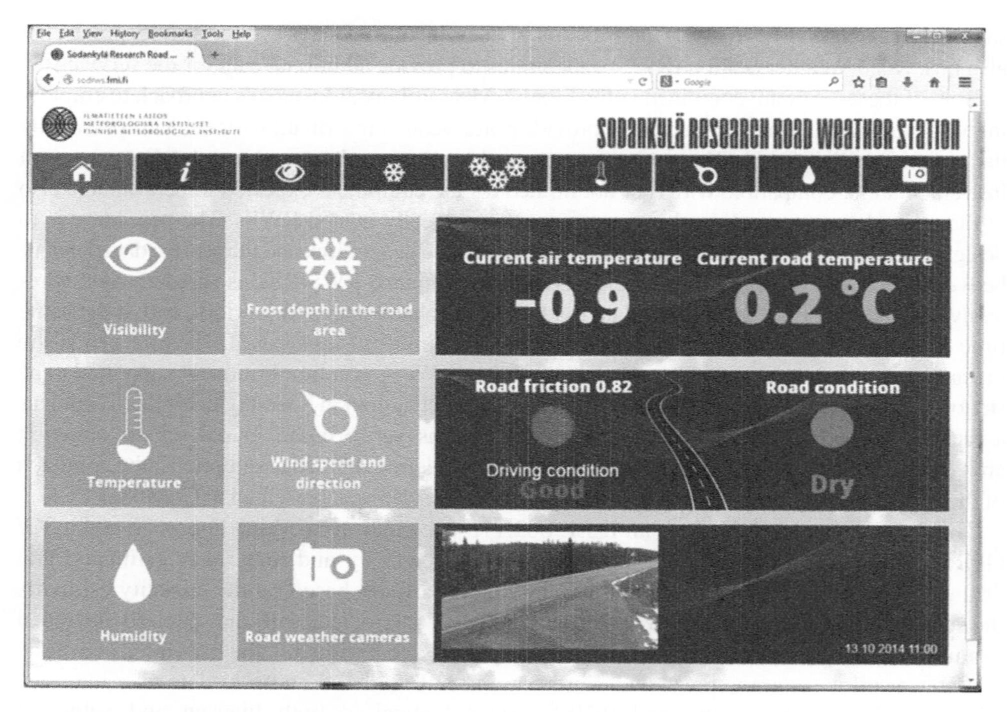

Figure 3.5 Road Weather Station user interface in the Internet (and vehicular networking).

Based on the experience gained from both the field measurements and the pilot deployment, a realistic system deployment strategy with simple scenarios has been proposed. The measurements proved that the IEEE 802.11p has clearly better general performance and behavior in the vehicular networking environment, compared to the traditional Wi-Fi solutions used for this purpose. The peak performance in terms of data throughput is lower with IEEE 802.11p, but still more than appropriate for the needs of vehicular networking.

The pilot system deployment proved that the new system operates also in practice, and we can provide defined pilot services properly. However, in the deployment, the 3G network plays an important role, and such a hybrid method could be an attractive solution for the ultimate commercial system. It has been shown that the solution presented in this chapter has a clear potential for a comprehensive heterogeneous vehicular communication entity, aiming at decreasing the amount of accidents and lives lost on the roads. The system deployment can be initiated in a cost-effective manner, relying purely on the existing 3G overlay network in the early deployment phase.

3.2 Vehicular Networking Approaches

Wireless networks refer to any type of network that does not have a physical connection using cables. The main features of the communication methods within the entities are listed in Table 3.1. The original motivation for wireless networks was to avoid the costly

Table 3.1 The main features of the communication methods related to vehicular networking.

Communication method	Theoretical data rate	Mobility support	Architecture	Connection delays	Theoretical range
Conventional WLAN; IEEE 802.11g	54 Mbps	Low	Local cells	Low	140 m
Conventional WLAN; IEEE 802.11n	600 Mbps	Very low[2]	Local cells	Low	250 m
V2V (vehicle-to-vehicle)[1]	3–54 Mbps	Good	Local cells	Very low	1 km
V2I (vehicle-to-infrastructure)[1]	3–54 Mbps	Good	Local cells	Very low	1 km
GPRS cellular data	56–114 kbit/s	Good	Cellular	Moderate	Unlimited[3]
3G cellular data	0.2 Mbps	Moderate	Cellular	Moderate	High[3]
LTE cellular data	300 Mbps	Moderate	Cellular	Moderate	Low[3]
Hybrid	0.2–54 Mbps	Good	Hybrid	Very low	Unlimited

[1] based on IEEE 802.11p networking.
[2] with maximum data rate mode.
[3] commercial cellular systems range is not defined as the range of one cell, but the coverage of operational systems in 2013.

process of introducing cables into office buildings, or as a connection between various equipment locations.

The commonly known term wireless local area network (WLAN) refers to a system that links two or more devices over a short distance using a wireless distribution method, usually providing a connection through an access point for Internet access. The major contribution for WLAN development has been produced through the IEEE (Institute of Electrical and Electronics Engineers), and more specifically through its standardization process, known as the IEEE 802.11 standard. The original standard, published in 1997, defines the wireless LAN Medium Access Control (MAC) and physical layer (PHY) specifications. The fundamental access method for the MAC realization is Carrier Sense Multiple Access with Collision Avoidance (CSMA/CA). The IEEE 802.11 architecture defines three different propagation modes. These are the 2.4 GHz FHSS (Frequency-Hopping Spread Spectrum), the 2.4 GHz DSSS (Direct Sequence Spread Spectrum) and the infrared system [11]. The basic version of the standard supports only 1 Mbps and 2 Mbps data rates, but there has been numerous amendments published since the original standard to update the data speed as well as other properties of the standard [11].

The first amendments for the standard were IEEE 802.11b and IEEE 802.11a. There is a fundamental difference between these amendments; while the objective in 802.11b was to maintain compatibility with the original standard, the 802.11a was aiming to increase capacity and efficiency by upgrading modulation, operating frequency and bandwidth, respectively. One can say that all the following amendments are inherited from these two, and therefore they are presented with details.

IEEE 802.11b is quite similar than the original 802.11 standard architecture. With this amendment, the name Wireless Fidelity was adopted to refer IEEE 802.11b and its subsequent amendments. 802.11b is operating in the same 2.4 GHz frequency band, and has the same MAC, CSMA/CA. It is also backwards compatible with the original standard, therefore supporting 1 Mbps and 2 Mbps data rates. As an extension to the original standard architecture, 802.11b also provides new data rates, 5.5 Mbps and 11 Mbps, respectively. The CCK (Complementary Code Keying) modulation method enables the possibility to achieve higher data rates. Otherwise, the IEEE 802.11b has only minor differences to the original standard architecture [12]. Basically IEEE 802.11b completely replaced the original 802.11 standard, due to the much higher capacity of extension b. 802.11b, which itself had the same destiny when it was later on replaced by IEEE 802.11g, and nowadays the de-facto standard for Wi-Fi communication is IEEE 802.11n [13].

IEEE 802.11a has very many differences compared to the original standard. The most significant differences are that the Physical Layer of 802.11a is based on OFDM (Orthogonal Frequency Division Multiplexing) modulation as the carrier system, and it uses 5.2 GHz frequency band. The underlying modulation schemes used are BPSK, QPSK (similar to the original standard) and different levels of QAM (Quadrature Amplitude Modulation). With these changes, 802.11a is able to achieve (from 6 Mbps) up to 54 Mbps data rates. Due to these major differences, 802.11a is not compatible with the original standard. However, the MAC architecture is the same CSMA/CA as in the original standard [14].

As stated before, the following Wi-Fi standard extension was IEEE 802.11g, providing an 802.11a type of architecture (with the same capacity, up to 54 Mbps), but operating still in the 2.4 GHz frequency [15]. The extension most commonly used nowadays is the aforementioned IEEE 802.11n. Its purpose was to significantly improve network throughput by combining elements of 802.11a and 802.11g, with the use of four spatial streams at a channel width of 40 MHz and with a significant increase in the maximum net data rate from 54 Mbit/s to 600 Mbit/s. This data rate can only be achieved when operating in the 5 GHz bandwidth, adapted from 802.11a. Therefore, IEEE 802.11n operates in two different bandwidths; in 2.4 GHz the downward compatibility is maintained with previous amendments but with relatively the same capacity, while in the 5 GHz band the ultimate improvements of capacity and efficiency are fully gained. Channels operating on a width of 40 MHz are the key feature incorporated into 802.11n; this doubles the channel width from 20 MHz in the previous 802.11 to transmit data, providing double data rate availability over a single 20 MHz channel. It can only be enabled in the 5 GHz mode, or within 2.4 GHz if there is knowledge that it will not interfere with any other 802.11 or non-802.11 (such as Bluetooth) system using the same frequencies [13,16].

The most recent proposed amendment of the standard is IEEE 802.11ac, aiming to gigabit-level throughput. In practice, the proposed approach has the goal of supporting at least 1 Gbps data rate in each of the multiple bands below 6 GHz (excluding 2.4 GHz band), meaning at least five times more capacity per single user compared to IEEE 802.11n. IEEE 802.11ac is expected to support 20, 40 and 80 MHz channels in its bands, with option to use 160 MHz channels [17]. 160 MHz channels in the highest band are partially using the same frequencies with IEEE 802.11p, therefore raising concerns and objections among IEEE 802.11p users. It remains to be seen what will be the ultimate channel and band allocation.

3.3 Vehicular Ad-hoc Networking

As mentioned earlier in this chapter, the main usage scenario in the Wi-Fi type of networking was originally the rather static office environment, with multiple communicating computers at a relatively small distance from each other, having only light physical walls and objects between them. Nowadays the concept has been expanded to the idea of a wireless home, with computers, printers, home multimedia entertainment systems, TVs, DVD players, tablet computers and mobile phones all connected to the same wireless network. The key concept in communications remains the same, communication units are located within rather short distances and are stationary or slowly moving. In this kind of scenario, the Wi-Fi works well; the capacity is high enough for even rather demanding usage scenarios, connection establishment time is not an issue, and even infrequent connection losses do not cause unbearable harm. However, this is not the case in a vehicular communication environment.

The first primitive experiments in vehicular networking were carried out already in 1989 [18], but more systematic research within the concept of vehicular networking was started in the early part of this millennium, when the maturity of needed technologies increased. Obviously, the starting point was Wi-Fi, as an existing and widely used wireless communication system. As expected, Wi-Fi networks were soon found to be rather inadequate for this purpose. Vehicular safety communication applications cannot tolerate long connection establishment delays before being enabled to communicate with centralized safety systems and/or other vehicles encountered on the road. Naturally, communication reliability all the time is also an important issue. Nonsafety applications also require fast and efficient connection setups with roadside stations that provide services (e.g., weather and road data updates) because of the limited time a car spends within the station coverage area. Additionally, rapidly moving vehicles and a complex roadway environment present challenges on the physical level. These problems typically arise when using Wi-Fi. The IEEE 802.11 standard body has created a new amendment, IEEE 802.11p, to address these concerns [7,19].

The primary purpose of IEEE 802.11p standard is to enhance public safety applications and to improve traffic flow by V2V and V2I communications. The underlying technology in this protocol is Dedicated Short-Range Communication (DSRC), which essentially uses the IEEE 802.11a standard OFDM-based physical layer and quality of service enhancements of IEEE 802.11e, adjusted for low overhead operations. The IEEE 802.11p uses an Enhanced Distributed Channel Access (EDCA) MAC sublayer protocol designed into IEEE 802.11e, with some modifications to the transmission parameters. DSRC is a short-range communication service designed to support communication requirements for enhancing public safety applications, to save lives and to improve traffic flow by V2V and I2V communications. Wireless Access to Vehicular Environment (WAVE) is the next generation technology, providing high-speed V2V and V2I data transmission. The WAVE system is built on IEEE 802.11p and IEEE 1609.x standards 0 operating at 5.850–5.9250 GHZ with data rates and supports between 3 and 27 Mbps with 10 MHz channel and 6–54 Mbps in 20 MHz channel, respectively. Up to 1.000 m range in a variety of environments (e.g., urban, suburban, rural) is supported, with relative velocities of up to 110 kph. Depending on usage needs, either 10 MHz or 20 MHz channel bandwidth can be chosen [7,19–23].

The development of vehicular communication networks has created a variety of emergency services and applications. The major contributions so far have been provided within the European Union EU IST 6th framework (FP6) and EU 7th framework (FP7) program for research and technology projects (main projects listed in [24]), in the Vehicle Safety Communication (VSC) project and Vehicle Infrastructure Integration (VII) supported by US DoT (Department of Transportation) in the USA and in the activities supported by Japanese Ministry of Land, Infrastructure and Transport (MLIT) in Japan. The pilot services developed so far contain different types of Cooperative Collision Warnings (CCW), (Post- and Pre-)

Crash Detection Systems (CDS) and Cooperative Intersection Safety Systems (CISS), among many others [24]. Examples of so-called "Day One" services to be deployed first are envisioned widely. For example ETSI standardization for the "day one set of services" [9] consists of warnings of approaching emergency vehicle, approaching roadworks area, a motorcycle approaching from "dead angle" and post-crash warning issued via vehicle emergency lights-initiated radio broadcast. These services have also been successfully demonstrated already in 2008 Car 2 Car Forum at Dudenhofen, Germany.

As stated above, a vehicular access network is often classified into V2V and V2I communications. As viewed in Figure 3.3, there are more subcategories of vehicular networking, but one can say that these two are the main subtypes, while the rest are some kind of special related cases. In this work, there are many special cases and scenarios dedicated purely to either V2I or V2V. Therefore it is important to consider the differences of these communication types in more details. In the following sections,V2I and V2V, as well as related hybrid combinations, are considered one by one.

3.3.1 Vehicle-to-infrastructure Communication

V2I communication means a Vehicular Ad-hoc Network (VANET) created between moving vehicles and a static infrastructure beside the road. The communication architecture is centralized, the roadside infrastructure acting as a central point for one or many vehicles. The communication is bidirectional, despite the fact that the term vehicle-to-infrastructure seems to refer to one direction only. However, in the V2I direction the communication is of the unicast type, while in the opposite direction, the communication type is both broadcast (while delivering general data) and unicast (while responding to the vehicles requests). The roadside unit/infrastructure, or simply RSU, is typically supplemented with a fixed power supply and a backbone network connection and therefore there is no need to consider the consumption of these resources in its operation. RSU can be equipped with multiple and/or directive antennas, making the downlink channel (from RSU) typically much stronger compared to the uplink. In some V2I applications, the uplink channel is meaningless or nonexistent, making the service more like a broadcast type. Nevertheless, V2I must not be mixed with broadcasting systems, the existence or at least preparedness for deployment of an uplink being an essential element when considering V2I.

V2I communication is usually employed to deliver information from road operators or authorities to the vehicles. Roadwork warning is a typical example of a V2I service; vehicular access network transceivers are deployed into the roadwork area, informing the vehicles approaching the area about the exceptional road operability. One particularly important advance is the ability for traffic signal systems to communicate the signal phase and timing (SPAT) information to the vehicle in support of delivering active safety advisories and warnings to drivers. One approach for traffic-light optimizing is the Shortest-Path-Based Traffic-Light-Aware Routing (STAR) protocol for VANETs [25]. Both of these services are broadcast-type, lacking the use of an uplink channel. On the contrary, RSU with a road weather station employed in [3] not only delivers the weather and warning data for the passing vehicles, but also gathers the weather and safety related observations from the vehicles to further update the data.

V2I communication has certain similarities to a wireless link between the mobile node and access point in a traditional wireless network. Just as an access point, RSU is a static element within moving vehicles, like the mobile nodes in a traditional wireless network. Due to its fixed nature, RSU possesses superior resources in terms of signal strength and therefore data capacity, just like access point. However, due to the temporary nature of V2I communication, RSU cannot provide continuous backbone network connectivity for the vehicles. Instead, RSU can merely act as a service hotspot, delivering a pre-configured high-band service data exchange between the vehicle and fixed network whenever in the vicinity area of an RSU. One example of such a data dissemination network is introduced in [26].

In some related work, there is discussion about Vehicle-to-Roadside (V2R) communications. V2R is a special case of V2I communications, in which the focus is strictly limited to roadside infrastructure, like roadwork and SPAT mentioned above. Nevertheless, V2R is a special case of V2I, and therefore it is not considered separately.

3.3.2 *Vehicle-to-vehicle Communication*

The V2V communication approach is mostly suited for short-range vehicular communications. The general idea is that moving vehicles create a wireless communication network between each other, in an ad-hoc networking manner and on a highly opportunistic basis. The communication architecture is distributed, as individual vehicles are communicating equally, in an ad-hoc manner. The data exchange between passing vehicles is typically of the unicast type, but also multicast (for example, in the case of a platoon of vehicles exchanging traffic information) and broadcast (in the case of accident warnings) transmissions are employed. A pure V2V network does not need any roadside infrastructure, making it fast and relatively reliable for sudden incidents requiring information distribution on the road. Therefore it is the primary communication candidate for real-time safety applications in vehicles.

One of the key motivations for V2V communications is the opportunity to enable cooperative vehicle safety applications that will be able to prevent crashes. Such cooperative collision-avoidance applications that are envisioned for initial deployment would be to (1) identify other vehicles in the immediate vicinity; (2) maintain a dynamic state map of other vehicles (location, speed, heading and acceleration); (3) perform a continuous threat assessment based on this state map; (4) identify potentially dangerous situations that require driver actions; and (5) notify the driver at the appropriate time and manner. In the long run, automatic vehicle intervention to avoid or mitigate crashes with these applications is envisioned, but it still needs much work on the validation of the required reliability in communications [1,27].

A special case of V2V communications is multihop dissemination (including broadcasting) with specific multihop protocols. Especially in the case of a traffic accident, the vehicle participating in or observing an accident will broadcast a warning message, which is forwarded by the vehicles receiving the message during a certain period of time, allowing others up to kilometers away to make smart driving decisions well ahead of time. In dense traffic conditions, there is a risk of a broadcast storm problem, where multiple vehicles are trying to transmit the message at the same time causing multiple packet collisions and in an extreme case total outage of the communication channel [28]. Several solutions exist to avoid the problem, most of them derived from the idea of forwarding the message with certain random, weighted or adjusted probability, instead of automatic "blind forward" [1,27].

The V2V communications entity is very challenging. In V2V, the connectivity between the vehicles may not be possible all the time since the vehicles are moving at different velocities, due to which there might be quick network topology changes. Without any roadside infrastructure, multihop forwarding must be enabled to propagate the messages or signals. The addresses of vehicles on highways are mainly unknown to each other. Periodic broadcasts from each vehicle may inform direct neighbors about its address, but the address-position map will inevitably change frequently due to relative movements among vehicles. It is the receiver's responsibility to decide on the relevance of emergency messages, and also decide on the appropriate actions.

Due to the crucial limitations presented above, V2V communications mainly focus on special cases of communications instead of a general "all-purpose" network. The most typical use cases are broadcasting of emergency or other critical data to all vehicles, exchanging data with bypassing vehicles and communication network between a platoon of vehicles moving into the same direction at the same speed. It is worth noting that a wireless ad-hoc network in trains can be seen as a special case of the last scenario. Furthermore, location information gathered from, for example, a GPS device, can be used to benefit V2V communication, allowing nearly continuous communication capabilities, especially when traffic is dense and multihop communications are used. Location based broadcast and multicast are also the proper communication methods for collision avoidance. In general, V2V communication is suitable for those roads with high vehicle density, and provides only small or even insignificant effort in rural areas.

3.3.3 Combined Vehicle-to-vehicle and Vehicle-to-infrastructure Communication

Combined V2V and V2I networking can be seen as plain V2V supplemented with V2I capabilities. V2V is the starting point, with applications defined in the previous section, and integrated V2I would enable an expanded range of vehicle crash-avoidance safety applications using the same wireless technology. One of the additional features enabled by V2I is intersection collision avoidance, whereby knowing the dynamic state map of all the vehicles, as well as the intersection geometry, the system could warn a driver about another potentially hazardous intersecting driver. From this perspective, hybrid V2V and V2I is often referred to as Vehicle-to-Vehicle/Infrastructure communications or simply V2X 0.

In this chapter, one of the essential issues is to consider the combined V2V and V2I as its own special case of communications. A similar kind of approach has been presented in [29]. RSU, infrastructure side of V2I usually has fixed power and can employ directive antennas especially tailored for RSU, often making the downlink signal from RSU to the vehicle dominant, compared to the uplink provided by the vehicle. Furthermore, RSU tends to communicate with all the vehicles, while the vehicle tries to optimize its use of communication resources by minimizing intervention with other vehicles. Finally, as RSU usually has a fixed network connection, it can also be seen as an access point of the wireless network in a special kind of vehicular wireless network.

As stated above, the combined V2V and V2I communications access network consists of vehicles and RSUs, with relatively different objectives. Vehicles are communicating between each other in a V2V manner whenever in the vicinity area of each other, basically exchanging their observations from the traffic or forwarding/broadcasting multihop messages, or possible wide-area data received earlier from RSU. However, when entering the vicinity area of an RSU, vehicles not only exchange data with RSU, but may also exchange data with services located in the fixed internet, through an access link provided by RSU (if such operability is employed). As the interaction time with RSU is very limited, such service hot-spot communication procedures must be pre-configured into the vehicle user profile, to be initiated automatically when entering into RSU vicinity. The vehicle should therefore initiate different operational procedures for vehicle and RSU interaction. On the contrary, RSU procedures are basically similar, regardless of whether the network is V2I or combined V2V and V2I.

3.3.4 Hybrid Vehicular Network

The concepts of V2V and V2I networking of VANET are based on local area networking, exploiting typically an IEEE 802.11p standard based access network, as stated above. Theoretically, an element of such a network can achieve up to a 1 kilometer communication range. In any case, it is not realistic to expect that such a local area networking system can be cost-effectively deployed to achieve complete coverage throughout the road network.

Mobile phone cellular networks provide (almost) complete geographical coverage, and nowadays they are also employed with a relatively high data capacity. A new and therefore very densely deployed LTE network goes up to a theoretical 100 Mbps throughput [30]. The widely deployed 3G cellular networking system allows data rates up to theoretical 2 Mbps and relatively good coverage, with underlying GPRS communication allowing very high coverage and typically around a 100 kbps data rate [27]. However, the mobile phone network, as the name states, is merely designed for supporting on-demand phone connections rather than continuous connectivity. This fact evidently leads to an unbearable response time in the case of accident warnings and related safety services expected to be delivered instantly to the vehicles approaching a brand-new accident location. Upcoming enhancements of mobile networks provide increasingly higher data rates, but as they move to a higher spectrum, coverage areas are getting smaller and smaller. However, services like WAZE 0 can be adequately supported by cellular networks.

The hybrid solution at the moment for the coverage/response time problem is to bind VANET and cellular networking into a hybrid vehicular networking system. Referring to Figure 3.3, this means that all the concepts presented in the figure are combined together. One approach for combining Wi-Fi and GPRS into a hierarchical hybrid network has been presented in [31]. The concept of a self-configurable heterogeneous radio network presents another approach to this topic [4]. A kind of general approach from the cellular networking perspective is presented in [32], more related to cellular communication. All of these approaches have a continuous networking perspective to this issue, as do the majority of existing approaches in general. From the continuous connectivity perspective, the handing over of the connection from one protocol to another plays a crucial role. For example, [4] presents several approaches for a smooth handover within different types of ad-hoc IP networks. However, in vehicular networking, the primary perspective is different. The continuous connectivity is not the main concern, but clearly more important is to ensure instant delivery of local vehicular safety data delivery for the vicinity area near the sending vehicle. Therefore, the straightforward approach for the handover in hybrid vehicular networking entity is to always promote VANET networking whenever available, and whenever arriving into the range of another vehicular networking unit, with the price of breaking up the ongoing cellular network data transfer.

3.3.5 LTE and Liquid Applications

In September 2014 Nokia Siemens Networks and HERE announced a new way of complementing Wi-Fi and using LTE more intensively for road safety applications [33]. Even though LTE is rather new, it is becoming universal faster than any previous network standard. The LTE network is being built "anyway" and thus it will be possible for road and car safety systems and applications to benefit from that.

It has been stated that with cellular systems the latency is a problem and therefore they are not suitable for safety applications. When a message is transmitted from a car to a base station and further to a data center it takes time, not to mention the fact that it also takes time to send the message back to cars. This is normally too much for at least some of the safety-related messages. As an example, the SME message from a sender to a receiver in the same room is not delivered directly and straightaway. There is at least a few seconds' gap, because of the roundabout it has to take. Connected cars would face the same issue, even with special rules to prioritize their communications. To avoid this, their applications have been and will be developed right to the base stations. Thus the base will no longer be just dumb transmitters, but powerful and fast mobile edge computers, able to adapt to the tasks they are asked to perform. This capability is called Liquid Applications. In the case of emergency, the base stations are capable of turning round the emergency messages in less than 100 milliseconds, in some circumstances in less than 50 milliseconds. This means that a car can get the emergency message from a car ahead in less time than it takes to draw breath.

Liquid Applications [34] fundamentally changes the role of the base station. Liquid Applications is unique in the sense that it takes data from inside the base station to create a personalized and contextualized mobile broadband experience. It also enables the creation of innovative new services that can use proximity and location to connect subscribers and local points of interest, businesses and events.

At the heart of Liquid Applications is the groundbreaking Nokia Radio Applications Cloud Server (RACS). RACS deploys the latest cloud technology and service creation capabilities, running on standard IT middleware, into the base station. RACS provides processing and storage capabilities, together with the ability to collect real-time network data, for example radio conditions, subscriber location, direction of travel and more. This data can be exploited by applications to offer context-relevant services that transform the mobile broadband experience and directly translate that experience into value.

Liquid Applications creates user experience enhancements through acceleration, primarily due to content in close proximity that can be delivered faster. This translates into significant throughput and time-to-content improvements that are only possible from the network edge. Liquid Applications also has the ability to extract and process real-time network insights, allowing operators to rapidly move from problem identification to precisely pinpointing the cause of service degradation. Real-time data from Liquid Applications enhances network operations as well as provides deeper insights into customer and network behavior. Liquid Applications is also the catalyst for a new services eco-system around the base station, leveraging from its direct connectivity or local breakout functionality. Direct connectivity is possible to local venues (e.g. stadiums) and private Enterprise networks, with customers connecting over LTE.

Close proximity to the mobile customer enables significant improvements in time-to-content and initial response: in excess of 100% throughput improvement is achievable, from content served from the base station, with download times improved by up to 80%.

References

1. T. Kosch, C. Schroth, M. Strassberger and M. Bechler (2012) *Automotive Interworking*. Chichester, UK: John Wiley & Sons Ltd.
2. WAZE application website, available at http://www.waze.com (last accessed April 24, 2015).
3. T. Sukuvaara and P.Nurmi (2009) Wireless traffic service platform for combined vehicle-to-vehicle and vehicle-to-infrastructure communications, *IEEE Wireless Communications* **16**(6): 54–61.
4. T. Sukuvaara (2004) The evaluation of a self-configurable IP-based heterogeneous radio network concept, Licentiate thesis, University of Oulu, Department of Electrical and Information Engineering, Oulu, Finland.
5. M. Kangas, M. Heikinheimo, M. Hippi, *et al.* (2012) The FMI road weather model, *Proceedings of 16th International Road Weather Conference, SIRWEC 2012, 23–25 May 2012, Helsinki, Finland*, pp. 117–23.
6. T. Sukuvaara, R. Ylitalo and M. Katz (2013) IEEE 802.11p based vehicular networking operational pilot field measurement, *IEEE Journal on Selected Areas in Communications* **31**(9): 409–17.
7. IEEE Std. 802.11p/D9.0: Draft Standard for Information Technology – Telecommunications and information exchange between systems – Local and metropolitan area networks – Specific requirements, Part 11: Wireless LAN Medium Access Control (MAC) and Physical Layer (PHY) specifications, Amendment 7: Wireless Access in Vehicular Environments (2009) New York: Institute of Electrical and Electronics Engineers Inc.
8. R. Baldessari, B. Bödekker, A. Brakemeier *et al.* (2007) Car 2 Car Communication Consortium Manifesto, Version 1.1. Available via http://www.car-to-car.org (last accessed April 24, 2015).
9. ETSI Standard; ETSI ES 202 663 V1.1.0 (2010-01) Intelligent Transport Systems (ITS); European profile standard for the physical and medium access control layer of Intelligent Transport Systems operating in the 5 GHz frequency band, European Telecommunications Standards Institute (2010).
10. T. Sukuvaara, K. Mäenpää, R. Ylitalo, *et al.* (2013) Interactive local road weather services through VANET-capable road weather station. *Proceedings of 20th World Congress on ITS, October 14–18, 2013, Tokyo, Japan.*
11. IEEE Std. 802.11: Wireless LAN Medium Access Control (MAC) and Physical Layer (PHY) specifications (1997) New York: Institute of Electrical and Electronics Engineers Inc.
12. IEEE 802.11b: IEEE Std. 802.11b: Wireless LAN Medium Access Control (MAC) and Physical Layer (PHY) specifications; Higher-Speed Physical Layer Extension in the 2.4 GHz Band (1999) New York: Institute of Electrical and Electronics Engineers Inc.
13. Y. Xiao (2005) IEEE 802.11n: enhancements for higher throughput in wireless *LANs, IEEE Wireless Communications Magazine* **12**(6): 82–91.
14. IEEE Std. 802.11a: Wireless LAN Medium Access Control (MAC) and Physical Layer (PHY) specifications; Higher-Speed Physical Layer Extension in the 5 GHz Band (1999) New York: Institute of Electrical and Electronics Engineers Inc.
15. IEEE Std. 802.11g, Further Higher-Speed Physical Layer Extension in the 2.4 GHz Band (2003) New York: Institute of Electrical and Electronics Engineers Inc.
16. IEEE Std. 802.11n, Wireless LAN Medium Access Control (MAC) and Physical Layer (PHY) specifications: Amendment 4: Enhancements for Higher Throughput (2006) New York: Institute of Electrical and Electronics Engineers Inc.
17. O. Bejarano, E.W. Knightly and M. Park (2013) IEEE 802.11ac: from channelization to multi-user *MIMO, IEEE Communications Magazine* **51**(10): 84–90.
18. J. Hellåker (2012) OEM commitment to deploy. *Proceedings of Car 2 Car Forum 2012, 13–14 November 2012, Gothenburg,* Sweden.
19. G.R. Hiertz, D. Denteneer, L. Stibor, *et al.* (2010) The IEEE 802.11 Universe, *IEEE Communications Magazine* **48**(1): 62–70.
20. IEEE P1609.0/D5, IEEE Draft Guide for Wireless Access in Vehicular Environments (WAVE) Architecture, (2012) New York: Institute of Electrical and Electronics Engineers Inc.
21. C. Han, M. Dianati, R. Tafazolli, *et al.* (2012) Analytical study of the IEEE 802.11p MAC sublayer in vehicular networks, *IEEE Transactions on Intelligent Transportation Systems*, **2012**: 873–86.
22. K. Dar, M. Bakhouya, J. Gaber, *et al.* (2010) Wireless communication technologies for ITS applications, *IEEE Communications Magazine* **48**(5): 156–62.
23. C. Suthaputchakun and Z. Sun (2011) Routing protocol in intervehicle communication systems: A survey, *IEEE Communications Magazine* **49**(12): 150–6.
24. F.J. Martinez, C-K Toh, J-C Cano, *et al.* (2010) Emergency services in future intelligent transportation systems based on vehicular communication networks, *IEEE Intelligent Transportation Systems Magazine* **2**(2): 6–20.

25. J-J. Chang, Y-H, Li, W. Liao and I-C Chang (2012) Intersection-based routing for urban vehicular communications with traffic-light considerations, *IEEE Wireless Communications Magazine* **19**(7): 82–8.
26. H. Liang and W. Zhuang (2012) Cooperative data dissemination via roadside WLANs, *IEEE Communications Magazine* **50**(4): 68–74.
27. M. Emmelmann, B. Bochow and C.C. Kellum (2010) *Vehicular Networking: Automotive Applications and Beyond*. Chichester, UK: John Wiley & Sons Ltd.
28. S. Ni, Y. Tseng, Y. Chen and J. Sheu (1999) The broadcast storm problem in a mobile ad hoc network. *Proceedings of the ACM International Conference on Mobile Computing and Networking (MOBICOM), 1999*, pp. 151–62.
29. J. Gozalvez, M. Sepulcre and R. Bauza (2012) IEEE 802.11p vehicle to infrastructure communications in urban environments, *IEEE Communications Magazine* **50**(5): 176–83.
30. Y. Kishiyama, A. Benjebbour, T. Nakamura and H. Ishii (2013) Future steps of LTE-A: evolution toward integration of local area and wide area systems, *IEEE Wireless Communications* **20**(1): 12–26.
31. A.K. Salkintzis, C. Fors and R. Pazhyannur (2002) WLAN-GPRS integration for next-generation mobile data networks, *IEEE Wireless Communications* **9**(5): 112–24.
32. M. Peng, D. Wei, W. Wang and H-H Chen (2011) Hierarchical cooperative relay based heterogeneous networks, *IEEE Wireless Communications* **18**(3): 48–56.
33. I. Delaney (2014) Milliseconds matter for accidents, HERE and Nokia Networks explain, *HERE News*, September 14, 2014, available at http://360.here.com/2014/09/10/milliseconds-matter-accidents-here-nokia-networks (last accessed April 24, 2015).
34. Nokia Liquid Applications website, available at http://nsn.com/portfolio/liquid-net/intelligent-broadband-management/liquid-applications (last accessed April 24, 2015).

4

The Case for Wireless Vehicular Communications Supported by Roadside Infrastructure

Tiago Meireles[1], José Fonseca[2] and Joaquim Ferreira[3]
[1]*Instituto de Telecomunicações, Universidade da Madeira, Portugal*
[2]*Instituto de Telecomunicações, DETI-Universidade de Aveiro, Portugal*
[3]*Instituto de Telecomunicações, ESTGA-Universidade de Aveiro, Portugal*

4.1 Introduction

Wireless vehicular networks for cooperative Intelligent Transport Systems (ITS) have raised widespread interest in the last few years, due to their potential applications and services. Cooperative applications with data sensing, acquisition, processing and communication provide an unprecedented potential to improve vehicle and road safety, passengers' comfort and efficiency of traffic management.

In order to support such visionary scenarios, applications running in the vehicles are required to communicate with other applications in other vehicles or with applications deployed in the back office of the emergency services, road operators or public services. These applications run unattended, reporting information and taking commands from counterpart applications in the vehicle or network.

The mobile units of a vehicular network are the equivalent to nodes in a traditional wireless network, and can act as the source, destination or router of information. Communication between mobile nodes can be point-to-point, point-to-multipoint or broadcast, depending on the requirements of each application. Besides the ad-hoc implementation of a network consisting of neighbouring vehicles joining up and establishing Vehicle-to-Vehicle (V2V) communication, there is also the possibility of a more traditional wireless network setup, with base stations along the roads in Vehicle-to-Infrastructure (V2I) communication that work as access points and manage the flow of information, as well as portals to external WANs. Devices

Intelligent Transport Systems: Technologies and Applications, First Edition. Asier Perallos, Unai Hernandez-Jayo, Enrique Onieva and Ignacio Julio García-Zuazola.

operating inside vehicles are called On Board Units (OBUs), while devices operating on the side of the road are Road Side Units (RSUs), and have different requirements and modes of operation.

Safety, efficiency and comfort ITS applications exhibit tight latency and throughput requirements, for example safety critical services require guaranteed maximum latencies lower than 100 ms while most infotainment applications require QoS support and data rates higher than 1 Mbit/s. Besides latency and throughput, safety applications also require deterministic communications (real-time). For example, a vehicle involved in an accident should be granted timely access to the wireless medium to transmit warning messages, even in congested road scenarios.

The IEEE 1609 family of standards for Wireless Access in Vehicular Environments (WAVE) defines an architecture and a standardized set of services and interfaces that collectively enable V2X wireless communications. Additionally, the IEEE 1609 standards rely on IEEE 802.11-2012 Amendment 6 [1], also known as 802.11p, and the equivalent European standard ETSI ITS G5 [2]. The physical layer is almost identical to IEEE 802.11a, using also orthogonal frequency-division multiplexing (OFDM) with BPSK, QPSK, 16QAM and 64QAM modulations, but with double timing parameters to achieve less inter-symbol interference due to the multipath propagation and the Doppler shift effect.

With double timing parameters, the channel bandwidth is 10 MHz instead of 20 MHz, and the data rate is half, i.e. 3…27 Mbit/s instead of 6…54 Mbit/s. The maximum range is 1000 m, with line of sight (close to 300 m in typical conditions) for vehicle speed below 200 kph.

The medium access control (MAC) layer adopts a carrier sense multiple access with collision avoidance (CSMA/CA), same as IEEE 802.11a, but with a new additional, non-IP, communication protocol, either Fast Network and Transport Protocol (FNTP) or Wave Short Message Protocol (WSMP). These non-IP protocols are essentially low overhead, port mapping protocols, designed to be small, efficient and tailored to the simple, single-hop broadcast over capacity constrained radio frequency channels. Due to the tight timing constrains, non-IP protocols do not perform channel scanning, authentication and association. The non-IP protocols coexist, in parallel, with IPv6. IEEE 802.11p adopts the QoS policy of IEEE 802.11e and defines a mechanism for the dynamic switching of channels.

Since the IEEE 802.11p medium access control is based on CSMA, collisions may occur indefinitely due to the nondeterminism of the back-off mechanism. So, native IEEE 802.11p MAC alone does not support real-time communications. Nevertheless, the probability of collisions occurring may be reduced if the load of the network is kept low, which is difficult to guarantee in vehicular communications, or if some MAC protocol restricts and controls the medium access to provide a deterministic behaviour. Strict real-time behaviour and safety guarantees are typically difficult to attain in ad-hoc networks, but they are even harder to attain in high-speed mobility scenarios, where the response time of distributed consensus algorithms, e.g. for cluster formation and leader election, may not be compatible with the dynamics of the system.

There are basically two main design choices to implement a deterministic MAC protocol for wireless vehicular communications. It could either rely on the roadside infrastructure (V2I) or it could be based on ad-hoc networks (V2V), without roadside units support. Hybrid approaches are also possible. There are two complementary ways to secure determinism (real-time), of the roadside network necessary for RSUs' coordination: use a real-time network technology, usually at Layer 2, and employ resource reservation protocols to extend the guarantees to multiple networks and to higher layers.

Our vision about the safety services is that they will firstly appear as safety warnings. As the user feels these warnings becoming reliable, he will start to adapt the driving accordingly. As happened with the ABS (Anti-lock Braking System) technology, the safety warnings eventually become safety critical as an incorrect operation can put in risk a driver relying on them. Even with a small number of probe vehicles, safety events tend to cause alarm showers, e.g. a hard braking or a car crash in intense traffic environments will generate a large number of other hard braking events. In some operational scenarios the IEEE 802.11p MAC may no longer be deterministic, possibly leading to unsafe situations. This calls for a reliable communication infrastructure with real-time, secure and safety properties, which is mandatory to support the detection of safety events and the dissemination of safety warnings. In this work, we argue that the presence of a backhauling network infrastructure, adds a degree of determinism that will very useful to enforce real-time and safety at the wireless end of the network. For this purpose, we present a proposal of a deterministic MAC protocol, the vehicular flexible time-triggered (V-FTT), which adopts a master multislave time division multiple access (TDMA), in which the roadside units act as masters to schedule the transmissions of the on-board units.

4.1.1 Rationale for Infrastructure-based Vehicle Communications for Safety Applications

Several vehicular safety applications have been devised in order to increase safety in road environments; some of these applications are based on vehicle-to-vehicle communications (V2V), others on vehicle to infrastructure communications (V2I), or both. In this subsection, we briefly analyse the advantages and disadvantages of using V2V communications or V2I communications for the deployment of safety applications. Please note that we will use V2I or I2V with the same meaning, since safety communication between infrastructure and vehicle is usually bidirectional.

Some advantages of deploying safety applications relying on V2V communications are:

- An infrastructure is not needed, which means it is cheaper and easier to deploy.
- In principle V2V offers lower latency than an infrastructure-based solution since the communication is directly from source to destination [3].
- V2V based networks are attractive for rural areas and developing countries, as it does not require roadside units and can be easily implemented.
- No specific protocol is required to coordinate different units.

However V2V communications present some strong disadvantages in what concerns safety applications [4–6].

- Proper work of V2V communications requires a certain market penetration before any effects or improvements can be shown. It was estimated that in order to make the network usable, at least penetration of 10% is needed [7,8]. It was projected for the US Market [9], that in the best case, it would take 3 years to achieve such threshold penetration level, once the full-scale deployment of V2X enabled vehicles starts. Considering this and the fact that other advanced automotive technology, e.g. drive-by-X, took over 10 years to become available in the market, it could realistically be estimated that the 10% penetration level of V2V enabled vehicles will take 7–10 years.

- A V2V system may be vulnerable to a badly intended user who can broadcast false information about safety events, which cannot be validated by the infrastructure (possibly using data from other sensors).
- OBUs will have a processing overhead in some applications. For example, Cooperative Collision Warning receives information about position, velocity, heading and more from several surrounding vehicles [10]. An OBU must then compute these values with the current data from the vehicle, in order to decide if there is a collision risk or not.
- When using V2V communications, alarm showers (also known as broadcast storm) can occur, overloading the medium, unless some protocol is enforced to avoid that situation [11].
- Because V2V communications are ad-hoc and totally distributed, there is no global vision of any zone.
- Protocols to enforce determinism in V2V communications, such as cluster membership and cluster leader election, are heavy in terms of the required communication rounds.
- Connectivity disruptions can occur due to quick topology network changes, vehicle speed, when the vehicle density is low or totally disconnected scenarios occur. As a consequence, vehicles are not always able to communicate with each other [12].
- Hopping might be needed in order to relay a message, increasing the end-to-end delay.
- V2V communications have privacy and security issues. In a pure V2V architecture, authentication and key management becomes extremely difficult to manage, as it requires a prior knowledge of each vehicle's public key in order to verify users' identity. However, having a fixed identity can in turn raise a lot of privacy concerns [7].

In summary, V2V communications might be a solution in rural or low to medium dense areas where the roadside unit has a higher cost per user. In urban or suburban areas, where traffic density and velocities are high and accidents are more likely to occur, it is better to deploy and maintain RSUs and use I2V communications, which can prevent the V2V issues already mentioned:

- In I2V, the roadside infrastructure can be seen as an instantiation of the wormhole metaphor [13], in which it is assumed that uncertainty is not uniform or permanent across all system components, i.e. some parts are more predictable than others. In this way, the more predictable parts of the systems can be seen as wormholes since they will execute certain tasks faster or more reliably than apparently possible in the other parts of the system. Thus, it can be argued that I2V communications could be made safer than pure V2V.
- Security is very important: in V2I communications the RSUs can behave as a broker, analysing and editing the received vehicle data, validating safety events by cross-examining with other sources of information such as cameras, induction loops, or other available data, therefore minimizing the vulnerability problem.
- Using an infrastructure-based approach solves the connectivity disruption problem and RSUs can also be used to improve positioning information, as their position is well known [14].

- Some vehicle manufacturers are developing proprietary solutions, which do not favour communication capabilities among vehicles. I2V communications can solve this by having RSUs that can function as gateways between different vehicle communication systems.
- The RSU can control the medium access in order to avoid the broadcast storm problem.
- RSUs (or an entity that coordinates RSUs) can have a global vision of the communication zone and therefore make potentially better decisions.
- To solve privacy and security issues a centralized key distribution agent can assign disposable temporary identities to vehicles OBUs. This centralized agency can (via RSUs) effectively verify the identities of the OBUs. Even in the case of a hybrid approach, where V2V and V2I communications co-exist, the need for a V2I infrastructure is critical [7].

Adding to all of the above, it is our strong belief that a long period of time is expected before all the circulating vehicles are factory equipped with IEEE802.11p/WAVE or other wireless communication system that allows inter-vehicle communication for the purpose of safety applications. In this transitory period RSUs will play a major role in implementing safety wireless applications, particularly if vehicles can be retrofitted with on-board units (OBUs) that are as inexpensive (or funded) as the current vehicle equipment used for electronic tolling. We also believe that users will place more trust on a safety system managed by the road infrastructure than a full ad-hoc V2V system.

4.2 MAC Solutions for Safety Applications in Vehicular Communications

This section briefly presents the main proposals found in the literature to overcome the medium access control (MAC) issues of IEEE802.11p and ETSI-G5, in what concerns real-time communications guarantees, i.e. to have a bounded delay. We will focus on infrastructure-based solutions, but since few infrastructure based solutions were found, we also mention V2V solutions that are relevant to our work.

The two main communication parameters that affect the performance of active traffic safety applications are reliability and delay. Reliability means packets should be correctly received at destination and it depends on the packet error rate. In active safety applications, most of the communication between vehicles relies on broadcast messages. Predicting the reliability of these messages is a hard task due to the absence of explicit acknowledgment. The major challenges in vehicular communication arise from the wireless communication channel and the inherent high mobility in the network. The wireless communication channel quality is not stable and it is prone to unpredictable time-varying changes in connectivity and potentially high bit error rates. Several effects such as shadowing and scattering, the effects of Doppler and multipath spread deteriorate the radio signals propagation and data sent over the wireless medium can easily get lost or corrupted. This signal propagation deterioration is also influenced by the surrounding environment (e.g. buildings, vegetation, large trucks) and in this context, vehicle communications are strongly affected by the vehicle's speed. Furthermore, it is really important to design a proper MAC scheme, which can help to reduce the interference by carefully scheduling the channel access and their power levels. Another important communication parameter in active traffic safety applications is predictable delay. This means data needs to be delivered to the destination in a predefined time window, i.e. in real-time.

A parameter of utmost importance for infotainment applications is throughput, which has less significance in real-time communications. Conversely, real-time communications do not require high data rate or a low delay, but need to exhibit deterministic delay, meaning that it should be possible to compute the worst-case transmission time, since a missed deadline may severely affect the system or it may degrade the performance temporarily. In case of real-time communication, a deadline miss ratio is a central performance parameter, which should be zero for the case of hard real-time systems.

In wireless broadcast communication systems, a missed deadline may be caused by two factors when looking from the MAC layer perspective. Either the packet has never granted channel access or the packet was not received correctly. The deadline miss ratio is the probability that a packet does not reach the intended destination before the deadline. Therefore, a missed deadline is closely related to the channel access delay, i.e. the total time it takes from channel access request to the actual channel grant at the MAC layer.

In ETSI G5, most of the active traffic safety applications rely on a message that is periodically broadcasted by every vehicle in a period of time: Cooperative Awareness Message (CAM). CAM messages are a really important part in active traffic safety applications because these messages are broadcast messages, which do not receive acknowledgements. CAM messages are defined in [15] and include several possible data elements (e.g. CrashStatus, Dimension, Heading, Latitude, Longitude, Elevation, Longitudinal Acceleration, Speed). CAM messages are transmitted periodically and have strict timing requirements. They are generated by the CAM Management and passed to lower layers according to the following rules [15]:

- Maximum time interval between CAM generations: 1 s.
- Minimum time interval between CAM generations: 0,1 s.
- Generate CAM when absolute difference between the current heading and the last is greater than 4°.
- Generate CAM when distance between current position and last CAM position is greater than 5 m.
- Generate CAM when absolute difference between the current speed and is greater than 1 m/s.

These rules are checked every 100 ms. A CAM message is dropped whenever the channel access request does not result in actual channel access before the next message is generated. This happens because more recent information is available, i.e. it means that the time to live is exceeded so it is dropped at the generated vehicle. There will be temporary reduction in the performance efficiency of the application, if a periodic message misses a time limit. If the channel access is denied for consecutive packets of same vehicle and is forced to drop them, this can become a critical problem.

The MAC scheme should be designed in such a way that it provides a fair channel access to all vehicles, so that the packet drops should be evenly distributed among all OBUs. Therefore, fairness and scalability are important parameters in vehicular networks, and consequently repeated deadline misses (packet drops) from the same OBUs should be taken into account. Even though packets are successful in the channel access, the packets may still not be received properly because of the unreliable physical communication channel, i.e. packets were never correctly decoded at the receiver because of an unreliable channel. Due to the broadcast nature of both event-driven and periodic messages, the performance measurement parameters like the deadline miss ratio should be redefined if the receiver side is to be taken into account. In active traffic safety applications, the throughput depends on the density of vehicles in the interest range. Furthermore, the interest range and communication range are not necessarily the same; hence some applications have a larger interest range than the communication range. Multihop communication schemes are used for solving these problems.

The broadcast nature of periodic messages, e.g. CAM in vehicular safety applications, requires scalability and fairness. The proper design of MAC protocols is a key factor to grant fairness, provide scalability and predictable channel access in VANETs.

IEEE 802.11p uses CSMA/CA as channel access; this mechanism can lead to unbounded channel access delays because of the potential random backoff algorithm. Moreover, the carrier sensing mechanism preceding each message transmission implies that there is a race for network resources, resulting in issues like scalability and fairness, e.g. some stations may have to drop several consecutive messages, because many stations simultaneously try to access the channel. Due to that, some stations never get access to the channel before the deadline, whereas other stations drop zero or a few messages. This problem becomes a concern in high-density networks. Finally, when the channel is occupied or busy, the vehicles in CSMA must perform a backoff procedure and during high-density periods this mechanism can cause several vehicles to transmit simultaneously within radio range of each other due to the limited discrete random numbers in the backoff procedure, impacting on scalability.

Due to very high-speed mobility, V2V and V2I communication links have a very short duration. One strategy for increasing the duration of communication links is increasing the transmission range in sparse traffic conditions, where only a few vehicles may be present on the road. However, increasing the transmission range may generate high levels of disruptive interference and high-network overhead in dense traffic conditions. It follows that dynamic adaptation of transmission power in response to changing traffic density is a critical requirement. Another strategy is assigning different priority levels to various traffic-related messages according to their urgency or delay requirements.

Most existing traffic models assume stationary distribution of vehicles on the road and use the Most Forward within Range (MFR) propagation model. That is, a packet is forwarded to the farthest vehicle within the transmission range of the source. However, vehicles located near the outer edge of the transmission range are more likely to cross the boundary and fail to successfully receive the packet. This will add more delay to the multi hop communication and challenge the performance of vehicle safety applications. Therefore, a scalable, fair, predictable and centralized MAC layer should be designed for achieving the real-time communication in traffic safety applications.

We will next review infrastructure based vehicle communications proposed solutions that attempt to deal with some of the problems mentioned above (a summary is shown in Table 4.1).

4.2.1 Infrastructure-based Collision-free MAC Protocols

Annette Böhm *et al.* [16] describe five different real vehicle traffic scenarios covering both urban and rural settings at varying vehicle speeds and under varying line-of-sight (LOS) conditions, discussing the connectivity that could be achieved between two test vehicles. The major conclusion from those tests is that connectivity is almost immediately lost with the loss of LOS. This limitation is a serious drawback for safety-critical applications usage.

This suggests the need for studies regarding the deterioration of signal propagation. The use of infrastructure to mitigate non-line-of-sight (NLOS) link failures is one solution to address the situations observed in these tests.

The IEEE 802.11p MAC method is based on 802.11e Enhanced Distributed Channel Access (EDCA) with QoS support, where four different access classes are provided. In IEEE 802.11e, time is divided into super frames, each consisting of a contention-based phase (CBF) and a collision-free phase (CFP). Unlike other 802.11 WLAN standards, *the 802.11p standard does not provide an additional, optional collision-free phase*, controlled centrally by an access point through polling.

Several authors 00 propose a deterministic Medium Access Control (MAC) scheme for V2I communication by extending the 802.11p standard with a collision-free communication phase controlled by an access point as provided in other 802.11 WLAN standards. Collision-free MAC protocols are considered deterministic as data collisions do not occur and a worst-case delay from packet generation to channel access can be computed. The collision-free phase needs support from a 'coordinator', in this case a Road Side Unit or a dedicated 'centralized' vehicle, which takes responsibility for scheduling the traffic and polling the mobile nodes for data. In this way the channel is assigned for a specific period of time to each vehicle equipped with an OBU without competition and safety-critical, real-time data traffic is scheduled in a collision-free manner by the RSU.

Böhm and Jonsson [5] assign each vehicle an individual priority based on its geographical position, its proximity to potential hazards and the overall road traffic density. This is done by introducing a real-time layer on top of the normal IEEE 802.11p. A super frame is created in order to obtain a Collision Free Phase (CFP) and a Contention Based Period (CBP). In the CFP the RSUs assume the responsibility for scheduling the data traffic and polls mobile nodes for data. Vehicles then send their heartbeats with position information and additional data (such as speed, intentions, etc.). 'A heartbeat message is a message sent from an originator to a destination that enables the destination to identify if and when the originator fails or is no longer available' [17]. The RSU sends a beacon to mark the beginning of a super frame, stating the duration of the CFP, so that each vehicle knows when the polling phase ends and when to switch to the regular CSMA/CA from IEEE 802.11p, which is used in the CBP, along with the random backoff mechanism which is similar to IEEE 802.11e.

The length of CFP and CBP is variable. Real-time schedulability analysis is applied to determine the minimum length of CFP such that all deadlines are guaranteed. The remaining bandwidth is used for best-effort services and V2V communications. In order for RSUs to start scheduling vehicle transmission, vehicles must register themselves by sending out connection setup requests (CSR) as soon as they can hear the RSU. This is done in the CBP, so a minimum risk exists of vehicles failing to register. They can, however, receive information from RSUs and communicate using the CBP. Böhm points out that vehicles might want to increase the number of heartbeats sent during lane change or in certain risk areas, but this is not clearly explained. Another interesting issue is that a proactive handover process is defined, based on the knowledge of road path and RSUs locations. Nothing is mentioned about RSU coordination and how it is done.

Böhm's protocol has many similarities with Tony Mak *et al.* [18], who proposed a variant to 802.11 Point Coordination Function (PCF) mode, so that it could be applied to vehicular networks. A control channel is proposed, in which time is partitioned into periodic regulated intervals (repetition period). Each period is divided into a contention free period, also named

CFP by the author (with the same meaning as Collision Free Phase used by Böhm), and an unregulated contention period. The scheme is similar to Böhm's, where each vehicle is polled by an RSU or Access Point (AP) during the CFP, similarly to the PCF of 'regular' IEEE 802.11.

Vehicles need to register and deregister so the polling list is kept updated. For this purpose a group management interval is created so that vehicles entering and leaving the region can notify the RSU. However, this beacon is sent in the CP and contends with other communications. The authors propose that the beacon is repeated to decrease probability of reception failure of the beacon. No schedulability analysis is made in [18] but the authors claim that the time between consecutive polls for vehicles in the RSU coverage area is bounded.

4.2.2 RT-WiFi – TDMA Layer

RT-WiFi [19] is a MAC protocol that aims to support real-time communications in IEEE802.11 networks in industrial environments. It allows a dynamic association of stations while supporting interference from non-real-time devices. Real-Time (RT) stations are interconnected by a central coordinator (in the case of vehicular communications it could be an RSU), which has a global vision of all the network traffic. All stations use EDCA, but RT stations use the Force Collision Resolution (FCR) mechanism, which aims to favour RT stations when collisions occur between RT and non-RT stations. This is done by simply deactivating the backoff mechanisms of RT stations. This means that whenever a collision occurs between one RT station and one or more non-RT stations, there is a higher chance that the RT station messages will be transmitted before the remaining ones. However, FCR does not solve the collision between RT stations. For that purpose, a TDMA layer is added so that RT stations can coexist in the same communication environment. In order to support interferences from other devices operating in the same frequency and coverage area, the slot size from the TDMA layer can be dimensioned to have the size of the maximum number of retransmissions that a station is allowed to perform, thus increasing the probability of message delivery. The order and size of the slots can be variable, in order to optimize the usage of the medium also giving some flexibility to the system. Another claimed advantage is that RT-WiFi is capable of supporting real-time communication service by controlling only a small group of stations (RT stations), without the need to update all devices that operate in the same frequency.

No reference is made about the applicability of RT-WiFi to vehicular environments, but it seems quite complex to implement due to the variable slot size. Besides that, the TDMA cycle grows linearly with the number of RT stations, which can be tricky if a large number of vehicles is considered.

4.2.3 Vehicular Deterministic Access (VDA)

Rezgui and Cherkaoui [20] proposed an adaptation of the Mesh Coordinated Channel Access (MCCA) standard (used in IEEE802.11s) for IEEE802.11p and named it VDA – Vehicular Deterministic Access. VDA aims at high-density scenarios and safety messages delivery within a two-hop range. The mechanism extends the typical 802.11p medium reservation procedure using schedule VDA opportunities (VDAops) within a two-hop neighbourhood. These VDAops are negotiated between neighbouring vehicles and then performed in multiples of time-slot units during the delivery traffic indication message (DTIM). Similarly to Böhm, the authors propose that the ratio of Contention Free Period and Contention Period can be adjusted dynamically. VDA provides better results than regular IEEE802.11p and offers a bounded delay. In order to integrate non-enabled vehicles, the authors suggest an extended VDA protocol.

4.2.4 *Self-organizing TDMA (STDMA)*

Although this protocol was not designed for I2V communications, it solves some IEEE802.11p MAC issues. In time slotted MAC approaches, the available time is divided into fixed length slots and further grouped into frames. STDMA is in commercial use in a collision avoidance system for ships. It is a self-organizing MAC method, using a nonblocking time-slotted MAC scheme. In most nonblocking time-slotted approaches a random access channel is used for slot allocation, where part of the frame is used for slot allocation but STDMA uses another method: nodes listen to the frame and determine the current slot allocation, based on what is perceived as free and occupied slots in the frame.

STDMA follows a distributed approach, where ships send their position message in the automatic identification system (AIS). The AIS frame length is 1 minute and has 2250 slots. The update rate of the position messages depends on the speed of the ship (the higher the speed, the higher the update rate). STDMA always grants channel access for all packets before a predetermined time, regardless of the number of competing nodes. It is scalable and the channel access delay is upper bounded. When no slots are available, simultaneous transmissions are allowed based on position information, a node that is forced to select an occupied slot will transmit at the same time as another node situated furthest away from itself.

Studies [21] have shown that STDMA can be well adapted to the vehicular environment for V2V communications, although it requires tight synchronization, e.g. using GPS. Simulations were made using a frame length of 1 second and the possibility of changing the number of slots from frame to frame. Obtained results show a lower probability of packet drop using STDMA when compared with CSMA/CA.

4.2.5 *MS-Aloha*

MS-Aloha is another slotted MAC protocol specifically designed for VANET, intended for V2V communications. Similar to STDMA, all nodes must synchronize using GPS and they share a common periodic frame structure divided into a variable number of slots. There is a Frame Information Field (FI) containing information on how each node perceives each slot (free, busy, collision). The FI is meant to propagate network information over three hops. Each node infers the state of each slot both by direct sensing and by the correlation among the received FIs. Based on them, it generates its own FI using the following mechanism:

- If node A receives a FI announcing slot J engaged by X, then A forwards it. If it receives two FI announcing the reservation by different nodes of the same slot J, A announces a collision in J.
- A node tries to reserve a slot simply by picking a free one, based on its direct channel sensing and on the FIs received.

- The reservation state of a slot is not forwarded more than two-hop far from the transmitter, in order to enable slot re-use.

The drawback of MS-Aloha is the overhead introduced by FI, which can be minimized by reducing the node identifier size to 8 bit and using a 'label swapping' algorithm in order to reuse the identifier geographically. resumes all the MAC protocols discussed in this section.

Analysing these protocols one may conclude that some critical aspects of I2V safety real-time communications were not addressed and there is room for improvement. Centralized protocols are not easily scalable; some protocols do not consider the interference of vehicles not compliant with a particular protocol, i.e. they assume that all vehicles run the same MAC protocol, and others do not consider the possibility of having RSUs overlapping transmission ranges, possibly causing a vehicle to receive inconsistent time slot allocation from different RSUs. These observations inspired the development of a novel deterministic MAC protocol, the vehicular flexible time-triggered (V-FTT), which adopts a master multislave time division multiple access (TDMA), in which coordinated roadside units act as masters to schedule the transmissions of the on-board units.

Table 4.1 Brief state of the art on MAC protocols for vehicular safety applications.

Protocol	V2V / I2V	Pros	Cons
Real-time I2V (Böhm)	V2I	• Provides bounded delay • Location based priority zones • Adaptable ratio between contention free phase and contention based phase.	• RSU uses polling mechanism. • Unspecified: vehicles change their warning message rate. • RSU coordination is not defined.
Multi-channel VANET	V2I / V2V	• No real-time analysis. • Basis for multi-channel WAVE proposal.	• RSU uses polling mechanism. • RSU Beacon must contend with other messages.
RT-WiFi	N/A	• Centralized mechanism. • Allows coexistence of RT stations and non-RT stations.	• It is not destined to vehicular environments. • The RT cycle grows with the number of RT stations. • No study yet on maximum number of RT stations it can allow.
Vehicle Deterministic Access (VDA)	V2V	• Provides bounded delay. • High-density scenarios. • Adaptable ratio between contention free phase and contention based phase.	• Two-hop range.
Self-organizing TDMA	V2V	• Provides bounded delay. • Lower probability of packet drop than regular CSMA/CA	• Requires GPS for synchronization
MS-Aloha	V2V	• Scalable with upper bounded delay	• Requires GPS for tight node synchronization • Overhead introduced by Frame Information Field

4.3 Vehicular Flexible Time-triggered Protocol

In terms of V2I communications, low traffic density scenarios have no MAC issues to solve, since all vehicles have the chance to communicate with the infrastructure. High traffic scenarios at low travelling speeds can cause some issues for nonsafety communications due to delays, but at these speeds, time critical safety events (in terms of maximum latency) have a lower probability of occurring. The particular scenario that can cause some problems for safety events dissemination occurs when a high number of vehicles travelling at high speeds need to communicate. This is the case for urban scenarios and motorways near urban areas. Urban scenarios will not be considered in this work since speeds are considerably lower than in motorways, and several studies have already been done in urban fields [10,22,23]. Furthermore, it can be assumed that the mechanisms adopted for urban motorways can be applied to urban scenarios.

We will consider the scenario of motorways near urban areas, since it is common for this type of motorway to have peak hours of large traffic with vehicles travelling at high speeds. This combination of high speed and high traffic means we have high probability of event occurrence and a large number of vehicles that need to be informed and/or to communicate an event. We start by presenting a possible model for RSU deployment in the following paragraphs.

4.3.1 Model for RSU Deployment in Motorways

In order to guarantee timely information about events that present a risk to driver safety, we will define a model for RSU deployment. We expect the deployment of RSUs to be expensive, assuming they are part of a wired backhauling network; therefore it is important to carefully choose their placement. In [24] an optimal RSU placement strategy was devised, based on vehicle density, average speed and accident probability. A more simple approach is to start RSU deployment in dense traffic areas (e.g. motorways near urban areas) and accident-prone zones such as dangerous curves or specific road sections such as tunnels or bridges. The road locations that have a record of a large number of crashes are also known as *blackspots* and we will use this term from this point on. In order to be effective, each blackspot zone must have total RSU coverage. In our work, we define specific and limited areas covered by RSUs which will be termed Safety Zones (S_z). Figure 4.1 depicts such a Safety Zone.

Whenever a vehicle enters the S_z (A), it must register itself in the infrastructure, so that RSUs know exactly how many vehicles are in the Safety Zone. In this process of registration each vehicle's OBU will be assigned a temporary identifier (t_{ID}). During its travel in the S_z, each OBU will be assigned to an RSU, i.e. each RSU shall be responsible for scheduling the

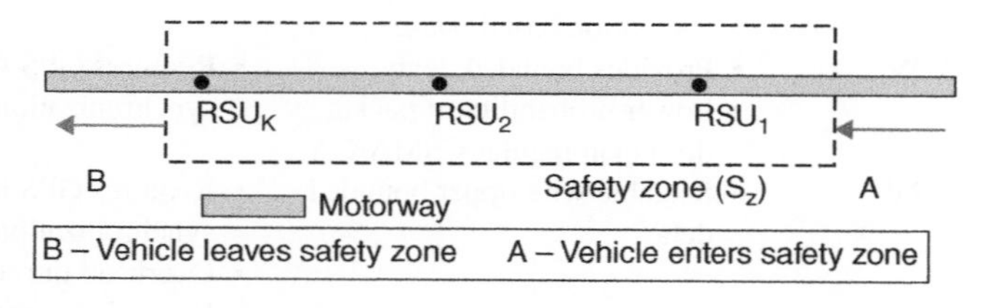

Figure 4.1 Definition of Safety Zone (S_z).

communications of the set of vehicles circulating in its coverage area. According to this scheme, RSUs can behave as a single entity having all the knowledge about the safety zone. All safety communications in the Safety Zone will therefore be controlled by the RSUs, which will process all the information received from the vehicles and, if needed, will cross-check it with their own information obtained from other sources (e.g. sensors, cameras). Whenever a vehicle leaves the S_z it will be deregistered from the system (B).

4.3.2 RSU Infrastructure Window (IW)

As RSUs cooperate to schedule OBUs safety communications, they must be able to coordinate their own transmissions, avoiding possible mutual interferences. To support RSU coordination, it is assumed that they are fully interconnected by a backhauling network. It is also assumed that RSUs are able to receive messages from vehicles travelling in both directions, RSUs' communication radius is considered to be circular and all RSUs are synchronized via GPS. Figure 4.2 depicts an example of such an infrastructured system.

In this RSU coordination proposal, RSUs transmit the OBUs scheduling in a reserved window, called the Infrastructure Window. Within this window, time slots are reserved for each RSU, as depicted in Figure 4.3. As RSUs are synchronized, they are able to respect the slot boundaries. The proposed mechanism to schedule RSU transmissions takes into consideration possible overlaps in the transmission range of each station to better mimic realistic propagation models and to add redundancy, thus increasing reliability. In nonurban areas RSUs will more likely be distributed across motorways. Therefore, consider that:

- R_z is the total number of RSUs in the Safety Zone (S_z);
- S_{IW} is the number of slots to be used in the RSU Infrastructure Window (one slot per RSU), corresponding to the maximum number of simultaneous RSU transmissions that an OBU can listen to.

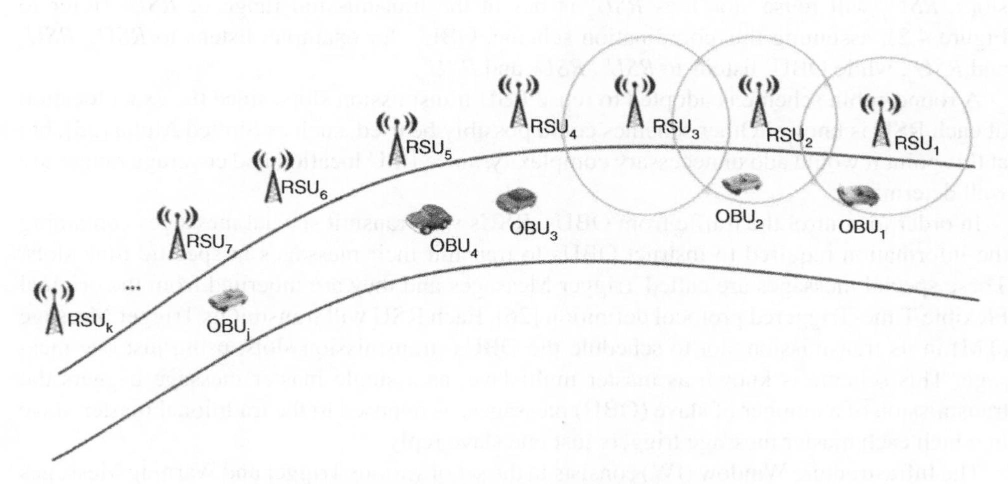

Figure 4.2 RSU distribution along the motorway.

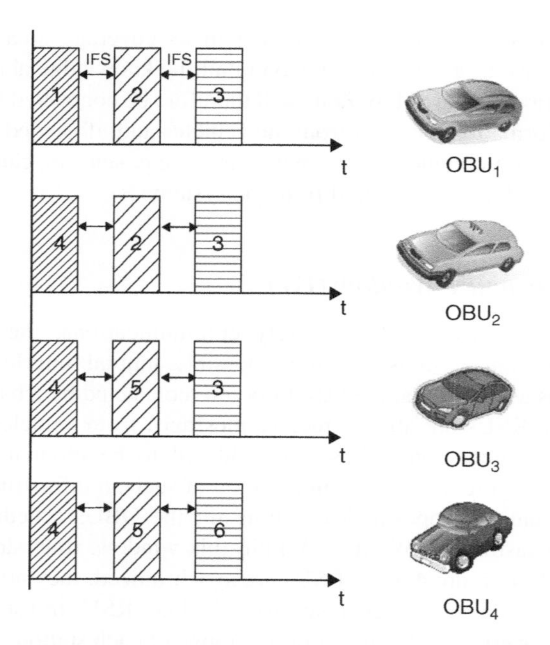

Figure 4.3 Round Robin scheme for RSU transmission slot in the IW (S_{IW}=3).

If RSUs are numbered according to their position, RSU_i with i=1 to R_z then RSU_i will transmit in a slot that can be computed by Equation (4.1).

$$Slot(RSU_i) = ((i-1)\%(S_{IW})) + 1 \tag{4.1}$$

As an example, if S_{IW} is 3 this means RSU_1 will transmit in slot 1, RSU_2 in slot 2, RSU_3 in slot3, RSU_4 will reuse slot 1 as RSU_4 is not in the transmission range of RSU_1 (refer to Figure 4.2): assuming this coordination scheme, OBU_2, for example, listens to RSU_2, RSU_3 and RSU_4, while OBU_3 listens to RSU_3, RSU_4 and RSU_5.

A round-robin scheme is adopted to reuse RSU transmission slots, since the exact location of each RSU is known. Other schemes could possibly be used, such as Slotted Aloha [25], but at this point it would add unnecessary complexity, since RSU location and coverage ranges are well determined.

In order to control the traffic from OBUs, RSUs will transmit special messages containing the information required to instruct OBUs to transmit their messages in specific time slots. These special messages are called Trigger Messages and they are inherited from the original Flexible Time-Triggered protocol definition [26]. Each RSU will transmit its Trigger Message (TM) in its transmission slot to schedule the OBUs' transmission slots, using just one message. This scheme is known as master multislave, as a single master message triggers the transmission of a number of slave (OBU) messages, as opposed to the traditional master-slave in which each master message triggers just one slave reply.

The Infrastructure Window (IW) consists in the set of various Trigger and Warning Messages transmitted by a configurable number of adjacent RSUs, depending on the desired redundancy level. The size of IW varies from (IFS + RSU_{slot}) to (S_{IW} * (IFS + RSU_{slot})), where IFS

represents an inter-frame space and S_{IW} is the maximum number of adjacent RSUs whose transmissions can be heard simultaneously by an OBU. All OBUs receive one or more TMs from the RSUs and search each TM for its temporary identifier (t_{ID}) in order to know if and when to transmit the OBU safety message.

4.3.3 V-FTT Protocol Overview

To ensure safety, it must be guaranteed that, in a worst-case scenario, all vehicles in the Safety Zone have the opportunity to transmit safety-related messages in a timely way, whether there are events to report or not.

The goal is to guarantee that information about events that can put at risk driver safety is transmitted in due time and, for this to happen, an infrastructure-based approach is proposed, where RSUs are coordinated among themselves and vehicles OBUs dynamically register and deregister from the system. As explained earlier in the model for RSU deployment, we assume that all OBUs register themselves in the Safety Zone, meaning that all registered OBUs will transmit their information (speed, position, any safety event) only in the instants determined by the RSUs.

The RSUs are responsible for two main operations:

- to schedule the transmission instants of the vehicle OBUs in what concerns the safety massages they have to broadcast during their stay in the Safety Zone;
- to receive information from the OBUs, edit that information and publish the edited safety information in the adequate places and instants.

From the communications point of view, the OBUs must:

- listen to the RSU transmissions (at least one RSU should be heard) and retrieve the safety information and dispatching information;
- always transmit its specific safety frame in the time window defined by the RSUs.

Figure 4.4 depicts the information flow diagram: each time an RSU receives any OBU safety event or information, it cross-validates it with its own sources of information (e.g. cameras, induction loops, infrared sensors, other vehicles). It is of the utmost importance that the information broadcast by the RSU is trustworthy. This is required in order to avoid possible intrusions where a badly intended user can try to cause a false alarm situation. RSUs must be very careful in validating OBU events and editing the information that is broadcasted to the vehicles in the Safety Zone. Consider, for example, that a malicious driver transmits a false Emergency Electronic Brake Light (EEBL) message. If no editing is made, several vehicles could receive a false alarm, which could lead to dangerous situations or even accidents. This edition operation must obviously be performed in bounded time so that the results can be transmitted to the OBUs in real-time. Due to the possibility of installing higher performance devices at the infrastructure as well as high throughput communication links with real-time operation capabilities, this operation, although complex, seems possible and with a controlled cost when compared with the construction and maintenance costs of the motorway.

Going back to the RSU to OBUs communications, note that nonregistered vehicle OBUs (or non-V-FTT-compliant OBUs) will also receive safety information from RSUs. They will, however, not be able to transmit information according to the proposed protocol, although they can

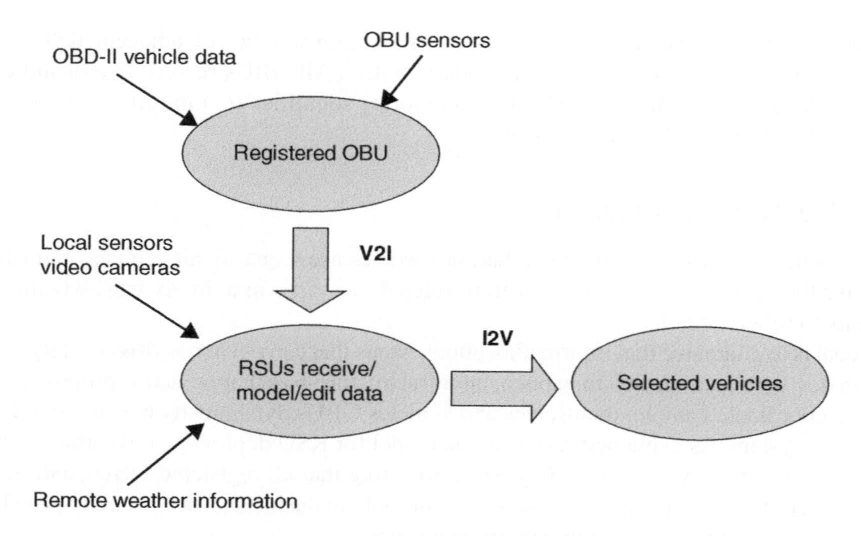

Figure 4.4 Vehicle information flow diagram.

still contend for transmission without any guarantees in the appropriate window. The V-FTT protocol timeline is cyclic and divided into elementary cycles (EC). Each EC has three windows:

- *Infrastructure Window (IW)* – based on the information received from the OBUs and some cross-validation with its own sources, the RSUs build the schedule for OBU transmissions. For that purpose each RSU periodically broadcasts a Trigger Message (TM) containing all identifiers (t_{ID}) of the OBUs allowed to transmit safety messages in the next period of OBU transmission, named Synchronous OBU Window. Based on OBU information and cross-validation, RSUs identify safety events and send warnings to OBUs belonging to vehicles affected by those specific safety events (protocol enabled and others). The warning messages (WM) have variable duration, depending on the number of occurred events. Each RSU will therefore transmit its TM and WM in its respective RSU transmission slot. Since each RSU slot will have a fixed size, care must be taken in order to fairly distribute slot time to TM and WM. There is no medium contention during the IW.
- *Synchronous OBU Window (SOW)* – this is where OBUs have the opportunity to transmit information to RSUs (V2I) without medium contention. Each OBU will have a fixed size slot (SM) to transmit vehicle information (speed, acceleration, etc.) and any safety event (e.g. EEBL). The SOW duration is variable. Each OBU will have a maximum of one slot per SOW, in order to ensure a fair access to the medium by all OBUs.
- *Free Period (FP)* – In the free period a contention period is ensured, where non-enabled OBUs are able to transmit safety messages and RSUs and OBUs are able to transmit non-safety short messages. Enabled OBUs may also transmit safety messages but without any guarantees since they have to contend for the medium. A minimum size for the FP must be guaranteed in order to reserve a contention period in the Elementary Cycle.

Figure 4.5 presents the Elementary Cycle and the transmission windows it contains.

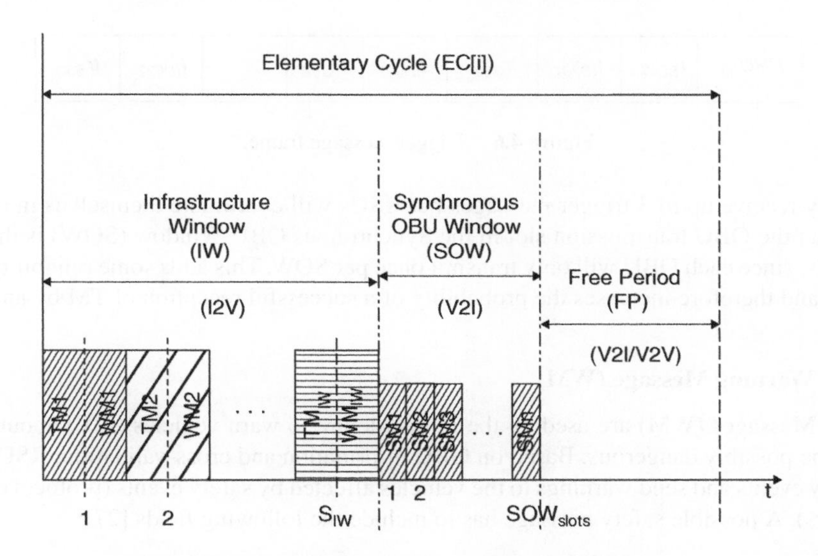

Figure 4.5 Proposed Vehicular FTT (V-FTT) protocol.

4.3.3.1 Trigger Message (TM)

A trigger message is a message broadcasted by each RSU, containing information about the OBUs that are allowed to transmit safety messages in the next period of OBU transmission (SOW). Each RSU is assigned a fixed size transmission slot, so that each RSU is able to know the exact position of its transmission slot. The RSU slot size limits the maximum TM length; therefore the choice of the maximum TM length is crucial, since, if a large enough length to accommodate all vehicles travelling inside a RSU coverage is chosen, bandwidth might be wasted when the number of vehicles served by an RSU is smaller than the TM number of slots. On the other hand, if a small maximum length for TM is chosen, it might not be enough to schedule all vehicles served by the RSU.

Figure 4.6 depicts the trigger message frame format, which starts with a field that identifies the RSU (RSU_{ID}), followed by an indication of when the Synchronous OBU window should start (t_{SOW}), and then by all temporary identifiers (t_{ID}) of the OBUs served by this RSU that are allowed to transmit in the next OBU Window. The transmission slot number (tr_s) in the OBU window is shown together with each t_{ID}, so that each OBU knows when to transmit.

- t_{SOW} – period of time between the beginning of this Trigger Message frame and the beginning of the Synchronous OBU window, measured in μs;
- RSU_{ID} – unique identifier for a Road Side Unit (RSU);
- t_{ID} [1 to N_{max}] – temporary OBU Identifier, from 1 to N_{max}, which is the absolute maximum number of vehicles that can be served simultaneously in the Safety Zone;
- tr_s [1 to SOW_{slots}] – OBU transmission slot, from 1 to SOW_{slots}, which is the maximum number of transmission slots allocated for the next Synchronous OBU window.

In Figure 4.6 an example is shown where OBUs with ID 207, 007 and 622 are allowed to transmit in slots 22, 87 and 33, respectively. As an example, for a redundancy level of 3, an

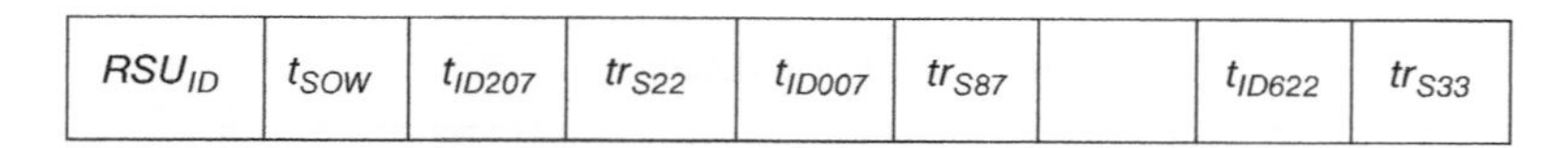

Figure 4.6 Trigger message frame.

OBU may receive up to 3 trigger messages but RSUs will coordinate themselves in order to ensure that the OBU transmission slot in the Synchronous OBU Window (SOW) is the same in all TMs, since each OBU will only transmit once per SOW. This adds some reliability to the protocol and therefore increases the probability of a successful reception of TM by an OBU.

4.3.3.2 Warning Message (WM)

Warning Messages (WM) are used by the infrastructure to warn vehicles (I2V) about events that can be possibly dangerous. Based on OBU information and cross-validation, RSUs identify safety events and send warnings to the vehicles affected by safety events (protocol enabled and others). A possible safety message has to include the following fields [27]:

- eventID, sourceID, transmitterID, location and additional info.

Most of the safety applications will only need the first four fields, while others (such as curve speed warning) need to send additional data. This means Warning Messages will have a variable size, contrary to Synchronous Messages. An OBU receiving the safety message will use the location data to find out if the event is 'in front' or 'behind' its travel path. This is done by comparing the event location with the current and recent vehicle locations.

4.3.4 Synchronous OBU Window (SOW)

The OBU window consists in the synchronous V2I window, divided into fixed size OBU slots (SM) where scheduled OBUs are able to transmit heartbeat messages, the Basic Safety Messages (BSM). The payload of the BSM includes, for example, vehicle speed, vehicle acceleration, vehicle position and any set of events detected by the vehicle (hard braking, malfunction, etc.). Consider the number of transmission slots in the Synchronous OBU window size (SOW_{slots}) to be:

$$SOW_{slots}\text{: 0 to}\left[\left(S_{IW} * N_{VRSU}\right)-\left(\left(S_{IW}-1\right) * N_{Vint}\right)\right] \quad \left(\text{number of OBU transmission slots}\right)\text{, where:}$$

- S_{IW} is the maximum number of adjacent RSUs which transmissions can be heard simultaneously by an OBU;
- N_{VRSU} is the maximum number of vehicles served by an RSU;
- N_{Vint} is the union of all sets of vehicles that can listen simultaneously to more than one RSU in a set of adjacent RSUs.

The number of transmission slots in the Synchronous OBU window (SOW_{slots}) is variable, in order not to waste bandwidth. It could even be zero in case there are no vehicles in the area covered by the RSUs. Its maximum size will not reach $S_{IW}*N_{VRSU}$ because OBUs will exist that can listen simultaneously to two or more RSUs, and each OBU will only have one opportunity to transmit per OBU window. This means RSUs must synchronize in case they serve common OBUs in order to assign them in the same slot in the OBU window.

Ideally each vehicle should have the opportunity to transmit its heartbeat data (speed, position and events) once per Elementary Cycle (EC), but in dense scenarios this might have to be re-evaluated. In that case a scheduling algorithm is required in order to allow fair transmission opportunities for all vehicles. The allocation of OBU slots must take into account the available bandwidth.

Each Synchronous Message (SM), where each OBU transmits important data and possible safety occurrences, is based on the standard messaging set for DSRC [28], and we will keep the same denomination: Basic Safety Message (BSM). We proposed some modifications, based on the dynamic data of the Cooperative Forward Collision Warning [10] and the Vehicle Safety Extension Data Frame referred by [29]. Notice that synchronous messages could also be Cooperative Awareness Messages (CAM).

4.4 V-FTT Protocol Details

After giving a V-FTT protocol overview, this section formalizes and further develops some concepts presented in the previous section.

4.4.1 *Trigger Message Size*

In the registration process, each OBU receives a temporary unique identifier (t_{ID}) to be used during its travel through the Safety Zone (S_Z). OBU MAC addresses could be used for the same purpose but a shorter-sized ID is more bandwidth efficient. The number of bits of t_{ID} (8 to n) depends on the absolute maximum number of vehicles that can travel simultaneously in the Safety Zone, $N_{\max}$, which depends on the Safety Zone characteristics (safety zone distance, number of lanes, etc.). t_{ID} is then related to $N_{\max}$, as is shown in Equation (4.2):

$$t_{ID} = \log_2\left(N_{\max}\right) \tag{4.2}$$

In order to calculate $N_{\max}$ it is necessary to know the safety zone characteristics:

• l_{Sz} 0 to x (m)	length of Safety Zone (m);
• n_{lanes} $[1, y]$	number of lanes for each travel path in motorway;
• V_{length}	average vehicle length (m);
• Tr_{length}	average Truck length (m);
• $v_{spacing}$	average spacing between two consecutive vehicles (m); the spacing between vehicles is a function of vehicle speed and traffic density;
• tr_{perct} [0, 1]	percentage of trucks among the total number of vehicles;
• n_{Sz} $[0, N_{\max}]$	number of vehicles in the safety zone, this number can vary depending on traffic conditions and presence of trucks.

The number of vehicles in the safety zone is then shown in Equation (4.3):

$$n_{S_Z} = \frac{l_{S_Z}}{\left(\left(V_{length} * \left(1 - tr_{perct}\right) + Tr_{length} * \left(tr_{perct}\right)\right) + v_{spacing}\right)} \times n_{lanes} \tag{4.3}$$

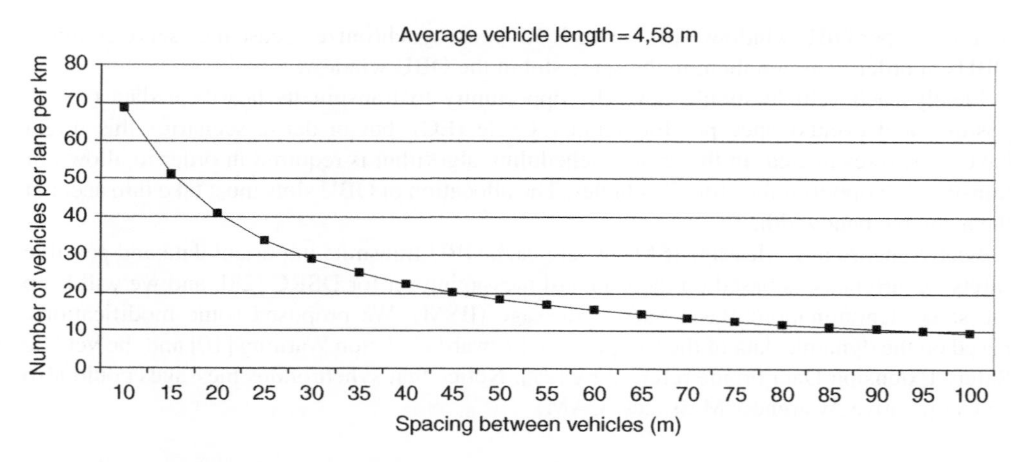

Figure 4.7 Number of vehicles per lane per km for an average vehicle length of 4,58 m.

N_{max} occurs when no trucks are present (tr_{perct}=0) and $v_{spacing}$ is minimum, which simplifies the previous equation into Equation (4.4):

$$n_{S_Z} = \frac{l_{S_Z}}{\left(V_{length} + v_{spacing}\right)} \times n_{lanes} \tag{4.4}$$

Figure 4.7 presents a graphical representation is shown where the number of vehicles per lane per km can be seen in function of the average spacing between vehicles.

The maximum possible number of vehicles in the safety zone (N_{max}) is shown in Equation (4.5):

$$N_{max} = \max\left(n_{S_Z}\right) \tag{4.5}$$

In order to determine the size of a Trigger Message, it is necessary to determine the following:

- number of bits of $t_{SOW} = \lceil \log_2(IW) \rceil$; where the maximum length of the Infrastructure Window is (SIW * (IFS + RSU_{slot}));
- number of bits of $RSU_{ID} = \lceil \log_2(n_r) \rceil$, where n_r is the total number of RSUs deployed in the Safety Zone; the number of RSUs depends on the length of the Safety Zone (l_{Sz}) and the coverage range of each RSU, C_r, as shown in Equation (4.6):

$$n_r = \left\lceil \frac{ls_z}{C_r} \right\rceil \tag{4.6}$$

- number of bits of $t_{ID} = \lceil \log_2\left(N_{max}\right) \rceil$;
- number of bits of $t_{rs} = \lceil \log_2(SOW) \rceil$, where SOW represents the maximum length of the Synchronous OBU Window.

Recall that the maximum number of OBUs that can appear in a TM is majored by the maximum number of vehicles served by an RSU, N_{VRSU}. In other words, ideally, an RSU to be able to include all vehicles in its coverage area in the Trigger Message. Bandwidth limitations will most likely pose a limit for the number of OBU slots in the OBU window and consequently limit the length of the TM. If N_{VTM} is the maximum number of OBUs that can appear in a TM, the size of a trigger message is given by as N_{VTM} we get in Equation (4.7):

$$TM = \lceil \log_2(IW) \rceil + \lceil \log_2(n_r) \rceil + N_{VTM} \times \left(\lceil \log_2(N_{max}) \rceil + \lceil \log_2(SOW) \rceil \right) \quad (4.7)$$

4.4.2 *Synchronous OBU Window Length (l_{sow})*

After determining the size of TM it is possible to compute the length of the Synchronous OBU window (l_{sow}). Starting by rewriting SOW_{slots} in function of N_{VTM}:

$$SOW_{slots} \, 0 \text{ to} \left[(S_{IW} * N_{VTM}) - ((S_{IW} - 1) * N_{Vint}) \right]$$

N_{Vint} is the union of all sets of vehicles that can listen simultaneously to more than one RSU in a set of adjacent RSUs and is shown in Equation (4.8):

$$N_{V\text{int}} = \bigcup_{i=1}^{(S_{ST}-1)} S_{RSU_i} \bigcap S_{RSU_{i+1}} \quad (4.8)$$

We conclude that the determination of N_{VTM} is crucial for this protocol, since this will influence the value of SOW_{slots} and consequently the length of synchronous window, l_{sow}, as shown in Equation (4.9).

$$l_{sow} = SOW_{slots} \times (IFS + BSM) \quad (4.9)$$

We recall that each elementary cycle (EC) is divided into an Infrastructure Window (IW), a Synchronous OBU Window (SOW) and an asynchronous free period (FP), as shown in Equation (4.10).

$$E = IW + SOW + FP \quad (4.10)$$

The Free Period (FP) corresponds to the remaining time in the Elementary Cycle after the Infrastructure Window and the OBU window. We must ensure that FP has a minimum guaranteed size in order to allow non-V-FTT communications to happen. This is shown in Equation (4.11).

$$FP_{\min} = \sigma \times E \text{ where } \sigma \in \,]0,1[\quad (4.11)$$

Since the vehicle density and the available bandwidth suffer strong variations, a scheduling mechanism will most likely be needed for the cases where the RSUs can not serve all OBUs in one EC. In some particular cases, the FP length can be reduced to zero, if emergency communications need to use the whole Elementary Cycle.

4.4.3 V-FTT Protocol Using IEEE 802.11p/WAVE / ITS G-5

This section describes how V-FTT can be supported by the IEEE 802.11p/WAVE standard for vehicular communications. In the IEEE802.11p/WAVE standard all vehicles must tune the Control Channel (CCH) during all Sync Intervals (refer to Figure 4.8), so this is the appropriate place for all short safety messages to be sent. The size of the CCH interval is 50 ms by default but it can have a maximum of 100 ms, considering that we are working in continuous mode [30]. In the European standard ITS G-5 0 all vehicles shall have two radio devices, which mean they will preferably work in continuous mode.

The CCH interval will then be the equivalent to our elementary cycle (EC):

- During the Infrastructure Window (IW), RSUs broadcast the scheduling table along with the safety messages in the beginning of the CCH interval, immediately after a Guard Interval (GI). To avoid contention with other devices, no IFS will be used in this case.
- OBUs have the opportunity to transmit important data to the RSUs during the Synchronous OBU Window (SOW).

Our approach assumes that:

- all registered OBUs in the Safety Zone are compliant with the protocol so that safety information is prioritized, meaning the IW and SOW are protected against any other type of communications;
- all OBUs and RSUs can hear other nodes transmission (no hidden node problem);
- care must be taken not to use the full remaining CCH interval (after IW) for OBU transmission of safety messages, since the CCH can also be used by other entities for Wave Service Announcements (WSAs). A 'free period' must be preserved so that OBU and/or RSUs can freely transmit WSAs in the CCH using the regular 802.11p MAC. Moreover, in certain cases of dense traffic, one CCH interval may not be enough to guarantee that every OBU has

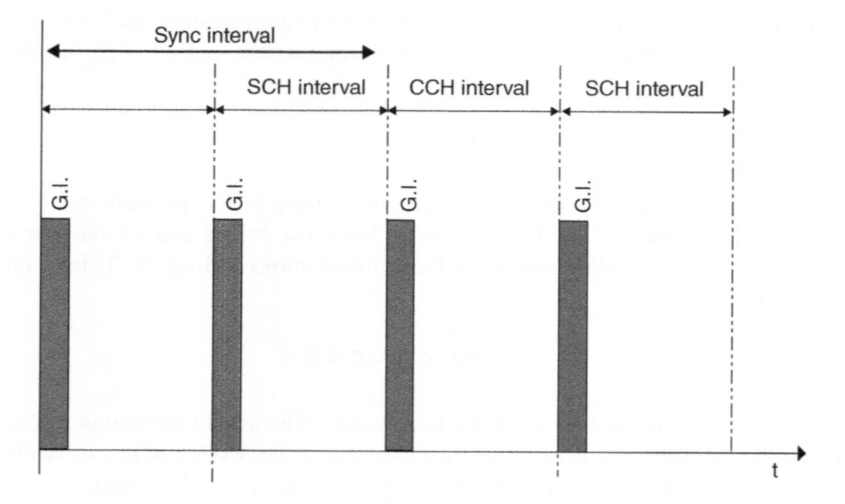

Figure 4.8 IEEE 802.11p/WAVE synchronization interval (adapted from [31]).

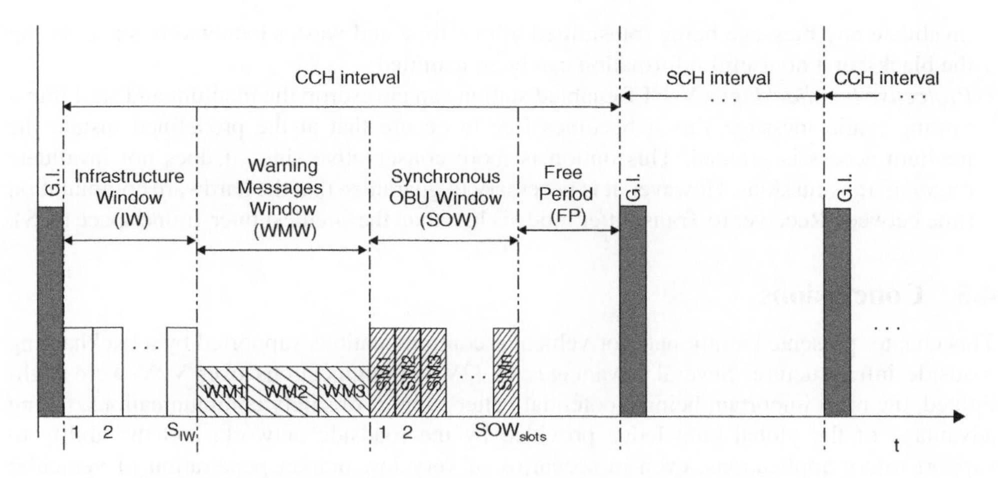

Figure 4.9 V-FTT protocol adaptation to IEEE802.11p/WAVE (normal mode).

the opportunity to transmit its data. This means that the maximum size of IW and SOW must be carefully chosen. A scheduling mechanism may also be introduced in order to guarantee delivering of high-priority OBU safety communications.

Figure 4.9 shows how the V-FTT protocol adapts to the WAVE Sync interval.

A relevant assumption is that V-FTT enabled OBUs will share the medium with non-V-FTT enabled OBUs. This implies that a noncompliant OBU could interfere with V-FTT TDMA scheduling, possibly compromising its timeliness, if V-FTT protection mechanisms are not put in place. Providing such protection mechanisms is an essential aspect to a V-FTT successful implementation. V-FTT protection mechanisms should enforce the periodicity of the trigger message transmission with low jitter, i.e. OBUs must not transmit when RSUs are close to begin the trigger message transmission. They should also guarantee that non-V-FTT compliant OBU could only transmit during the Free Period.

As was seen earlier, the carrier sense mechanism of IEEE 802.11p evaluates if the medium is free before starting a transmission. If the medium is not free, the message transmission is postponed for a later time according to the backoff algorithm. Otherwise, the message is transmitted immediately. A drastic approach can be used in order to gain access to the medium. A modified station having the ability to transmit a long enough noise sequence (black-burst), without performing the carrier sense procedure, will eventually force the remaining stations to evaluate the channel as occupied. Therefore, if the modified station is able to transmit immediately after the end of the noise sequence, violating the Inter-Frame Space (IFS), it gains access to the shared medium. This technique, called bandjacking [32], is a medium access control scheme that provides determinism, even in the presence of other contention-based technologies, as long as the channel capture is performed during the shortest IFS. In this sense, bandjacking enables a station to 'forcefully gain access' to a communication channel. There are two types of bandjacking:

- *Destructive bandjacking*: Transmit the black-burst, ignoring all the information that exists in the medium, with a length equal to the longest message available. This possibility would invalidate any message being transmitted at that time and wastes bandwidth, since during the black-burst no useful information can be transmitted.

- *Protective bandjacking*: a V-FTT enabled station can eavesdrop the medium and start transmitting (valid messages) as it becomes free to ensure that at the predefined instant the medium access is granted. This option is more conservative since it does not invalidate ongoing transmissions. However, it is necessary to guarantee that the hardware commutation time between Receiver to Transmitter mode is less than the smallest inter-frame space (IFS).

4.5 Conclusions

This chapter presented a rationale for vehicular communications supported by a backhauling roadside infrastructure. Several advantages of I2V communications over V2V were highlighted, the most important being a potential better support of safety communications, taking advantage of the global knowledge provided by the roadside network, and the ability to support safety applications, even in scenarios of very low market penetration of vehicular communications. For this purpose, related work was analysed and a new protocol was proposed, the Vehicular Flexible Time Triggered Protocol (V-FTT), which is an adaptation of the FTT protocol to wireless vehicular communications. V-FTT protocol is based on a roadside infrastructure in order to guarantee road safety, data privacy and safety events timeliness delivery in high vehicle density scenarios. A model where RSUs are deployed near motorway blackspots and are responsible for scheduling OBU communications as well as broadcasting safety events was presented. Then, V-FTT protocol was detailed and formalized, including the definition of the Basic Safety Message (BSM) that every OBU must periodically transmit.

An important issue is how to ensure that OBU information is credible. Security is very important, so RSUs must have mechanisms to check OBUs identity and cross-validate the received data with other means (cameras, induction loops, even information from other OBUs or RSUs). Data privacy is also very important, so all communications should be encrypted, in order to protect information. OBU certificates management in order to guarantee identity is out of the scope of this work but various works can be referenced on this subject [33–36]. Note that security is important to the V-FTT protocol since cryptographic operations need to be time bounded.

In case an RSU does not have enough OBUs in its coverage area to fill the Trigger Message, the space can be used by the RSUs to broadcast safety warnings (WM), so that the medium keeps busy to OBUs that are noncompliant to the protocol. This ensures that non-V-FTT compliant stations can only transmit during the Free Period If an RSU is responsible for all OBUs in its coverage area, a handover process must be considered, in order for an RSU to pass information and responsibility for an OBU to the following RSU in the motorway. The fact that vehicles follow a known path (motorway) and that the RSUs know the speed and positions of their (under control) OBUs can be very useful for the handover process [4].

Care must be taken in determining the RSU coverage area. A compromise needs to be made between coverage area and terminal (vehicle) capacity. More power can augment the area but will most likely increase channel congestion, while lower transmission power implies fading and loss of packets, which is not acceptable for safety critical applications.

Finally, the initial OBU registration process in the Safety Zone was not defined but since motorways have access ramps, a possible simple solution is to install RSUs in all entries and exits of the motorway, to keep track of all vehicles [37]. Vehicles could use any non-V-FTT MAC protocol to register themselves in the Safety Zone using the Free Period.

References

1. IEEE. 802.11-2012 – IEEE Standard for Information Technology – Telecommunications and information exchange between systems. Local and metropolitan area networks – Specific requirements Part 11: Wireless LAN Medium Access Control (MAC) and Physical Layer (PHY) Specifications (2012).
2. ETSI ITS-G5 standard – Final draft ETSI ES 202 663 V1.1.0, Intelligent Transport Systems (ITS); *European Profile Standard for the Physical and Medium Access Control Layer of Intelligent Transport Systems Operating in the 5 GHz Frequency Band* (2011).
3. A.M. Vegni and T.D.C. Little (2011) Hybrid vehicular communications based on V2V-V2I protocol switching, *International Journal of Vehicle Information and Communication Systems* **2**(3/4): 213–31.
4. A. Böhm and M. Jonsson (2007) *Handover in IEEE 802.11p-based Delay-Sensitive Vehicle-to-Infrastructure Communication*, Research report IDE-0924, School of Information Science, Computer and Electrical Engineering (IDE), Halmstad University, Sweden.
5. A. Böhm and M. Jonsson (2011) Real time communications support for cooperative, infrastructure-based traffic safety applications, *International Journal of Vehicular Technology*, 2011, Article ID 54103.
6. V. Milanes, J. Villagra, J. Godoy, *et al.* (2012) An intelligent V2I-based traffic management system, *IEEE Transactions on Intelligent Transportation Systems* **13**(1): 49–58.
7. G. Chandrasekaran (2008) *VANETs: The Networking Platform for Future Vehicular Applications*, cs.rutgers.edu, Department of Computer Science, Rutgers University.
8. Matheus, K., Morich, R. and Lübke, A. (2004) Economic background of car-to-car communication, *Proceedings of the 2nd Braunschweiger Symposium Informationssysteme für mobile Anwendungen. Braunschweig, October 2004.*
9. M. Emmelman, B. Bochow and C. Kellum (2010) *Vehicular Networking, Automotive Applications and Beyond.* New York: John Wiley & Sons, Inc.
10. US Department of Transportation, National Highway Traffic Safety Administration, Vehicle Safety Communications Project (Task 3 Final Report) – *Identify Intelligent Vehicle Safety Applications – Enabled by DSRC*, March 2005.
11. Y. Ni, Y. Tseng, J. Chen and S. Sheu (1999) The broadcast storm problem in a mobile ad hoc network. *ACM Mobicom (The Annual International Conference on Mobile Computing and Networking), Seattle, Washington, USA, 15–20 August 1999*, pp. 151–62.
12. Moustafa H. and Zhang Y. (2009) *Vehicular Networks: Techniques, Standards and Applications*. Boston, MA: Auerbach Publishers.
13. P. Veríssimo (2003) Uncertainty and predictability: can they be reconciled? In A. Schiper, A.A. Shvartsman, H. Weatherspoon and B.Y. Zhao (eds), *Future Directions in Distributed Computing*. Berlin: Springer-Verlag, pp. 108–113.
14. E. Schoch, F. Kargl, M. Weber and T. Leinmuller (2008) Communication patterns in VANETs, *IEEE Communications Magazine*, **46**: 119–25.
15. ETSI Technical Specification 102 637-2: *Intelligent Transport Systems (ITS); Vehicular Communications; Basic Set of Applications; Part 2: Specification of Cooperative Awareness Basic Service*, v.1.2.1 (March 2011).
16. A. Böhm, K. Lidström, M. Jonsson and T. Larsson (2010) Evaluating CALM M5-based vehicle-to-vehicle communication in various road settings through field trials, *The 4th IEEE LCN Workshop on User MObility and VEhicular Networks (On-MOVE), Denver CO USA, Oct. 2010*, pp. 613–20.
17. D.W. Brown, J.W. Leth and J.E. Vandendorpe, *et al.* (1987) *Fault Recovery in a Distributed Processing System*, US Patent 4 710 926.
18. T. Mak, K. Laberteaux and K. Sengupta (2005) A multi-channel VANET providing concurrent safety and commercial services, *VANET05, Proceedings of the 2nd ACM International Workshop on Vehicular Ad hoc Networks, September 2005, Germany*, pp. 1–9.
19. R. Costa (2013) RT-WiFi, *Um Mecanismo para Comunicação de Tempo-Real em Redes IEEE802.11 Infraestruturadas*, PhD dissertation, Faculdade de Engenharia da Universidade do Porto (FEUP).
20. J. Rezgui and S. Cherkaoui (2014) About deterministic and nondeterministic vehicular communications over DSRC/802.11p, *Wireless Communications and Mobile Computing* **12**(15): 1435–49.
21. K. Bilstrup, E. Uhlemann, E.G. Ström and U. Bilstrup (2009) On the ability of the 802.11p MAC method and STDMA to support real-time vehicle-to-vehicle communication, *EURASIP Journal on Wireless Communications and Networking* 2009, Article ID 902414.
22. K. Macek, D. Vasquez, T. Fraichard, R. Siegwart (2008) Towards safe vehicle navigation in dynamic urban scenarios. *Proceeding of the IEEE Conference on Intelligent Transportation Systems, Beijing, China, October 2008*, pp. 482–9.

23. E.D. McCormack and B. Legg (2000) *Technology and Safety on Urban Roadways: The Role of ITS for WSDOT*, Washington State Transportation Center (TRAC).
24. B. Aslam and C.C. Zou (2011) Optimal roadside units placement along highways. *8th Annual IEEE Consumer Communications and Networking Conference*, Work in Progress Paper.
25. R. Scopigno and H.A. Cozzetti (2009) Mobile slotted Aloha for Vanets, *70th IEEE Vehicular Technology Conference Fall (VTC 2009-Fall)*, pp. 1–5.
26. L. Almeida, P. Pedreiras and J.A.G. Fonseca (2002) The FTT-CAN protocol: why and how. *IEEE Transactions on Industrial Electronics* **49**(6): 1189–1201.
27. B. Hu and H. Gharavi (2011) A joint vehicle-vehicle/vehicle-roadside communication protocol for highway traffic safety, *International Journal of Vehicular Technology*, 2011, Article ID 718048, Hindawi Publishing Corporation.
28. *Dedicated Short Range Communications (DSRC) Message Set Dictionary*, SAE Std. J2735, SAE Int., DSRC Committee, November 2009.
29. J.B. Kenney (2011) Dedicated Short-Range Communications (DSRC) standards in the United States. *Proceedings of the IEEE* **99**(7): 1162–82.
30. IEEE Std 1609.4 2010, *IEEE Standard for Wireless Access in Vehicular Environments (WAVE) – Multi-channel Operation*, February 2011.
31. IEEE Std 1609.4 2006, *IEEE Trial-Use Standard for Wireless Access in Vehicular Environments (WAVE) – Multi-channel Operation*, November 2006.
32. P. Bartolomeu, J. Ferreira and J. Fonseca (2009) Enforcing flexibility in real-time wireless communications: A bandjacking enabled protocol. *IEEE Conference on Emerging Technologies Factory Automation, September 2009*, pp. 1–4.
33. G. Calandriello, P. Papadimitratos, J.P. Hubaux and A. Lioy (2007) Efficient and robust pseudonymous authentication in VANET. *Proceedings of the Fourth ACM, International Workshop on Vehicular Ad hoc Networks, Montreal, September 2007*, pp. 19–28.
34. X. Lin X., X. Sun, P.H. Ho and X. Shen (2007) GSIS: a secure and privacy preserving protocol for vehicular communications. *IEEE Transaction on Vehicular Technology* **56**(6): 3442–56.
35. M. Raya and H.P. Hubaux (2007) Securing vehicular ad hoc networks, *Journal of Computer Security* **15**(1): 39–68.
36. A. Wasef, Y. Jiang and X. Shen (2008) ECMV: Efficient Certificate Management Scheme for Vehicular Networks. *Proceedings of IEEE Globecom '08, New Orleans, USA, December 2008*, pp. 1–5.
37. A. Böhm and M. Jonsson (2009) Position-based data traffic prioritization in critical, real-time vehicle-to-infrastructure communication, *Proceedings of the IEEE Vehicular Networking and Applications Workshop (VehiMobil 2009) in conjunction with the IEEE International Conference on Communications (ICC), Dresden, Germany, June 14, 2009*, pp. 1–6.

5

Cyber Security Risk Analysis for Intelligent Transport Systems and In-vehicle Networks

Alastair R. Ruddle and David D. Ward
MIRA Ltd, Nuneaton, UK

5.1 Introduction

In the past, road vehicles were independent and largely mechanical systems. However, modern vehicles are increasingly reliant on internal networks that link sensors, actuators and control systems in order to achieve higher levels of functionality than can be provided by stand-alone subsystems. Many of these functions are safety-related, such as anti-lock braking, advanced emergency braking, electronic stability control, and adaptive cruise control. In the automotive environment, therefore, cyber security attacks could have significant safety implications for vehicle occupants and other road users, in addition to the privacy and financial implications that are more commonly associated with attacks on information systems. Such safety issues could be unintentional side effects of some cyber security attacks, but it is conceivable that causing death or injury could even be a primary attack goal for some potential attackers.

At the same time, cellular telephones [1] and other 'nomadic' devices now interact with vehicle systems, and maintenance facilities such as remote diagnosis and wireless 'flashing' of software are being developed for automotive applications [2,3]. In addition, in-vehicle wireless networks are commonly implemented using Bluetooth, and on-board devices are increasingly able to access the internet. These features provide possible opportunities for unauthorized access to vehicle systems. Furthermore, considerable research effort is devoted to the development of V2X communications (i.e. vehicle-to-vehicle, or vehicle-to-infrastructure) in order to allow vehicles to participate in ad-hoc mobile networks. Roadside infrastructure and

Intelligent Transport Systems: Technologies and Applications, First Edition. Asier Perallos, Unai Hernandez-Jayo, Enrique Onieva and Ignacio Julio García-Zuazola.

intelligent transport systems (ITS), which are also expected to be integrated with vehicle networks via V2X communications, are also potential targets for attack.

Thus, the opportunities for malicious individuals and groups to attempt to gain access to vehicle systems and information could widen significantly if suitable counter-measures are not implemented. Roadside infrastructure and intelligent transport systems (ITS), which are expected to be integrated with vehicle networks via V2X communications, are also potential targets for attack. Furthermore, a modified vehicle could become a 'weapon' for the attacker, by transmitting false warnings and messages in order to subvert the operation of vehicle networks, ITS and telematics services. Ensuring adequate security against such threats will therefore be critical for the successful deployment of V2X and ITS technologies.

The unified approach to security and safety risk analysis that is outlined here is based on work carried out during the EVITA project [4], which aimed to prototype a toolkit of techniques and components to ensure the security of in-vehicle systems, including security hardware, software, and analysis methods, as well as an evaluation of the associated legal aspects [5]. The EVITA project was focused on in-vehicle networks as other projects (e.g. SeVeCom [6]) had addressed the security of V2X communications. Nonetheless, it is not possible to investigate cyber security for in-vehicle networks without considering the external networks that the vehicle will need to interact with for ITS and telematics purposes.

5.2 Automotive Cyber Security Vulnerabilities

Published investigations of possible cyber security attacks against vehicle systems include the use of wireless relay techniques to extend the range of keyless entry systems and so achieve physical access to the vehicle [7–9]. Simulations of attacks on the FlexRay protocol [10] and CAN bus [11] have also been reported. A conclusion of the experiments outlined in [12] is that achieving access to almost any subsystem can provide access to almost all of the other subsystems. Attacks against vehicle systems by exploiting security vulnerabilities in the wireless communications used by a tyre pressure monitoring system (TPMS) are described in [13]. The vulnerability of TPMS is notable as the installation of TPMS has been mandated for all new light motor vehicles since 2007 in the USA [14]. Furthermore, TPMS have become mandatory equipment in the EU for passenger cars and similar vehicles since November 2012 [15], and similar measures are envisaged in other parts of the world.

Ongoing changes in automotive requirements, technology and legislation will have an impact on the evolution of the automotive security landscape, as new safety systems, which are increasingly likely to exploit wireless communications, become mandatory equipment. In Europe, it has been proposed that e-Call (automated emergency calls for road accidents) should become mandatory from October 2015 [16]. Furthermore, both the US National Highway Traffic Safety Administration (NHTSA) and the EC are actively promoting the development of V2X communication systems, but have not yet specified timescales for their implementation. Possible threats to V2X communications are discussed in [17].

Cyber security is concerned with maintaining the availability, integrity and confidentiality of system data. Methods for attacking vehicle systems could be broadly categorized in terms of direct attacks on system information, and remote attacks achieved via electromagnetic methods.

5.2.1 Information Security

Future vehicles, as well as the transport management systems with which they interact, will be subject to all of the cyber security threats that currently afflict computer networks and other computer-based systems, or will in the future. These include:

- malware – malicious software (e.g. worms, Trojans, key loggers) that may be used to steal information or disrupt normal system operation;
- unauthorized control of system assets – gain control of on-board hardware, often achieved via malware;
- backdoors – methods for bypassing authentication measures in order to achieve access to a computer system, often achieved via malware;
- social engineering – psychological manipulation of people into divulging confidential information or performing actions that compromise cyber security.

A further emerging technology of relevance is 'cloud computing', which is currently of considerable interest for applications such as cooperative ITS, but potentially challenging from the cyber security point of view because of its distributed and dynamic nature.

5.2.2 Electromagnetic Vulnerabilities

The mobile nature of vehicles and their increasing reliance on wireless services result in some electromagnetic vulnerabilities in addition to the cyber security threats that more commonly afflict other computer-based systems. A further trend of significance is that, in order to minimize weight and thereby maximize the range of electric vehicles (or the fuel efficiency of vehicles using internal combustion engines), vehicle construction is moving away from traditional steel bodyshells. Current alternatives include aluminium or nonconducting panels fixed to a steel frame, although the use of alternative materials such as carbon-fibre is also being investigated to implement light-weight chassis members, as well as for vehicle body panels. These changes will have implications for the electromagnetic properties of vehicles, including security aspects.

Electromagnetic security vulnerabilities of vehicles that have already been demonstrated include:

- GNSS jamming – many proposed ITS functions rely on accurate and reliable position data, but the inherently low power GNSS signals are vulnerable to interference from simple low-cost jamming systems [18] (i.e. compromising data availability);
- microwave weapons – pulsed microwave signals can be used to induce engine stalls in vehicles that use electronic engine management systems [19] (i.e. compromising data integrity);
- eavesdropping – radio transmissions [20], or even unintentional electromagnetic emissions from electronic circuits, may be monitored in order to acquire information (i.e. compromising data confidentiality).

These kinds of vulnerabilities may be exploited in direct attacks, or indirectly as part of the development of more complex attack sequences. For example, information obtained by eavesdropping on valid radio messages may be of immediate use to the attacker, but could also be used to construct 'spoof' messages for use as part of a later attack. In the latter, possible attack objectives could include falsifying hazard warnings propagated through vehicle networks, or changing traffic light priorities (e.g. impersonating an emergency services vehicle).

5.3 Standards and Guidelines

For safety-related control systems, of which vehicles are perhaps the most widely encountered examples, it is possible that some cyber security threats may also have safety implications. This possibility is now reflected in IEC 61508 [21], the functional safety standard for safety-related electronic control systems in the process industries. However, security threats that are not safety-related, such as those affecting privacy or financial security, are beyond the scope of IEC 61508.

Standards and guidelines have already been developed to help ensure the security of information technology (IT) systems. However, as some cyber security threats to cyber physical systems may also impact on functional safety, there is a need to ensure that their treatment meets the requirements of safety engineering processes. Standards and guidelines have already been developed for functional safety of safety-related control systems, including vehicles. There is also a need to adapt IT security evaluation approaches in order to address the particular issues of automotive applications, such as the possibility that a security threat may also have safety implications. This section therefore summarizes a number of key concepts from these two disciplines.

5.3.1 Risk Analysis Concepts

In functional safety engineering a 'safety hazard' is a source of possible 'harm' to life, health, property or the environment. In the security domain the source of harm is usually described as a 'security threat' and has other potential impacts, such as unauthorized access to data or loss of privacy. The 'severity' is a measure of the expected degree of harm that may result from a hazard/threat in a specific situation. The associated 'risk' is then a combination of the 'probability' of occurrence of harm and the severity of that harm, such that the risk increases with greater probability and/or severity. In some applications it is possible to quantify the probability and severity, with the result that the derived risk can also be quantified. In other applications, however, it is only possible to rank these measures in a qualitative manner. Such qualitative rankings often employ a classification based on order of magnitude differences.

The purpose of risk analysis is to identify the hazards/threats, assess their associated severity and probability, and hence evaluate their relative risks. This allows the analyst to identify and prioritize requirements for mitigating the risks, and thus ensure that the residual risks are acceptable. These activities can only be undertaken when the system characteristics and intended operating environment have been defined. Nonetheless, it is important to note risk analysis activities can be initiated at the concept stage, and should be in order to ensure that risks are considered from the outset of the development lifecycle. The preliminary risk analysis can then be progressively refined as the design matures and the details become better defined.

5.3.2 *Functional Safety Standards*

Risk analysis is a key element of the functional safety standard IEC 61508, which reflects the following views on safety risks:

- zero risk is unachievable;
- safety must be considered from the outset;
- where risks are considered unacceptable, measures must be employed to reduce the risks to a level considered 'broadly acceptable'.

The latter leads to safety requirements that are described in terms of 'safety function requirements' (i.e. what the function should do) and 'safety integrity requirements' (i.e. the likelihood that the safety function will be carried out satisfactorily). The safety integrity requirements associated with the safety functions are specified in terms of a number of discrete levels, known as 'safety integrity levels' (SILs), which are related to the risk level and range from SIL1 to SIL4. The SILs reflect requirements for increasingly rigorous processes to be applied for more critical safety functions in a wide variety of development activities, including specification and design, through configuration management, testing, validation, verification, and independent assessment. This is intended to provide greater confidence in the reliability of complex systems that are not amenable to exhaustive testing.

Although IEC 61508 originated from the process control industry, it provides a basic functional safety standard for other sectors. It is also intended to provide a basis for the development of sector-specific safety standards, such as the automotive interpretation ISO 26262 [22]. This addresses the functional safety aspects of the entire system development process and describes a risk-based approach for determining 'automotive safety integrity levels' (ASILs) that are analogous, but not directly equivalent, to the IEC 61508 SILs. The main difference between the ISO 2626 ASILs and IEC 61508 SILs is that the latter employ quantitative target probability values, while the ASILs are based on qualitative measures. Related guidance regarding safety analysis for vehicle based programmable systems [23] has also been developed by the Motor Industry Software Reliability Association (MISRA).

For automotive applications an additional parameter should be taken into account that represents the potential for the driver to influence the severity of the outcome. In the MISRA guidelines and ISO 26262 this possibility is taken into account by means of a qualitative measure known as 'controllability' (i.e. avoidance of a specified harm or damage through timely reactions of the persons involved), which was originally developed by the project DRIVE Safely [24].

5.3.3 *IT Security Standards*

Security evaluation for IT products is described ISO/IEC 15408 [25], but this has some limitations for use in automotive applications. In particular, it does not explicitly address the possible safety implications of security breaches for safety-critical control systems, and it does not provide a framework for risk analysis. Methods for evaluating the efforts required for a successful attack (described as 'attack potential') are described in ISO/IEC 18045 [26], but the probability of a successful attack and the severity of the impact are not evaluated to allow risk to be assessed.

As with safety, the achievement of zero security risk is an unachievable goal. Thus, a risk-based approach is also needed in order to evaluate potential security threats and to identify and prioritize security requirements. Risk analysis in an IT security context is outlined in ISO/IEC TR 15446 [27] and described in more detail elsewhere (e.g. ISO/IEC 13335 [28], NIST IT Security Handbook [29]).

In ISO/IEC 15408 the concept of 'evaluation assurance levels' (EAL) has a similar role for security to the SIL and ASIL categories used in the safety context. The EALs are similarly associated with graded levels of increasing development rigor, ranging from functional testing where the security risk is not deemed to be serious (EAL1), through to formally (i.e. mathematically rigorous) verified design and testing for cases where the risks are judged to be extremely high (EAL7).

5.3.4 Combining Safety and Security Analysis

The MISRA consortium identified the need to protect vehicle software from unauthorized access that could compromise the performance of safety-related systems, as well as the need to detect such tampering [30]. The similarities between the EAL and SIL/ASIL concepts suggest the potential for developing a unified approach for automotive safety and security [31]. Similar observations have also been made in other fields, such as mobile ad-hoc networks [32] and defence applications [33]. Unifying safety and security engineering processes offers potential benefits in terms of reduced development costs through sharing of evidence and re-use of risk analysis where security may also have possible safety implications.

5.4 Threat Identification

In order to define security and safety requirements it is first necessary to have an understanding of both the intended behaviour of the system and the intended operating environment. This can be achieved through analysis of a range of representative use cases. Although the EVITA use cases themselves revealed some security-related user requirements, they also provided the basis for investigating so-called 'dark-side scenarios', which aimed to identify possible cyber security attacks. In order to facilitate this, it was first necessary to identify the various 'security actors' and their interests, including both the 'threat agents' (i.e. the attackers) and the 'stakeholders' (i.e. their potential victims).

5.4.1 Use Cases

The EVITA use cases [34] assumed a generic vehicle network architecture (see Figure 5.1) that was derived from the EASIS project [35]. A total of 18 use cases were developed, which can be broadly grouped into six classes:

- V2X – involving wireless communication between vehicles and with roadside infrastructure;
- eToll – relating to toll transactions;
- eCall – relating to emergency assistance calls;

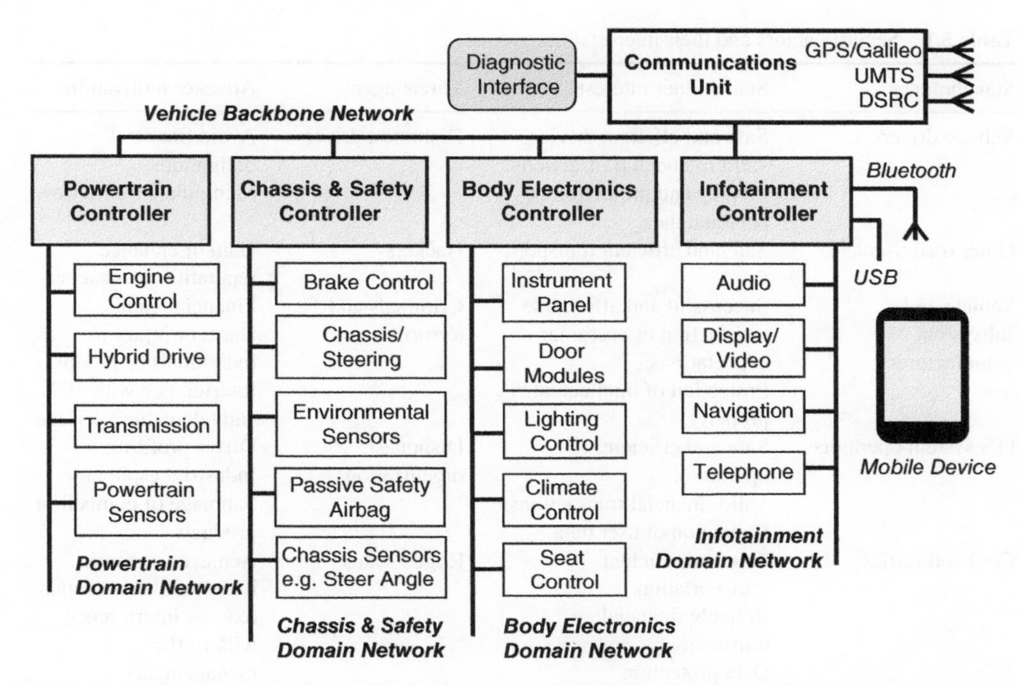

Figure 5.1 Generic vehicle network architecture assumed in the development of the EVITA use cases.

- nomadic devices – involving in-vehicle wireless communications links or temporary wired connections such as USB devices;
- aftermarket – installation of aftermarket modules or replacement of defective modules;
- diagnosis – including both diagnostic and software maintenance activities.

The various on-board sensors, actuators, data buses, and electronic control modules constitute the 'assets' of the in-vehicle network.

5.4.2 *Security Actors*

In assessing possible cyber security threats it is necessary to consider the security actors that may be involved, including both the 'threat agents' (i.e. the attackers) and their motivations, as well as the 'stakeholders' (i.e. their potential victims) and their interests.

For networked vehicles participating in ITS and telematics applications, unauthorized access to personal and/or vehicle data may become possible, while the corruption of data and/or software could result in anomalies in vehicle functionality, telematics services or traffic behaviour. Potential threat agents therefore range from dishonest drivers seeking personal advantage through to 'rogue states' perhaps aiming to achieve large scale harm to other societies. Consequently, the financial and technical resources that may be available to the various threat agents in mounting cyber security attacks may vary enormously between them.

Table 5.1 Security actors and their interests/motivations.

Stakeholders	Stakeholder interests	Threat agents	Attacker motivations
Vehicle drivers	Safe and efficient driving, valid financial transactions. Privacy and protection of personal data.	Dishonest drivers	Avoid financial obligations. Manipulate traffic flow.
Other road users	Safe and efficient transport.	Hackers	Gain or enhance reputation as a hacker.
Vehicle and subsystem manufacturers	Successful and affordable satisfaction of customer expectations. Protection of intellectual property.	Criminals and terrorists	Financial gain Harm or injury to individuals or groups. Interference with ITS and telematics services.
ITS system operators	Safe and efficient operation. Valid financial transactions. Protection of user data.	Dishonest organizations	Driver profiling. Industrial espionage. Sabotage of competitor products.
Civil authorities	Safe and efficient transportation. Reliable financial transactions. Data protection compliance.	Rogue states	Achieve economic harm to other societies (e.g. by interference with traffic management).

Future visions of road transportation include networked vehicles and intelligent transport systems (ITS) that will enhance the safety of drivers and other road users, minimize pollution and maximize the efficiency of travel. Thus, the stakeholders involved in future road transport systems range from road users with a simple need to travel through to civil authorities responsible for ensuring efficient traffic flow.

The nature and interests of the various stakeholders and threat agents are outlined in Table 5.1.

5.4.3 Dark-side Scenarios and Attack Trees

The dark-side scenarios were developed from the use cases and assumptions about the security actors by brainstorming. This activity was focused primarily on in-vehicle networks, but also considered the possibility that a vehicle might be modified to act as a weapon for the attacker (e.g. used to transmit false warnings and other messages).

The main objectives of constructing the dark-side scenarios were to identify possible security threats and to provide a basis for assessing their relative risk. Attack trees [36], which are similar to the fault trees that are often used for identifying causes of safety hazards, provided a convenient mechanism for documenting the dark-side scenarios. Subsequent analysis and rationalization of these attack trees to extract common elements and to develop a method to support risk analysis [37] suggested a generic model, which is illustrated for their interpretation.

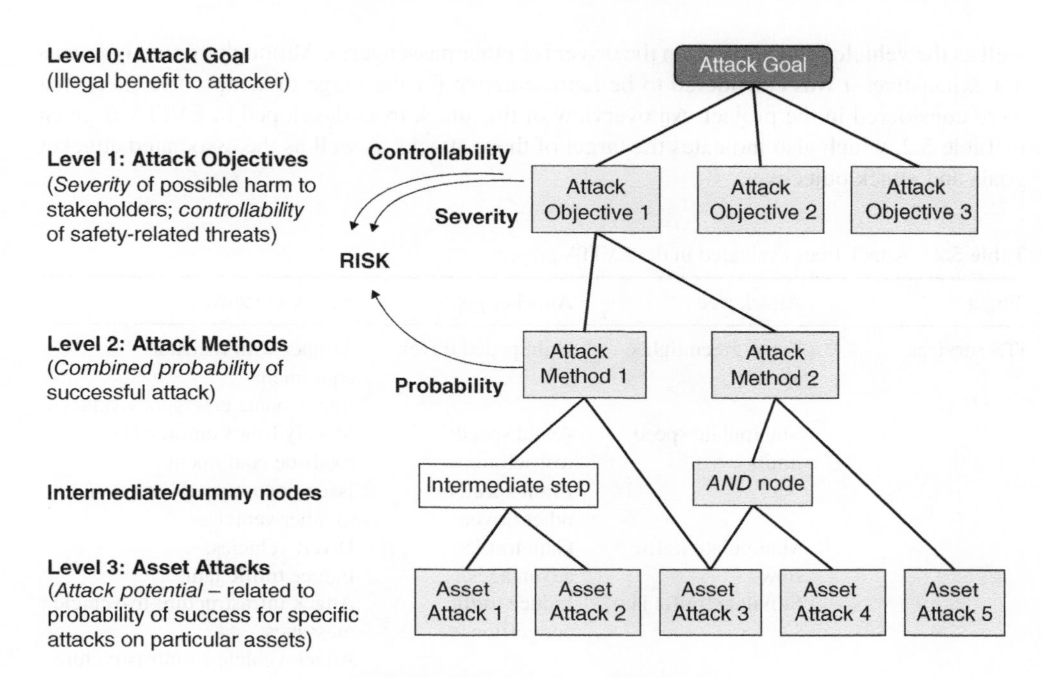

Figure 5.2 Generic attack tree model.

In Figure 5.2, the root of the attack tree (Level 0) represents an abstract 'attack goal' that is associated with some kind of benefit to the attacker. Its child nodes (Level 1) represent one or more 'attack objectives' that could satisfy this attack goal. The importance of the attack objectives is that they can be associated with negative impacts for the stakeholders, thus allowing the severity of the outcome to be estimated at this level.

The attack objectives may be further decomposed into one or more 'attack methods' (Level 2) that could be employed in order to achieve the attack objective. Each attack method is described in terms of a logical combination of lower level attacks, thus populating the branches of the attack tree. The terminal nodes of the tree, described here as 'asset attacks' (Level 3), represent the level at which the probability of successful attack can be estimated for specific attacks against particular system assets. Intermediate nodes may be added as needed for describing the logic of combining different asset attacks in order to implement a particular attack method.

The individual asset attack success probabilities can then be combined using the attack tree logic in order to assess the overall probability (which is described here as the 'combined attack success probability') for each of the attack methods. Thus, attack trees provide a convenient mechanism for identifying the probability and severity measures required in order to assess the risks associated with possible attacks, as well as for documenting the threat identification and analysis processes.

The 18 EVITA use cases suggested 10 attack trees [37], in which 44 different asset attacks were identified involving 16 of the system assets. This activity was focused on in-vehicle networks, but the targets of the attacks investigated also included ITS and telematics services, as well as the vehicle systems or even the driver (or other passengers). Although this analysis was not exhaustive, it was considered to be representative for the range of sample use cases that were considered in the project. An overview of the attack trees developed in EVITA is given in Table 5.2, which also indicates the target of these attacks as well as the associated attacker goals and attack objectives.

Table 5.2 Attack trees evaluated in the EVITA project.

Target	Attack tree	Attacker goals	Attack objectives
ITS services	Force green lights ahead	Unimpeded travel.	Tamper with roadside equipment. Impersonate emergency vehicle.
	Manipulate speed limits	Avoid speed restrictions. Limit speed of other drivers.	Modify limits enforced by roadside equipment. Issue bogus speed limit notices to other vehicles.
	Manipulate traffic flow	Gain traffic advantages.	Divert vehicles. Induce traffic jam.
	Simulate traffic jam	Induce traffic congestion.	Attack infrastructure-to-vehicle messages. Attack vehicle-to-infrastructure messages. Tamper with roadside equipment.
Telematics services	e-Call attack	Disrupt e-Call service.	Trigger spurious e-Call. Degrade e-Call service quality. Denial of service for e-Call.
	e-Toll attack	Obtain driver data. Avoid financial obligations. Harm driver.	Access victim's private data. Reduce toll bill. Increase toll bill. Prevent vehicle/driver from passing toll.
	Tamper with warning message	Harm/irritate driver.	Delay warning message. Prevent warning message. Display wrong warning message.
Driver/vehicle systems	Active brake function attack	Harm driver.	Delay active braking. Prevent active braking. Degrade active braking.
	Unauthorized braking	Harm driver.	False brake demand on Chassis and Safety Domain Bus.
	Engine denial of service	Damage reputation of vehicle manufacturer. Harm/irritate driver.	Engine controller not reachable. Powertrain controller does not receive demand. Engine controller receives warning. Essential component out of order.

5.4.4 Identifying Security Requirements

The security requirements that are identified for systems often tend to be lists of general security features that are perhaps more accurately described as implementation mechanisms that are intended to satisfy security requirements that have not been explicitly stated [38]. Furthermore, if the likely threats to the system, which should take account of its intended operating environment, are not analysed then the attacker perspective is not considered. Thus, without a detailed threat analysis the security requirements will not actually be specific to the system, and are therefore unlikely to be complete.

Methods for identifying security requirements are beyond the scope of this discussion, but the approaches used in EVITA are outlined in [37]. However, threat identification is certainly an essential element in helping to identify specific security requirements for a particular system in its intended operating environment. The attack trees described above are also helpful in supporting the identification of possible requirements for security countermeasures. Risk analysis is an essential element of the requirements engineering process, as it provides an objective basis for judging whether implementation of the security features that could address the security requirements identified from the threat analysis is actually justified in terms of the perceived risks.

5.5 Unified Analysis of Security and Safety Risks

Risk analysis is essential for cost-effective security engineering, as it provides an objective basis for judging whether the implementation of security features to address the security requirements identified from the threat analysis is justified in terms of the perceived risks. Consequently, the functional safety analysis techniques and relevant IT security concepts outlined above were adapted and merged to derive a unified approach to risk analysis that takes account of both security and security-related safety aspects.

5.5.1 Severity Classification

In safety engineering the focus is on physical injuries that might be sustained as the consequence of a safety hazard. However, in considering security threats in the context of networked vehicles and ITS services, physical safety is only one of a number of aspects that may be subject to 'harm'. Other types of harm could include loss of privacy or fraudulent financial transactions, as well as compromised intellectual property and even sabotage threats to competitors. Furthermore, the impact of security threats may be more widespread than a single vehicle, and there is a wider range of stakeholders to be considered who may be influenced by the consequences of security hazards.

In order to accommodate this more complex situation, the severity classification proposed in Table 5.3 separates and categorizes different aspects of the potential consequences of possible security breaches. The starting point for this scheme is the safety severity classification of ISO 26262, which can be characterized using a scale such as the Abbreviated Injury Scale [39]. These severity categories have been adapted and augmented to consider both the greater numbers of vehicles that may be involved and the possible implications for aspects other than safety.

Table 5.3 Proposed severity classification for automotive security threats.

Severity class	Classes of harm to stakeholders			
	Safety (S_S)	Privacy (S_P)	Financial (S_F)	Operational (S_O)
0	**No injuries.**	No unauthorized access to data.	No financial loss.	No impact on operational performance.
1	**Light or moderate injuries.**	Anonymous data only (no specific driver or vehicle data).	Low-level financial loss (~€10^1).	Operational impact not discernible to driver.
2	**Severe and life-threatening injuries (survival probable).** Light/moderate injuries for multiple vehicles.	Identification of vehicle or driver. Anonymous data for multiple vehicles.	Moderate financial loss (~€10^2). Low losses for multiple vehicles.	Driver aware of performance degradation. Indiscernible operational impacts for multiple vehicles.
3	**Life threatening (survival uncertain) or fatal injuries.** Severe injuries for multiple vehicles.	Driver or vehicle tracking. Identification of driver or vehicle, for multiple vehicles.	Heavy financial loss (~€10^3). Moderate losses for multiple vehicles.	Significant impact on operational performance. Noticeable operational impact for multiple vehicles.
4	Life threatening or fatal injuries for multiple vehicles.	Driver or vehicle tracking for multiple vehicles.	Heavy financial losses for multiple vehicles.	Significant operational impact for multiple vehicles.

Note: Entries highlighted in bold font in the Safety column derive from ISO 26262.

In future road transport scenarios, breaches in the security of vehicle information or functions could lead to possible concerns for the various stakeholders in four main areas:

- Safety – unwanted or unauthorized interference with on-board vehicle systems or V2X communications that may impact on the safe operation of vehicles and/or ITS systems.
- Privacy – unwanted or unauthorized acquisition of data relating to vehicle/driver activity, vehicle/driver identity data, or vehicle/subsystem design and implementation.
- Financial – fraudulent commercial transactions, or access to vehicle.
- Operational – unwanted or unauthorized interference with on-board vehicle systems or V2X communications that may impact on the operational performance of vehicles and/or ITS systems (without affecting physical safety).

An important consequence of considering these aspects collectively is that the severity associated with an attack objective becomes a 'severity vector' **S** with four components (S_S, S_P, S_F, S_O), relating to safety, privacy, financial and operational threats, respectively. Each of these components may have different ratings, depending on their impact, rather than a single severity parameter for all aspects. For example, it is possible that an attack could have little or no impact on safety, but still present significant risks in terms of compromised driver privacy.

5.5.2 *Probability Classification*

Attack potential [25] is a measure of the effort required to mount a successful attack against an IT system. Some attackers may be willing to exert greater efforts if they have the necessary resources and they consider that the benefits justify the effort required. If the attack potential of the attacker exceeds the attack potential that the system is able to withstand, then the system will not withstand the attack. The factors to be considered in assessing attack potential [26] include the time needed to identify a potential vulnerability and develop a method to exploit it, the specialist expertise, equipment, software and system knowledge that may be required, and the window of opportunity. In many cases these factors are not independent, but may be substituted for each other to varying degrees. For instance, expertise or equipment may be a substitute for time. The approach adopted in the EVITA project is summarized in Table 5.4, which is based on [26] and [40].

Attack potential has been suggested as a probability measure for security risk analysis [41], although it in fact describes the difficulty, rather than the probability, of mounting a successful attack. The 'basic' attack potential suggests a high probability of successful attack, since many possible attackers will have the necessary resources. Conversely, a 'high' attack potential is assumed to correspond to a lower probability of successful attack, since the number of attackers with the necessary resources is expected to be comparatively small. Consequently, Table 5.5 proposes a scale that maps the relative probability of success to the attack potential in a more intuitive manner. In this scheme, the numerical ranking (P) for 'attack success probability' is higher for easier attacks that are associated with lower attack potentials, and lower for the more difficult attacks that are characterized by higher attack potentials.

5.5.3 *Controllability Classification*

Where the severity vector includes a nonzero safety component, the risk assessment should include an additional probability parameter that represents the potential for the driver to influence the severity of the outcome. In the MISRA guidelines and ISO 26262 this possibility is taken into account by means of a qualitative measure known as 'controllability' (i.e. avoidance of a specified harm or damage through timely reactions of the persons involved).

The concept of 'controllability' for automotive applications was originally developed by the EU project 'DRIVE Safely' [42], and is now used as a qualitative probability measure in safety risk analysis methods that are applied to vehicle functional safety engineering (i.e. ISO 26262 [22]). The controllability classifications used in ISO 26262 are outlined in Table 5.6.

5.5.4 *Risk Classification*

In general the severity associated with a security threat is a 4-component vector (**S**), reflecting possible harm relating to safety, privacy, financial loss and operational performance (see Section 5.5.1). Consequently, the risk level (**R**) is also a 4-component vector, with

Table 5.4 Rating of aspects of attack potential.

Factor	Level	Comment	Value
Elapsed time	≤1 day	Time required by an attacker to identify that a particular potential vulnerability may exist, to develop an attack method and to sustain effort required in mounting the attack.	0
	≤1 week		1
	≤1 month		4
	≤3 months		10
	≤6 months		17
	>6 months		19
	Impracticable	Not exploitable within a timescale that would be useful to an attacker.	∞
Expertise	Layman	With no detailed knowledge of the underlying principles, product types or attack methods.	0
	Proficient	Familiar with the security behaviour of the product or system type.	3
	Expert	Knowledge of the underlying algorithms, protocols, hardware, structures, security behaviour, principles and concepts of security employed, techniques and tools for the definition of new attacks, cryptography, and classical attack methods for the product type.	6
	Multiple experts	Requiring involvement of several Experts from different fields in order to mount an attack.	8
Knowledge of system	Public	Available from the internet or in readily available publications.	0
	Restricted	Knowledge that is controlled by the originating organization and shared with other organizations only under a nondisclosure agreement.	3
	Sensitive	Knowledge that resides with discreet teams within the developer organization, access to which is limited to team members only.	7
	Critical	Knowledge that is known to only a few individuals, access to which is very tightly controlled on a strict need-to-know basis.	11
Window of opportunity	Unnecessary/ unlimited	There is no risk of being detected during access to the target and there is no difficulty in accessing the necessary number of targets for the attack.	0
	Easy	Access is required for ≤ 1 day and the number of targets is ≤ 10.	1
	Moderate	Access is required for ≤ 1 month and the number of targets is ≤ 100.	4
	Difficult	Access is required for > 1 month or the number of targets is > 100.	10
	None	Access to the target is too short to perform the attack, or a sufficient number of targets is not accessible to the attacker.	∞
Equipment	Standard	Readily available to the attacker.	0
	Specialised	Not readily available to the attacker, but acquirable without excessive effort. This could include purchase of moderate amounts of equipment or development of more extensive attack scripts or programs.	4
	Bespoke	Equipment that is not readily available because it is so specialized that its availability is restricted, or is very expensive, or may need to be developed by experts.	7
	Multiple bespoke	Several different types of bespoke equipment are required for different steps of an attack.	9

Table 5.5 Proposed mapping between attack potential and attack success probability.

Attack potential, reflecting effort required to mount a successful attack		Attack success probability, reflecting relative likelihood of a successful attack	
Value	Description	Ranking (P)	Description
0–9	Basic	5	Likely
10–13	Enhanced-basic	4	Possible
14–19	Moderate	3	Unlikely
20–24	High	2	Remote
≥25	Beyond high	1	Very remote

Table 5.6 Controllability of automotive safety hazards.

Controllability class (C)	Meaning
0	Controllable in general
1	Simply controllable
2	Normally controllable
3	Difficult to control or uncontrollable

components (R_S, R_P, R_F, R_O), since it is determined from the severity vector that is associated with a specific attack objective. However, it should be noted that there may be several risk vectors associated with each attack objective where a number of attack methods with differing combined attack success probabilities (A, a scalar) could be employed in order to achieve the attack objective.

The relative risk level is identified using a 'risk graph' approach [43], in which the 'security risk level' increases with increasing severity and/or combined attack success probability. Furthermore, the security risk level for safety-related threats also increases for those threats that are judged to have potential consequences that are less likely to be controllable for the driver. This is illustrated in Table 5.7, in which a range of risk levels are mapped to combinations of severity, probability of successful attack, and controllability. The portion of the risk graph corresponding to controllability class $C = 0$ also provides the mapping to risk levels for those aspects of security that are not related to safety, and which therefore have no controllability dimension.

The risk levels of Table 5.7 range from $R_i = 0$ (i.e. no security risk) to $R_S = 7+$, which represents safety-related security hazards with the highest severity classes and probability levels, coupled with low levels of driver controllability. Any safety hazards that are judged to correspond to risk levels $R_S = 7+$ are unlikely to be considered to be acceptable.

5.5.5 *Evaluating Risk from Attack Trees*

For evaluating the risk levels, it is convenient to describe the attack tree structure of using a table that can be augmented with the risk analysis parameters (see Table 5.8). The aim of this is to achieve a more compact representation of the attack tree information by focusing on the

Table 5.7 Tabular representation of security risk graph mapping risk level to severity, combined attack success probability and controllability.

Controllability class (C)	Severity (S_i)`	Combined attack success probability (A)				
		$A = 1$	$A = 2$	$A = 3$	$A = 4$	$A = 5$
$C = 0$	$S_i = 1$	0	0	1	2	3
	$S_i = 2$	0	1	2	3	4
	$S_i = 3$	1	2	3	4	5
	$S_i = 4$	2	3	4	5	6
$C = 1$	$S_S = 1$	0	1	2	3	4
	$S_S = 2$	1	2	3	4	5
	$S_S = 3$	2	3	4	5	6
	$S_S = 4$	3	4	5	6	7
$C = 2$	$S_S = 1$	1	2	3	4	5
	$S_S = 2$	2	3	4	5	6
	$S_S = 3$	3	4	5	6	7
	$S_S = 4$	4	5	6	7	7+
$C = 3$	$S_S = 1$	2	3	4	5	6
	$S_S = 2$	3	4	5	6	7
	$S_S = 3$	4	5	6	7	7+
	$S_S = 4$	5	6	7	7+	7+

Note: Nonsafety automotive security threats correspond to C=0.

contributing 'asset attacks' (denoted by a–g in Table 5.8), for which the attack potential (and hence the associated probabilities *Pa–Pg*) can be evaluated, and how they may be combined to contribute to the achievement of 'attack objectives' (denoted by X and Y in Table 5.8). At the attack objective level the severity of the attack consequences ($\mathbf{S}_X$ and $\mathbf{S}_Y$) and the controllability of safety-related security threats (C_X and C_Y) can be assessed. In this process it is useful to retain the construction of the 'attack methods' (X1, X2, Y1 and Y2 in Table 5.8), since these are described in terms of logical combinations of the asset attacks, whilst suppressing the details of any intermediate steps. The construction of the attack method provides the information that is required to work up the 'combined attack success probability' measure (i.e. A_{X1} etc.) that is needed in order to determine the associated risk level (i.e. $\mathbf{R}_{X1}(\mathbf{S}_X, A_{X1}, C_X)$ etc.).

If an attack method can be implemented using any one of a number of possible asset attacks (i.e. an OR relationship), then the combined attack success probability is taken to be the highest of the attack success probabilities for the possible asset attack options (i.e. it is as easy as the easiest option). Where the attack method requires a conjunction of asset attacks (i.e. an AND relationship), then the combined attack success probability is taken to be the lowest of the attack success probabilities associated with the contributing asset attacks (i.e. it is as hard as the hardest of the essential steps).

When the attack method involves asset attacks combined using both AND/OR relationships the combined attack success probability can then be built up using these simple rules, as illustrated in Table 5.8 (where the OR relationship is represented by separate lines in the 'Attack method construction' column). The numerical probability ranking that has been associated with the attack potential rating (see Table 5.5) therefore provides a convenient basis for

Table 5.8 Compressed tabular representation for attack trees augmented with risk analysis parameters.

Attack objective	Severity (**S**)	Controllability (C)	Attack method	Attack method construction	Asset attack success probability (P)	Combined attack success probability (A)	Risk level (**R**)
X	$\mathbf{S}_X$	C_X	X1	a &	Pa	$A_{X1} = \min\{Pa,Pb\}$	$\mathbf{R}_{X1}(\mathbf{S}_X,A_{X1},C_X)$
				b	Pb		
			X2	d	Pd	$A_{X2} = \max\ \{Pd,Pe,Pf\}$	$\mathbf{R}_{X2}(\mathbf{S}_X,A_{X2},C_X)$
				e	Pe		
				f	Pf		
Y	$\mathbf{S}_Y$	C_Y	Y1	a &	Pa	$A_{Y1} = \max[\min\{Pa,Pb,Pc\}$, $\min\{Pc,Ph\}]$	$\mathbf{R}_{Y1}(\mathbf{S}_Y,A_{Y1},C_Y)$
				b &	Pb		
				c	Pc		
				c &	Pc		
				h	Ph		
			Y2	g	Pg	$A_{Y2} = Pg$	$\mathbf{R}_{Y2}(\mathbf{S}_Y,A_{Y2},C_Y)$

evaluating the overall success probability for a given attack method, based on the logical combination of asset attacks that are required to implement the attack and their individual probabilities of success (see 'Combined attack success probability' column of Table 5.8).

In practice the same asset attacks often appear at different points in a given attack tree, as well as in other attack trees. Nonetheless, they may contribute to very different risk levels, depending on the particular security aspects (i.e. safety, privacy, financial and operational), stakeholders and severity ratings and that are associated with different attack objectives.

5.5.6 Prioritizing Security Functional Requirements

Specific subsets of security function requirements that are identified to mitigate particular asset attacks can, through the asset attack categorization, be mapped to security risks. Thus, the risk analysis results can be used to rank the relative importance of these subsets of security function requirements. This is illustrated in Table 5.9, which shows a few examples of the results obtained during the EVITA project [37].

From the illustrative examples shown in Table 5.9 it can be seen that protecting the wireless communications against corrupt or fake messages is a high priority, since several instances of high risk levels were identified from the attack trees in relation to this asset attack. By contrast, only a few instances of low risk levels were identified in relation to denial of service attacks on the chassis safety controller, which is therefore perceived to be a lower priority for protection.

The attack trees also provide a convenient basis for the systematic evaluation of possible attack methods and 'cutting branches' from the tree is a possible mechanism for identifying requirements for specific countermeasures. The selection of branches to be cut can be prioritized based on risk levels and attack success probabilities, in the following manner:

- Where a number of possible attack objectives may achieve the attack goal, the attack objective with the highest perceived risk level is the priority for countermeasures to reduce the overall risk.

Table 5.9 Mapping asset attacks to risks and security requirements.

Identified threats		Risk analysis results		Security functional requirements
Asset	Attack	Risk level	Number of instances	
Chassis safety controller	Denial of service	1	3	Subset 1
		2	1	
	Exploit implementation flaws	4	1	Subset 2
		5	1	
Wireless communications	Corrupt or fake messages	2	5	Subset 3
		3	5	
		4	4	
		5	1	
		6	4	
		7	3	
	Jamming	4	3	Subset 4
		5	2	

- Where a number of possible attack methods may lead to the same attack objective, the attack method with the highest combined attack success probability is the priority for countermeasures to reduce the risk level for the attack objective.
- Where a number of asset attacks may lead to the same attack objective, the asset attack with the highest perceived success probability (i.e. lowest attack potential) is the priority for countermeasures to reduce the risk level associated with the attack objective.

Introducing countermeasures to protect against those asset attacks that are judged to have the highest success probability (i.e. lowest attack potential) reduces the threat level for the associated attack method, and if the combined attack success probability for this attack method dominates the risk level for the associated attack objective then the risk level will also be reduced.

An important result of mapping the risk analysis results to the system assets is that this naturally leads to a 'defence in depth' approach [44]. The asset-based security architecture and counter-measures developed in EVITA [45] provide for a distributed and multilevel approach that is intended to ensure robust security in a cost-effective manner.

5.5.7 Security Assurance and Safety Integrity Requirements

It is also necessary to identify safety integrity and security assurance requirements that are appropriate to the safety and security functional requirements. In order to achieve this, Table 5.10 proposes a mapping from the risk levels of Table 5.7 to the ASILs of ISO 26262 [22] and the EALs of ISO/IEC 15408 [25], as well as to the SILs of IEC 61508 [21] and MISRA [23,30].

The recommended approach for establishing the safety of complex electronic control systems, based on experience in safety-related applications found in the aerospace, defence, nuclear, rail and off-shore oil industries, is to create a safety argument to show that the system is acceptably safe for the intended application and operating environment. The important points to note here are that:

- complete safety is recognized as unachievable, although mitigation measures must be implemented as necessary in order to ensure that any residual risks are deemed to be acceptable;
- the safety argument only applies to the intended application and operating environment.

Table 5.10 Proposed mapping of EVITA risk levels to EAL, ASIL and SIL (IEC 61508 and MISRA).

EVITA Risk Level	IEC 15408 EAL	MISRA SIL	IEC 61508 SIL	ISO 26262 ASIL (C>0)
0	0	0	Not applicable	Quality Management
1	1	1	1	A
2	2	1	1	A
3	3	2	2	B
4	4	2	2	B
5	5	3	3	C
6	6	3	3	D
7	7	4	4	Not applicable
7+	Risks deemed to be beyond normally acceptable levels			

The safety argument and supporting evidence should be documented in the form of a 'safety case', which should provide the following features [46]:

- make an explicit set of claims about the system properties;
- identify the supporting evidence;
- acknowledge and refute 'counter evidence';
- provide a set of safety arguments that link the claims to the evidence;
- state the underlying assumptions and judgments;
- provide different viewpoints and levels of detail.

The safety case is generally subject to assessment and audit by an independent third party. Constructing the safety case in a hierarchical form makes it easier to understand the main arguments and to partition the safety case development activities. Claims can be made more robust by using independent evidence and more than one safety argument to support them. The safety argument for the achievement of a particular SIL [47] is then that the processes described in the standard as appropriate to the required SIL have been applied, and evidence generated in so doing has been independently assessed to confirm this.

For networked vehicles, ITS and telematics services the operating environment is known to include hackers, criminals and terrorists, who are already actively engaged in cyber security attacks against existing computer networks and can be expected to turn their attention to vehicles and roadside infrastructure in future. Thus, safety cases for vehicle and road transport applications should also take account of safety-related security threats.

As the safety case concept has been widely adopted in many safety-related industrial sectors, it seems logical to consider developing an analogous 'security case' [48] to present the security argument for security-related applications, particularly for cases where security threats may also have potential safety implications. Evidence and arguments from the security case can then be shared with the safety case, and subjected to independent third party assessment and audit, as with the safety case.

5.6 Cyber Security Risk Management

Achieving zero security risk is impracticable, as the cost would be excessive and the nature of the threats is constantly changing. Consequently, a more pragmatic 'risk management' approach is recommended, in which:

- security must be considered from the outset;
- evaluated risks are considered unacceptable, measures must be employed to reduce these risks to a level that is considered to be 'broadly acceptable';
- security analysis must be an on-going process throughout the product lifecycle.

The proposed approach to analysing security risks and prioritizing security functional requirements includes the following:

- description of system under investigation and its environment;
- definition of relevant use cases;

- listing of the assets to be protected (e.g. electronic control units, applications, processes, sensors, data, communication links);
- identification of possible threat agents and their objectives, as well as the relevant stakeholders;
- investigation of the potential threats posed to each asset, using methods such as 'attack trees';
- identification of safety-related security threats, and assessment of their controllability;
- evaluation of the probability of successful attack, severity of outcome, and resulting risk classification;
- prioritization of security functional requirements for critical threats identified in the risk analysis.

This kind of approach mirrors the measures that are currently employed in order to achievable acceptable levels of functional safety in the automotive industry, based on ISO 26262. A number of international standards relating to information security have already been developed, and the US Society of Automotive Engineers (SAE) has recently initiated work on an automotive cyber security guidebook [49]. At present, however, there are no standards specifically addressing automotive cyber security.

5.7 Conclusions

The IEC 61508-based safety analysis techniques of ISO 26262 and MISRA have been merged with relevant IT security concepts from ISO/IEC 15408 and ISO/IEC 18045 to develop a unified approach for assessing security threats in automotive applications, including those with possible safety implications. The objectives of this were to ensure that the assessment of safety risks arising from security threats would comply with automotive functional safety engineering processes, and to introduce risk analysis into the evaluation of security threats that do not have safety implications.

To achieve this, the notion of severity as used in automotive functional safety analysis was extended to encompass nonsafety aspects of security threats for a wide range of stakeholders. The attack potential concept was used to assess the probability of success for attacks. In addition, the impact of vehicle controllability is taken into account in assessing the safety-related security risks. A threat model based on attack trees was used to document possible attacks and to evaluate their associated risks. As with safety, security risk analysis will need to be an on-going process in order to take account of the possible evolution of security threats over the anticipated operational lifetime of the vehicle, as well as during the vehicle development lifecycle.

The EVITA project did not aim to develop specific vehicle systems with particular levels of security, or to enhance the security of existing systems. Instead, the project was concerned with prototyping a 'toolkit' of analysis methods and security measures (software, hardware, and architectural) that could be selected for further development and implementation in future systems. Consequently, the requirements analysis activity was based on a representative range of use cases and a generic vehicle network architecture, with the aim of identifying what kind of security requirements may arise, as well as their likely prevalence and distribution amongst the vehicle assets.

Nonetheless, it is expected that the security engineering process developed in EVITA could be adapted to support future vehicle development processes, as well as the development of safe and secure ITS and telematics services. In particular, the risk analysis approach proposed here could be used, in combination with the developer's security policy, in order to decide whether to accept or transfer the security risks identified, or to take measures to reduce or avoid specific risks where this is deemed necessary.

Given that cyber security attacks are unavoidable, the aim should be to develop vehicles and transport management systems that are able to maintain resilience against such threats over their lifecycle in an environment that will continue to change. The most important factor in this is to recognize that cyber security threats exist, and to implement counter-measures that address the most significant of the foreseeable threats. However, it should be recognized that the cyber security environment is evolving very rapidly, and unforeseen threats will emerge. Design measures to detect intrusion and tampering, to limit the opportunities for unauthorized access, and to revert to 'fail-secure' operating modes, will be required in order to achieve the necessary resilience.

Acknowledgements

Part of the research leading to these results received funding from the European Community's Framework Programme (FP7/2007–2013) under grant agreement FP7-ICT-224275 (EVITA). The authors are grateful for the contributions of the other EVITA project participants, from BMW Group (Germany), Continental Teves (Germany), escrypt (Germany), EURECOM (France), Fraunhofer Institute for Secure Information Technology (Germany), Fraunhofer Institute for Systems and Innovation Research (Germany), Fujitsu (Sweden, Austria, and Germany), Infineon Technologies (Germany), Institut Telecom (France), Katholieke Universiteit Leuven (Belgium), MIRA (UK), Robert Bosch (Germany), and Trialog (France).

References

1. R. Bose, J. Brakensiek, K.Y. Park and J. Lester (2011) Morphing smart-phones into automotive application platforms, *IEEE Computer* **44**(5): 53–61.
2. K. Amirtahmasebi and S.R. Jalalinia (2010) Vehicular Networks – Security, Vulnerabilities and Countermeasures, MSc Thesis, Chalmers University of Technology.
3. K. Bjelkstal (2008) *Exchange of Diagnostic Information between Car and Centralized Functions*, VINNOVA Information 2008-04l, ISSN 1650-3120, Vehicle-ICT Sweden.
4. EVITA project overview. Available online at: http://www.evita-project.org (last accessed 27 April 2015).
5. J. Dumortier, C. Geuens, A.R. Ruddle and L. Low (2011) *Legal Framework and Requirements of Automotive On-board Networks*, EVITA Deliverable D2.4, 19 September 2011. Available on-line at: http://www.evita-project.org (last accessed 27 April 2015).
6. SeVeCom project overview. Available online at: http://www.transport-research.info/web/projects/project_details.cfm?id=46017 (last accessed 8 May 2015).
7. A. Alrabady and S. Mahmud (2003) Some attacks against vehicles' passive entry security systems and their solutions, *IEEE Transactions on Vehicular Technology* **52**(2): 431–9.
8. A. Alrabady and S. Mahmud (2005) Analysis of attacks against the security of keyless-entry systems for vehicles and suggestions for improved designs, *IEEE Transactions on Vehicular Technology* **54**(1): 41–50.
9. A. Francillon, B. Danev and S. Capkun (2011) Relay attacks on passive keyless entry and start systems in modern cars, *Proceedings of 18th Annual Network and Distributed Systems Security Symposium, San Diego, CA, USA, February 2011*.

10. D.K. Nilsson, U.E. Larson, F. Picasso and E. Jonsson (2008) A first simulation of attacks in the automotive network communications protocol FlexRay, *Proceedings of 1st International Workshop on Computational Intelligence in Security for Information Systems (CISIS)*. London: Springer, pp. 84–91.
11. T. Hoppe and J. Dittmann (2007) Sniffing/replay attacks on CAN buses: a simulated attack on electric window lift classified using an adapted CERT taxonomy, *Proceedings of 2nd Workshop on Embedded System Security, October 2007*, pp. 1–6.
12. K. Koscher, A. Czeskis, F. Roesner, *et al.* (2010) Experimental security analysis of a modern automobile, *Proceedings of 31st IEEE Symposium on Security and Privacy*, Oakland, CA, USA, May 2010, pp. 447–62.
13. I. Rouf, R. Miller, H. Mustafaa, *et al.* (2010) Security and privacy vulnerabilities of in-car wireless networks: a tire pressure monitoring system case study, *Proceedings of 19th USENIX Security Symposium, Washington DC, USA, August 2010*, pp. 11–13.
14. US Department of Transportation, National Highway Traffic Safety Administration, Federal Motor Vehicle Safety Standard FMVSS No. 138, 49 CFR, Parts 571 & 585: Tire Pressure Monitoring Systems.
15. Commission Regulation (EC) No. 661/2009 of 13 July 2009 concerning type-approval requirements for the general safety of motor vehicles, their trailers and systems, components and separate technical units intended therefor, *Official Journal of the European Union*, L 200, July 2009, pp. 1–24.
16. COM(2013) 315, *Proposal for a Decision of the European Parliament and of the Council on the Deployment of the Interoperable EU-wide eCall*, 13 June 2013.
17. D.K. Nilsson and U.E. Larson (2009) A defense-in-depth approach to securing the wireless vehicle infrastructure, *Journal of Networks* **40**(7): 552–64.
18. R. Bauernfeind, T. Kraus, A. Sicramaz Ayaz, *et al.* (2012) Analysis, detection and mitigation of incar GNSS jammer interference in intelligent transport systems, *Deutscher Luft- und Raumfahrtkongress 2012, Berlin, 10–12 September 2012*, Paper 281260.
19. C. Vallance (2013) RF Safe-Stop shuts down car engines with radio pulse, BBC radio 4, 3 December. Available online at: http://www.bbc.co.uk/news/technology-25197786 (last accessed 27 April 2015).
20. S. Checkoway *et al.*, Comprehensive experimental analyses of automotive attack surfaces, *Proceedings of 20th USENIX Security Symposium, San Francisco, 10–12 August 2011*.
21. IEC 61508, *Functional Safety of Electrical/Electronic/Programmable Electronic Safety-related Systems*, 2nd Edition, April 2010.
22. ISO 26262, *Road Vehicles – Functional Safety* (9 parts), November 2011.
23. MISRA (2007) *MISRA Guidelines for Safety Analysis of Vehicle Based Programmable Systems*, MIRA Ltd.
24. *Towards a European Standard: The Development of Safe Road Transport Informatics Systems*, Draft 2, DRIVE Safely (DRIVE I Project V1051), March 1992.
25. ISO/IEC 15408, *Information Technology – Security Techniques – Evaluation Criteria for IT Security* (3 parts), December 2009.
26. ISO/IEC 18045, *Information Technology – Security Techniques – Methodology for IT Security Evaluation*, August 2008.
27. ISO/IEC TR 15446, *Information Technology – Security Techniques Guide for the Production of Protection Profiles and Security Targets*, Technical report, July 2004.
28. ISO/IEC 13335-1, *Information Technology – Security Techniques – Management of Information and Communications Technology Security*, November 2004.
29. NIST Special Publication 800-12, *An Introduction to Computer Security: The NIST Handbook*, October 1995.
30. MISRA (1994) *Development Guidelines for Vehicle Based Software*, MIRA Ltd, November, p. 43.
31. P.H. Jesty and D.D. Ward (2007) Towards a unified approach to safety and security, *Proceedings of 15th Safety-Critical Systems Symposium, Bristol, UK, February 2007*. London: Springer.
32. A. Clark, H.R. Chivers, J. Murdoch and J.A. McDermid (2007) *Unifying MANET Safety and Security*, International Technology Alliance in Network-Centric Systems, Report ITA/TR/2007/02, V. 1.0.
33. S. Lautieri, D. Cooper and D. Jackson (2005) SafSec: commonalities between safety and security assurance, *Proceedings of 13th Safety Critical Systems Symposium, Southampton, UK, February 2005*. London: Springer, pp. 65–75.
34. E. Kelling, M. Friedewald, M. Menzel *et al.* (2009) *Specification and Evaluation of e-Security Relevant Use Cases*, EVITA Deliverable D2.1, 30 December 2009. Available online at: http://www.evita-project.org (last accessed 27 April 2015).
35. EASIS project overview. Available online at: http://cordis.europa.eu/result/rcn/45019_en.html (last accessed 8 May 2015).

36. B. Schneier (2000) *Secrets and Lies – Digital Security in a Networked World*. New York: John Wiley & Sons, Inc., Chapter 21.
37. A. Ruddle, D. Ward, B. Weyl, *et al.*, Security requirements for automotive on-board networks based on dark-side scenarios, *EVITA Deliverable D2.3, 30 November 2009*. Available online at: http://www.evita-project.org (last accessed 27 April 2015).
38. N.R. Mead (2010) *Security Requirements Engineering*. Available online at: https://buildsecurityin.us-cert.gov/bsi/articles/best-practices/requirements/243-BSI.html (last accessed 27 April 2015).
39. *Abbreviated Injury Scale*, Association for the Advancement of Automotive Medicine, Barrington, IL, USA, 2005.
40. *Application of Attack Potential to Smartcards*, Common Criteria Supporting Document – Mandatory Technical Document, Version 2.5, Revision 1, April 2008, CCDB-2008-04-001.
41. M. Scheibel and M. Wolf (2009) Security risk analysis for vehicular IT systems – a business model for IT security measures, *Proceedings of 7th Embedded Security in Cars Workshop (escar 2009), Düsseldorf, Germany, November 2009*.
42. *Towards a European Standard: The Development of Safe Road Transport Informatics Systems*, Draft 2, DRIVE Safely (DRIVE I Project V1051), March 1992.
43. DIN V 19250, *Control Technology: Fundamental Safety Aspects to be Considered for Measurement and Control Equipment*, May 1994.
44. R. Anderson (2001) *Security Engineering: A Guide to Building Dependable Distributed Systems*. Chichester, UK: Wiley Computer Publishing, p. 296.
45. O. Henniger, L. Apvrille, A. Fuchs, *et al.* (2009) Security requirements for automotive on-board networks, *Proceedings of 9th International Conference on Intelligent Transport System Telecommunications (ITST 2009), Lille, France, October 2009*, pp. 641–6.
46. P. Bishop and R. Bloomfield (1998) A methodology for Safety Case development, In F. Redmill and T. Anderson (eds), *Industrial Perspectives of Safety-Critical Systems: Proceedings of 6th Safety-Critical Systems Symposium, Birmingham, UK, February 1998*. London: Springer, pp. 194–203.
47. F. Redmill (2000) Understanding the use, misuse and abuse of Safety Integrity Levels, *Proceedings of 8th Safety-Critical Systems Symposium, Southampton, UK, February 2000*.
48. G. Despotou and T. Kelly (2004) Extending the Safety Case concept to address dependability, *Proceedings of 22nd International System Safety Conference, Providence, RI, USA, August 2004*, pp. 645–54.
49. SAE Vehicle Electrical System Security Committee, J3061 – Cybersecurity Guidebook for Cyber-Physical Automotive Systems. Available online at: http://www.sae.org/works/documentHome.do?docID=J3061&inputPage=wIpSdOcDeTaIlS&comtID=TEVEES18 (last accessed 27 April 2015).

6

Vehicle Interaction with Electromagnetic Fields and Implications for Intelligent Transport Systems (ITS) Development

Lester Low and Alastair R. Ruddle
MIRA Ltd, Nuneaton, UK

6.1 Introduction

The modern vehicle incorporates numerous electronic devices to provide communication, information and entertainment services for the driver and passengers [1]. All wireless devices communicate over a radio frequency (RF) channel via antennas. In the context of a vehicle environment, antennas are often found within the cabin or mounted to the exterior of the vehicle chassis. As externally mounted antennas are covered in another chapter of this book, this chapter focuses on the study of electromagnetic (EM) fields generated by antennas that are found mainly within the vehicle cabin.

For antennas placed within the confines of the vehicle cabin it is important to understand how the vehicle structure interacts with the EM field, as the vehicle interior is a partial cavity that introduces multiple reflections and indirect coupling that may degrade the quality of the communication channel [2]; for example, for real-time intra-vehicle control applications wireless data transmission must satisfy stringent requirements that include low latency and high reliability. The objective of this chapter is therefore to outline the behaviour of EM fields within such resonant environments and possible implications for intelligent transport systems (ITS). Related issues such as installed antenna performance and electromagnetic

Intelligent Transport Systems: Technologies and Applications, First Edition. Asier Perallos, Unai Hernandez-Jayo, Enrique Onieva and Ignacio Julio García-Zuazola.

compatibility (EMC) will also be discussed in this chapter. In order to emphasize how ITS functions are related to wireless devices that are used in the automotive environment, a review of existing and potential future technologies will be introduced in this section.

The rapid development of information and communication and technologies has changed consumers' expectations of future in-car technologies. Drivers and passengers often carry portable connected devices such as smartphones, tablets and internet media communication devices, and they increasingly expect access to connected services in their vehicles [3]. Examples of such services include the ability to operate a smartphone using controls on the steering wheel while driving, and the streaming of music from their brought-in devices to the music system on board the vehicle [4]. In order to enhance the driving and ownership experience, some vehicles already have built-in devices that enable internet browsing via a monitor, and the ability to read and dictate emails whilst driving [5]. In situations where a vehicle manufacturer provides fixed placement locations for a portable device or a repeater to enhance wireless communications within the car, a good understanding of the EM field behaviour within the cabin may allow an optimized antenna placement location to be found. The optimization could be performed for the communication link between the portable device and an external base-station, or for on-board transceivers. Connected vehicles play a pivotal role in ITS, and can provide advanced services such as remote diagnostics and vehicle security [6]. This is a fast-growing market that is now a strategic priority for the automotive industry [7,8].

In addition to the demand for connected devices, the number of on-board sensors is also increasing. It is forecast that there will be about 200 sensors per vehicle by 2020 [9]. Most of these sensors are used in engine management and advanced driver assistance systems (ADAS), for which road conditions, vehicle status and even driver fatigue may be monitored. These event- or time-driven sensors communicate with on-board electronic control units (ECUs) via an intra-vehicle network. Present intra-vehicle communications are based on wired solutions, such as CAN (Controller Area Network), LIN (Local Interconnect Network) and MOST (Media Oriented Systems Transport). With increasing numbers of sensors, the number of cables required to connect the sensors to the ECU will increase accordingly. This can add significant weight to the vehicle [10]. The vehicle wiring harness already contributes up to 50 kg to the vehicle sprung mass and is the heaviest, single most complex and expensive electrical component in a vehicle [11,12]. Replacing the cables with wireless technology could reduce the weight of the vehicle and at the same time enable a more flexible ad-hoc network architecture, enabling sensors to be added without the physical constraints of wired interconnect technologies. The reduction in vehicle weight may contribute to reducing carbon emissions by improving fuel efficiency in a conventional internal combustion engine. For electric vehicles, the weight reduction may translate to increased driving range. At present, wireless sensors are only found in locations that are inaccessible by cables, such as the tyres and steering wheel. One example of such a wireless system is the tyre pressure monitoring [13], which has become mandatory equipment in the EU for passenger cars and similar vehicles since November 2012. At present, however, wireless communications are used mainly in noncritical applications because vehicle control and monitoring systems have stringent requirements relating to latency and reliability. Due to the complex communications channels present in the automotive environment, intra-vehicle wireless communications cannot provide the same performance and reliability characteristics as wired communications [14]. However, with significant research and development efforts it is considered that the latency and reliability properties could be

Table 6.1 European frequency bands for wireless technologies found in the automotive environment.

Wireless technology	Frequency (MHz) – Europe
Dynamic Spectrum Access (DSA) – TV White Space	470–790
Passive Radio Frequency Identity (RFID) Tag	865–868
GSM	890–1880
Bluetooth	2402–2470
Zig Bee	868, 915, 2400
DSRC	5725–5875
Ultra-Wideband	3100–10 600
60 GHz millimetre wave	57 000–64 000

improved through having a better understanding of the channel characteristics, improved network protocols and the introduction of redundancy [15].

Identification of the type of wireless technologies is required in order to narrow down the frequency bands for which optimization of the communications channels is required. Typical operating frequencies of wireless technologies that are used in the automotive environment are listed in Table 6.1, for Europe. The frequency range is rather wide. For this reason, the investigation presented in this chapter is focused on three important communications frequencies: 900 MHz, 1800 MHz and 2400 MHz.

The outline of this chapter will be as follows: Section 6.2 will briefly describe the present state of art of channel characterization and in-vehicle EM field investigations; Section 6.3 will introduce software tools and techniques used to perform in-vehicle field simulations; Section 6.4 will describe the use of a novel low disturbance scanner used for experimental validation; Section 6.5 will investigate the effects of field distribution within the cabin of the vehicle due to different placement locations of the source antenna; and Section 6.6 will discuss human exposure issues and possible EM field mitigation techniques. Finally, Section 6.7 will provide a summary of the work described in this chapter.

6.2 In-vehicle EM Field Investigation and Channel Characterization

Studies of various EM field investigations and propagation characterizations for transmitting and receiving antennas placed within the vehicle cabin can be found in the published literature. This section summarizes relevant literature that may have implications for ITS.

Channel characterization experiments are generally performed to obtain a better understanding of the propagation characteristics. With basic knowledge of channel parameters, communication links can be improved and the design of RF front ends may be simplified [16]. Usually, measurements [17] or simulations based on deterministic ray tracing methods [18] are used to obtain these important parameters. Full-wave simulations can also be used to obtain detailed three-dimensional (3D) EM field within the vehicle cabin, but require greater computational resources. Full-wave methods are more commonly used for investigations relating to EMC phenomena, human exposure to EM fields and optimization of antenna placement locations [19]. However, they can also be used to extract important communication channel parameters such as small- or large-scale fading characteristics [20]. The above-mentioned analysis may impact ITS performance. As field external to the vehicle is covered in another chapter of this book, this chapter will focus more on the use of full-wave electromagnetic solver and measurements to obtain EM fields within the vehicle cabin. For completeness, studies not involving 2D/3D in-vehicle field distributions will also be briefly described.

Propagation models based on statistical models can provide initial estimates of signal characteristics without the resources required by ray-tracing or full-wave simulation. They are also less sensitive to the geometry of the environment if the propagation domain is sufficiently large. The main parameters used for characterizing wireless channels include:

- path loss;
- fading;
- power-delay profile;
- time-delay spread.

There are many other parameters of possible interest. An excellent summary of the definitions, terminology and measurement methods for characterizing various parameters of a propagation channel is provided in [21]. The above-mentioned characteristics are often used to describe the propagation behaviour between a pair of points with minor variations in the input parameters. The changes to the input parameters may be the position of the RF source, or changes occurring in the environment. For optimizing placement of antennas for ITS systems, and also for EMC problems, statistical measures such as the field amplitude probability distribution functions within the entire spatial volume of interest can be useful. This involves the collection of a large sample of field values from different spatial points within the vehicle, followed by calculation of mean values, standard deviations and plotting of probability distribution functions. A large data set is required in order to obtain sufficiently accurate estimates for the statistics of the probability distribution function. For EMC analysis, the extrema are also of importance, as overestimation of the likelihood of a high field level could lead to over-engineering. For occupant field exposure analysis, an insufficient number of samples may underestimate the average field level, resulting in a lower margin of safety.

A number of studies have been carried out to derive in-vehicle propagation models from measurements or simulation data. In one example, the radiation mechanism was found by evaluating surface currents obtained using a full-wave field solver [22]. In another, results from ray tracing simulations were combined with empirical measurements to obtain a directional path loss model at 1.8 GHz which required minimal knowledge of the detailed vehicle geometry [23]. The resulting model is analytical, and based on an approximation to physically interpretable edge diffraction. A comparison between ray tracing models, full-wave simulations and measurements for investigating path losses within vehicular environments is provided in [18], for frequencies of 866 MHz and 5.8 GHz. The different level of path loss observed between the lower and higher frequencies showed that the channel performance is dependent on both the frequency of the wireless system and the location of the transceiver. To optimize the communication link, appropriate wireless protocols should be chosen.

As listed in Table 6.1, a range of different wireless protocols such as DSA, RFID, Bluetooth, ZigBee and UWB are used for wireless communications within vehicles. Detailed channel characterization studies using RFID devices have been performed at 915 MHz [15,24]. In these studies, some transceivers were placed outside the cabin to communicate with transceivers placed inside the cabin. The presence of the vehicle structure resulted in high transmission losses. However, the authors concluded that the channel is still usable as the coherence times are large at the operating frequency of the RFID under investigation (915 MHz). Many experiments were conducted on the feasibility of using Bluetooth and ZigBee for intra-car wireless network. Channel models suitable for vehicular environment in the industrial, scientific and medical (ISM band) have been proposed [15,25]. ZigBee and Bluetooth coexistence, as well as ways to optimize the channel, were also investigated [26]. However, most of these experiments have not evaluated the channel capacity. The authors of Refs. [27,28] attempted to understand the time variation of the channel within the vehicle cabin in order to evaluate the channel capacity. It was observed that the Doppler spread increases with carrier frequency, but the type of channel variation and interior contents of the vehicle have little effect on the Doppler spread due to the presence of windows and other apertures in the vehicle shell. This is an important property as optimal communication depends on the channel having a well-defined coherence time.

Ultra-wideband (UWB) is another wireless protocol that can provide reliable and robust wireless communications in harsh environments such as vehicle interiors. Various in-vehicle channel measurements have been performed in order to evaluate the potential of UWB technology for use in vehicles [29–31]. Small- and large-scale statistics of the vehicular UWB channel were analysed with transceivers located beneath the chassis, in the vicinity of the wheel arches, and within the vehicle cabin [14]. It was found that the channel characteristics for transceivers placed under the wheel arches were quite different from those for transceivers placed under the chassis. This study showed that it is important to investigate the harsh communication channel presented by the vehicle in more detail. Hence, the paragraphs below will survey available studies, which have focused on the characteristics of spatial field distributions within the cabin space of a vehicle.

The spatial field distribution within a vehicle may be obtained with numerical models and from measurements [32]. Most of these analysis are conducted near mobile phone frequency bands as they are needed for EMC compliance tests and for the assessment of human exposure to EM fields. However, investigations have also been conducted using full-wave simulations for the TETRA (Terrestrial Trunked Radio) and WiFi (Wireless Fidelity) frequency bands [33,34].

The electric field distribution produced within an empty vehicle cabin due to a micro-strip patch antenna has been described using a scaled vehicle-like object [35]. Comparison between the measured and simulated data was only available on a single 2D plane. The size of the measurement equipment and connectors cannot be scaled, hence the field sampling resolution is limited. However, the paper showed that it is possible to obtain reasonable correlation between measurements and simulation on a 2D plane when an optical probe was used for the EM field measurements.

In Refs. [36,37], a comparison between simulations and measurements of EM field created by a mobile phone antenna mounted on the vehicle roof and a mobile phone placed inside a vehicle cabin was investigated. Both investigations concluded that dielectric furnishings significantly alter the EM field distributions within a vehicle. Hence, the interior contents of the vehicle should be included in the simulation model when computing EM field distributions at and above mobile phone frequencies.

To verify that fields emitted by mobile phones do not exceed a specific level for EMC compatibility testing, a fine positioning apparatus that can perform 3D measurement of EM fields both inside and out the vehicle was used [38]. This scanning system was constructed from reinforced plastic, with the scanning mechanism and supporting frames placed outside the vehicle cabin. Consequently, the windows had to be opened to allow the robotic arm to be inserted into the vehicle. However, opening the windows for vehicles that has solar shields on the windows may alter the EM field level and distribution. In order to measure the EM field within the cabin in cases where the window has solar shields, it is necessary to use a low disturbance positioner that can be mounted within the vehicle. A novel positioner that was designed to carry out the task will be described in Section 6.4.

6.3 Field Simulation Tools and Techniques

The evaluation of electromagnetic field distributions within a vehicle is a complex task. This is due to the geometrical complexity and highly reflective surfaces of the vehicle structure. For experimental assessment of electromagnetic fields within a vehicle, the introduction of RF measurement equipment into the semi-resonant structure may alter the Q-factor and spatial distribution of the electromagnetic field emitted by a transmitter placed within the vehicle cabin. Furthermore, space restrictions limit the volumes that can be measured and the presence of significant movable parts (i.e. front seats and steering wheel) make it difficult to achieve repeatable measurements.

A more practical way to evaluate the electromagnetic field behaviour and placement of in-vehicle wireless devices is to use numerical simulation. Numerical models can provide an initial estimate of the field levels that could be present within a vehicle. In addition, simulation has the advantage of not being affected by movement of the internal cabin furnishings or changes in Q-factor resulting from the introduction of measurement equipment or personnel performing the measurements.

Numerical modelling methods, based on the full-wave numerical solution of Maxwell's equation, can be performed in either the time or frequency domain. Some examples of time domain techniques include the finite integration technique (FIT) method, finite differences in the time-domain (FDTD), and the transmission-line matrix method (TLM). Frequency domain methods include the method of moments (MoM), the multilevel fast multipole method (MLFMM), and the finite element method (FEM). Commercial software suites such as CST, FEKO and HFSS offer these kinds of tools together with a convenient interface for importing computer-aided design (CAD) data. Hence, these products are well-suited for use in vehicle EM field simulation. The list of simulation techniques given here is not exhaustive and there are many other commercial software products that are suited to perform the required task. The CST toolset [39] offers a variety of solvers (e.g. FIT, TLM, and frequency domain solvers), whereas FEKO [40] and HFSS [41] offer only frequency domain solvers. The frequency domain approach may seem better suited for resonant structures with high Q-factors, but as the vehicle contains apertures in the body shell and lossy internal furnishings the resonances are damped and time-domain solvers may be used to perform broadband simulation. Typical Q-factors of the order 10^2–10^3 are expected in vehicles at the frequencies considered here [42]. The Q-factors of an empty cavity of the same volume, surface area and conductivity but with no window apertures would be significantly higher.

Vehicle CAD data is highly detailed, down to the level of welds and fixings such as screws and bolts. In addition, it often contains defects such as intersecting surfaces, gaps between surfaces or overly complex definitions. These defects may not be apparent until mesh generation is attempted, or may result from translations between CAD different formats. Thus, before a numerical simulation can be performed, the vehicle geometry will generally have to be de-featured to remove unnecessary elements, and the remaining CAD data repaired and simplified to ensure that the CAD model comprises topologically coherent and unambiguous 3D geometry. The resulting simplified CAD model should then provide a suitable basis for generating a mesh for simulation purposes. A typical work flow for creating vehicle models suitable for numerical EM modelling is shown in Figure 6.1. A representative example of a complex vehicle requiring de-featuring to create a viable EM model is shown in Figure 6.2. The resulting simplified model is illustrated in Figure 6.3.

For commonly used time-domain methods, such as FIT, TLM and FDTD, the geometry is usually discretized into hexahedral volume cells as shown in Figure 6.4, whereas in the frequency domain a triangular surface mesh (see Figure 6.5) or a tetrahedral volume mesh are used. For meshing, a typical discretization level of 10 cells/segments/triangle-edges per wavelength is recommended, but in regions where there are fine features or high spatial field

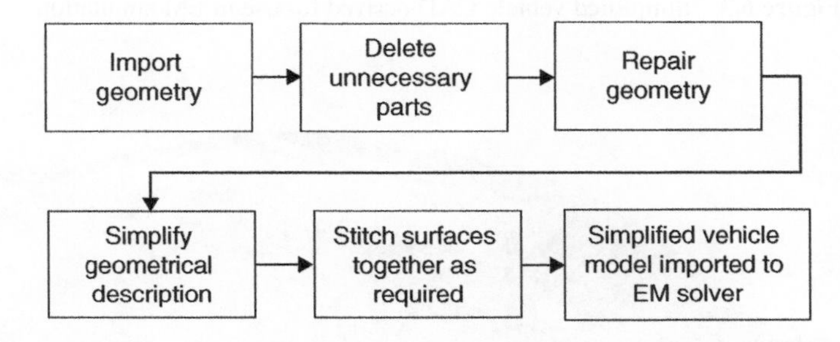

Figure 6.1 Typical work flow for vehicle CAD processing.

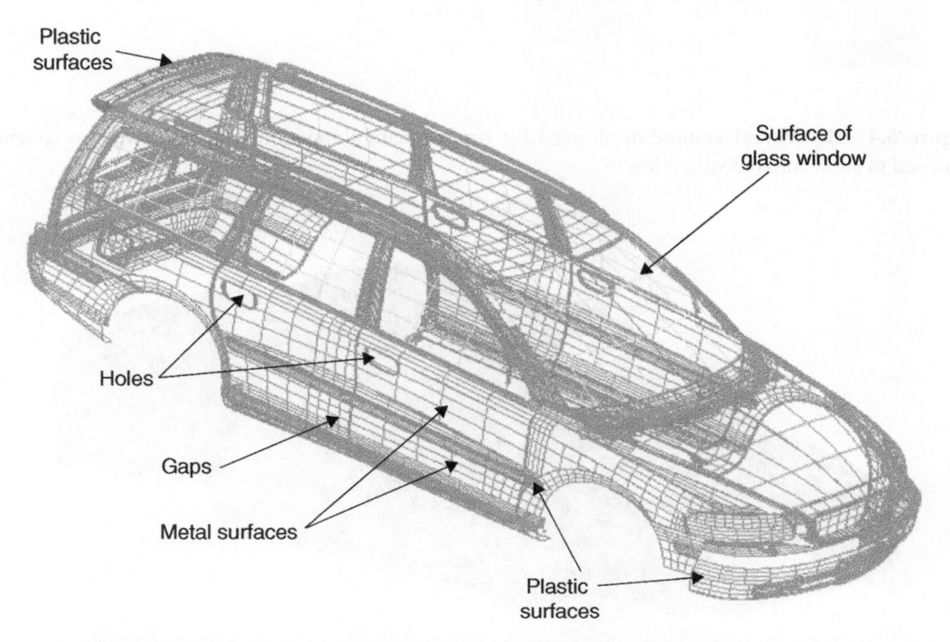

Figure 6.2 Example of complex CAD geometry for the outer surfaces of a vehicle.

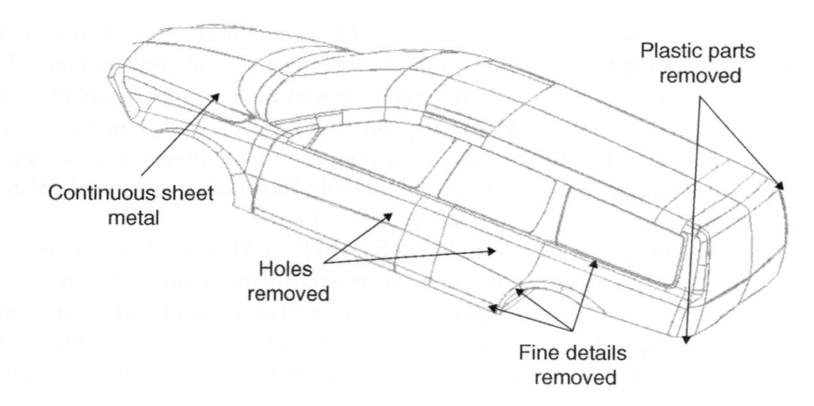

Figure 6.3 Simplified vehicle CAD derived for use in EM simulation.

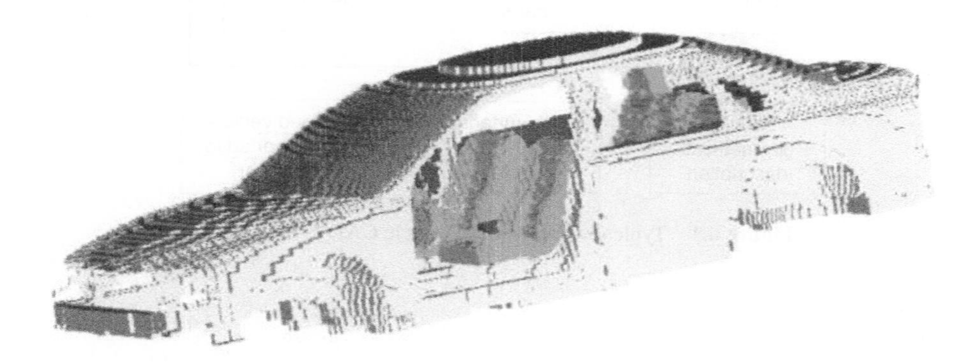

Figure 6.4 Hexahedral volume mesh used for time-domain simulation, with side window glazing removed to show internal structures.

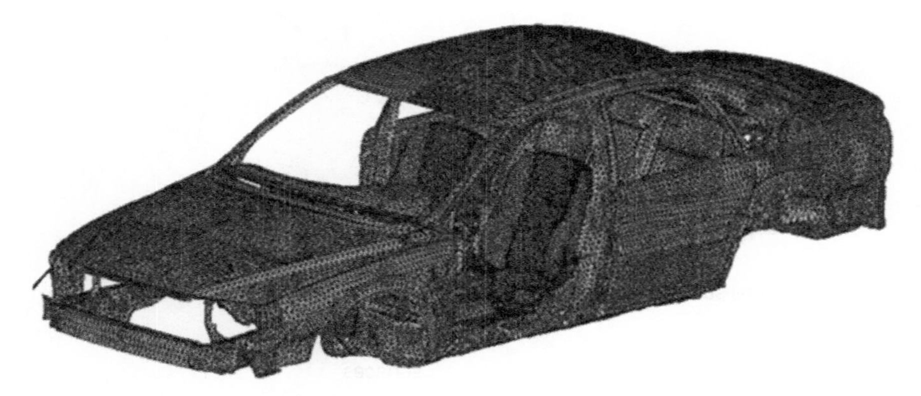

Figure 6.5 Triangular surface mesh used for frequency domain simulation.

gradients a higher resolution will be required to ensure the accuracy of the simulation. The structural complexity of the numerical model required to carry out a satisfactory simulation is often dependent on the frequency of interest. In the lower frequency range (e.g. 400 MHz), features such as door gaps, thin lossy internal trims and furnishings may not have significant impact on field levels within the vehicle cabin. However, at higher frequency (e.g. 2.4 GHz), features such as tiny apertures, window heater arrays and lossy internal furnishings can have a significant effect on field levels. For this reason, it is important to include the appropriate geometry to obtain satisfactory simulation data. However, due to limitations on the available computer memory and processing speed, the complicated geometry of internal structures of a vehicle may have to be simplified or eliminated in order to obtain practicable run-times.

Electrical properties of materials that are indicative of those commonly found in a vehicle are shown in Table 6.2. The material measurements were performed at the University of Sheffield either using the waveguide method [43] or circular resonant cavity method [44]. It should be noted that material properties can vary between different vehicles. In addition, due to the difficulties in engineering and placement of the dielectric sample in the measurement apparatus, the loss tangent measurements for fibrous materials (e.g. foam/fabric) are considered to be less reliable than those made on more rigid materials.

From Table 6.2 it can be seen that the frequency usage of vehicular communications could be higher, and may span up to up to 60 GHz. At these frequencies, full wave vehicular simulation which includes the entire vehicle body may not be feasible in terms of the numerical problem size. Asymptotic ray tracing approaches, such as the uniform theory of

Table 6.2 Electrical properties (at 3 GHz) of representative dielectric materials used in vehicles.

Vehicle part	Material type	Relative permittivity	Loss tangent
Headrest/seats	High density polyethylene	2.31	0.01
	Thermoplastic rubber (PP4)	2.15	0.0006
	Foam	1.12	0.001
	Polyurethane	3.12	0.04
Seat covers	Fabric	1.74	0.02
	Leather	2.73	0.05
Sun visor	Polyvinyl chloride	3.22	0.04
	Acrylonitrile butadiene styrene	2.46	0.006
	Foam	1.05	0.00005
Door trims	Polyvinyl chloride (grey)	2.86	0.006
Pillar trims	Polypropylene	2.12	0.0007
Roof lining	Fibre board	1.31	0.007
	Black insulation	1.41	0.01
	Fabric	1.53	0.006
Carpet	Fabric	1.19	0.02
Dashboard	Polycarbonate	3.90	0.04
	Polypropylene	2.12	0.0007
	Nylon	3.1	0.01
Airbag	Fabric	1.87	0.01
	Polybutylene terephthalate	4.00	0.002
Window glazing	Glass	6.5	0.03

diffraction (UTD) and geometrical optics (GO), may be required to evaluate channel performance. However, the use of these method for vehicular communications will not be covered in this chapter.

6.4 In-vehicle EM Field Measurement

The previous section has described simulation techniques used for modelling intra-vehicle field distribution. To investigate if the simulation model is able to predict EM field levels to a satisfactory level in a vehicle, measurement will have to be performed to validate the data. A measurement system that can measure 3D electric field distribution with minimum field disturbance will be described in this section. 1D and 2D results are usually sufficient for validation activities in simple environments such as an empty rectangular cavity [45]. However, in a complex environment such as a vehicle, field distribution validation will required a larger dataset to provide more reliable field population statistics which may be useful for propagation studies in ITS, EMC analysis and assessment of human exposure to EM fields.

An automated position system which can provide sufficient tensile strength and rigidity to support the field probe and be mounted in a vehicle was developed in [46]. As the use of a relatively large structure could cause significant electric field disturbance while measuring field level within a vehicle, the scanning frame was mostly made out of Perspex and has a total of 12 small stepper motors for automated probe positioning. The electric field probe used for the measurement was laser-powered and controlled by an optical cable to minimize potential field disturbance. The scanning frame without the vertical positioner is shown in Figure 6.6a. The low disturbance properties of the scanning frame were obtained by using a sandwich structure (see Figure 6.6b), which is made out of two thin pieces of Perspex separated by a spacer placed at an optimized location, thus creating an air gap that reduce the RF visibility of the structure. The sandwich structure was 11 mm thick in order to maintain rigidity. When compared to a solid Perspex block of the same thickness, the field disturbance cause by the scanner with the sandwich structure is in the range 0.6–1.4 dB at 2 GHz, whereas the solid Perspex scanner created a field disturbance of 1.2–2.6 dB. The scanning frame mounted in the cabin of the vehicle is shown in Figure 6.6c, with the vertical arm and optical probe attached. The number of points that were sampled totalled 12,852 and 3402 at 1800 MHz and 900 MHz, respectively.

In order to compare the correlation of field data obtained by simulations with measurements, the cumulative distributions of the measured and simulated fields for a source monopole placed on the vehicle storage compartment are plotted in Figure 6.7 and Figure 6.8. Two different time-domain techniques were used for the simulations, each using different models. The model used with CST Microwave Studio included dielectric furnishings as well as the major metallic parts of the vehicles, whereas the Microstripes (CST TLM solver) model was limited to the metal parts only.

At 900 MHz, Figure 6.7 shows that the simulated cumulative amplitude distribution curve obtained from the CST Microwave Studio data closely follows that obtained from the measured data and is within the uncertainty bounds of the DARE RSS1006A laser-powered probe and the expanded uncertainty for the measurement as a whole [47]. The TLM results were slightly poorer but still within the uncertainty bounds. As the source antenna on the storage compartment was placed very close to large dielectric structures such as the driver's and front passenger's

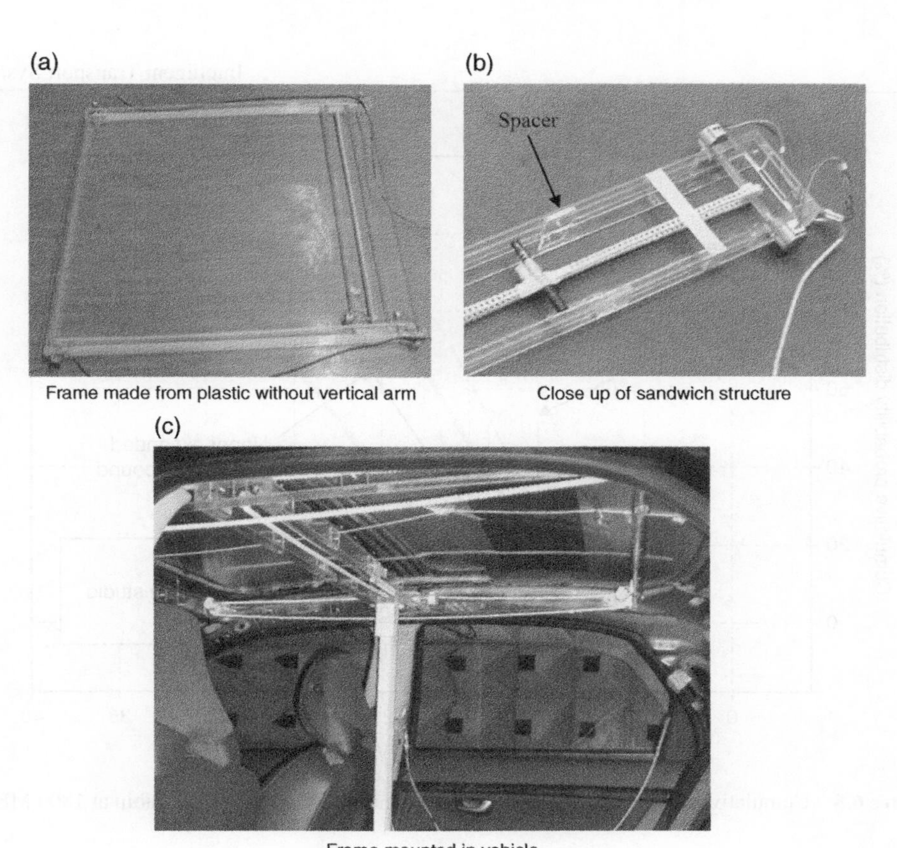

Figure 6.6 Low disturbance scanning frame.

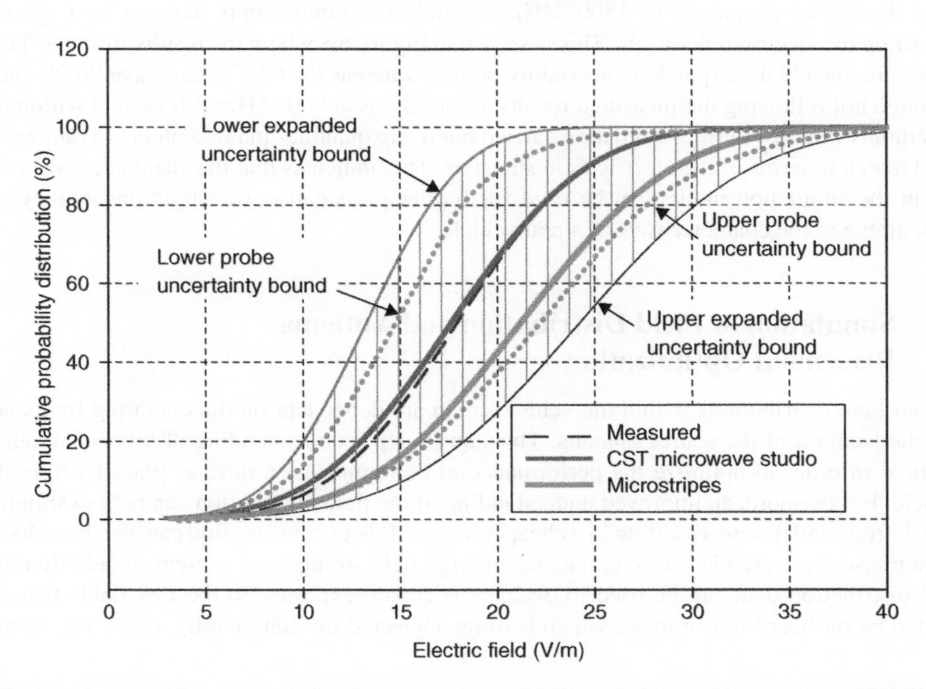

Figure 6.7 Cumulative distribution of measured and simulated field in vehicle cabin at 900 MHz.

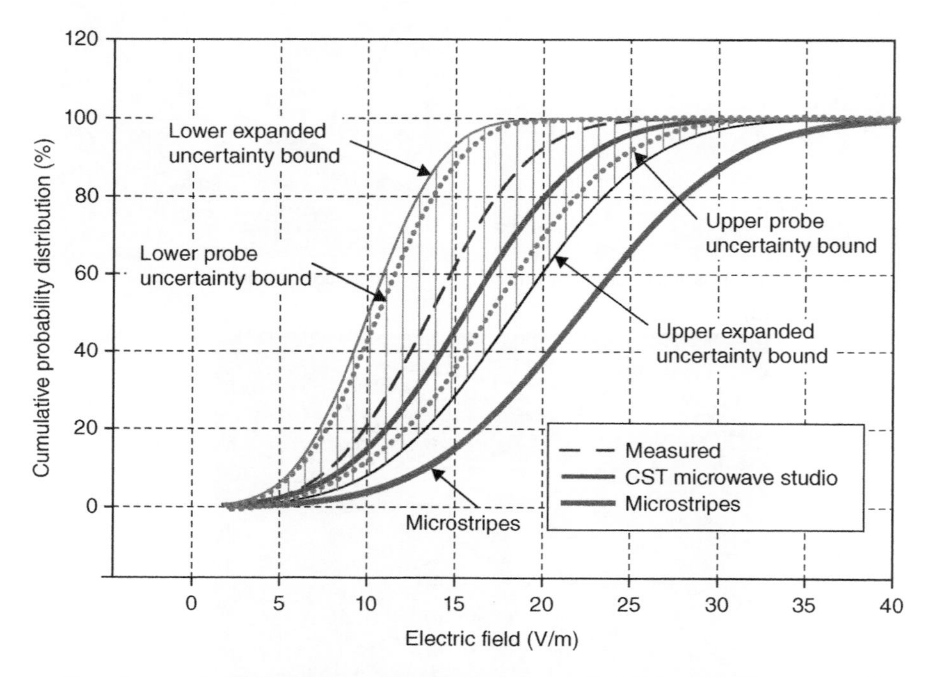

Figure 6.8 Cumulative distribution of measured and simulated field in vehicle cabin at 1800 MHz.

seats, the absence of the dielectric materials in the TLM model may have been the reason for the deviation of the simulated field from the measured data.

At the higher frequency of 1800 MHz the dielectric components have an even greater impact on the electric field levels. This is shown in Figure 6.8 where the results from the TLM solver are outside the expanded uncertainty bounds whereas the CST Microwave Studio data, although not following the measured results as closely as at 900 MHz, still remain within the uncertainty bounds. Further validation carried out using quantile-quantile plots [46] showed a good match in terms of the electric field statistics. This indicates that the channel characteristics in the simulation model can be used for further propagation investigations as they are comparable to the characteristics of a real vehicle.

6.5 Simulation of Field Distribution and Antenna Placement Optimization

Spatial field distributions within the vehicle cabin are dependent on the operating frequency and the location of the source antenna. Thus, an in-depth understanding of this behaviour is required in order to optimize the performance of communication devices placed within the vehicle. Furthermore, an improved understanding of the field distribution can help to simplify the RF front-end design, resulting in a cheaper device. Power consumption can also be reduced if the transceivers are placed in regions where high field strength are present. In addition, the field distribution data can be used to evaluate occupant exposure to the EM fields that are created by on-board transmitters, which is often neglected in such investigations. The results

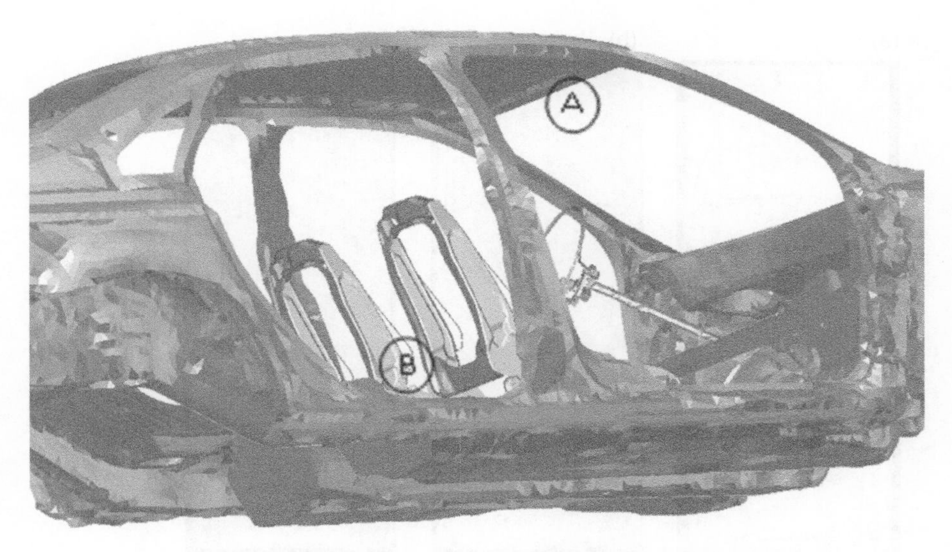

Figure 6.9 Location of source antenna within vehicle.

outlined in Section 6.4 demonstrate the potential for using numerical simulation to obtain field population statistics that are comparable to those of measured fields. This section illustrates the use of simulation to assess the changes in spatial field distribution for different source antenna locations at different frequencies.

Major metallic parts of a vehicle are used in the 900 MHz simulation model to investigate field distribution within the cabin from source antennas found in two typical locations. At 1800 MHz and 2400 MHz, dielectric parts are included as they have an effect on field distribution. The location of the source antennas are labelled in Figure 6.9, where location 'A' is at the top of the windscreen aperture and location 'B' is represents mounting on a storage box located between the two front seats. For simplicity, quarter-wave monopole antennas were used to represent the RF source.

Plotting 2D field distributions can be used to visualize and locate field levels at specific points and regions of high or low field levels within the vehicle cabin at a specific height. Figures 6.10 and 6.11 show the field distribution on a 2D single horizontal cut-plane at frequencies of 900 MHz, 1800 MHz and 2400 MHz, for source antennas placed at locations A and B, respectively. The optimal location to place an RF source or a receiving communication device can be obtained by comparing field plots. For example, at 900 MHz, an antenna located near the top of the windscreen will produce average electric field levels within the cabin which could be lower than for a source antenna located on the storage compartment. This can be quickly visualized by the smaller number of high field peaks observed in the cabin in Figure 6.10a. At 1800 MHz, the numbers of peaks and troughs are significantly larger and it is more difficult to determine which source antenna location will give a higher average field level within the cabin. However, regions where field levels are higher can still be clearly identified. Comparing Figures 6.10a and 6.11b, placing the RF source at the storage compartment will still give a large choice of receiver placement locations due to the wider regions of high field level. At 2400 MHz, visual comparison of

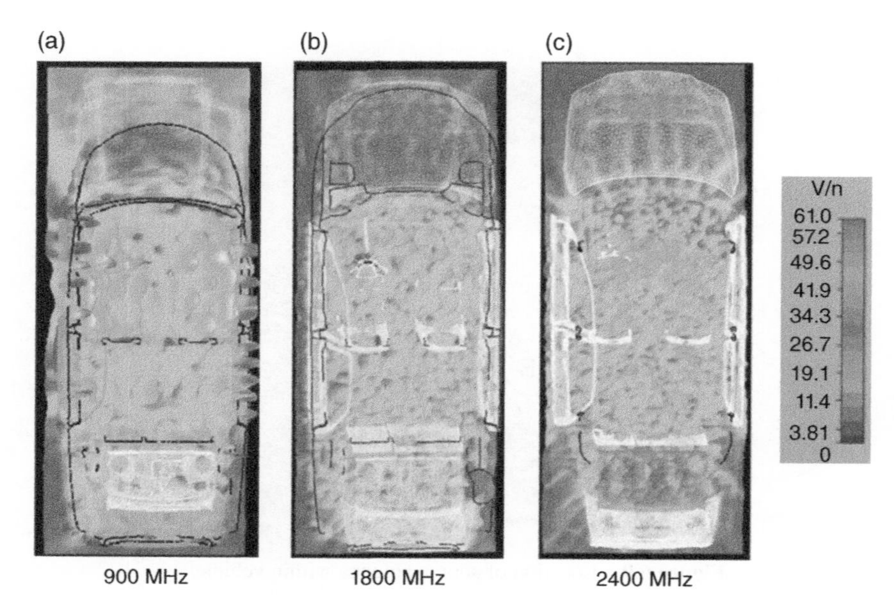

Figure 6.10 Field distribution within vehicle with source antenna at location A.

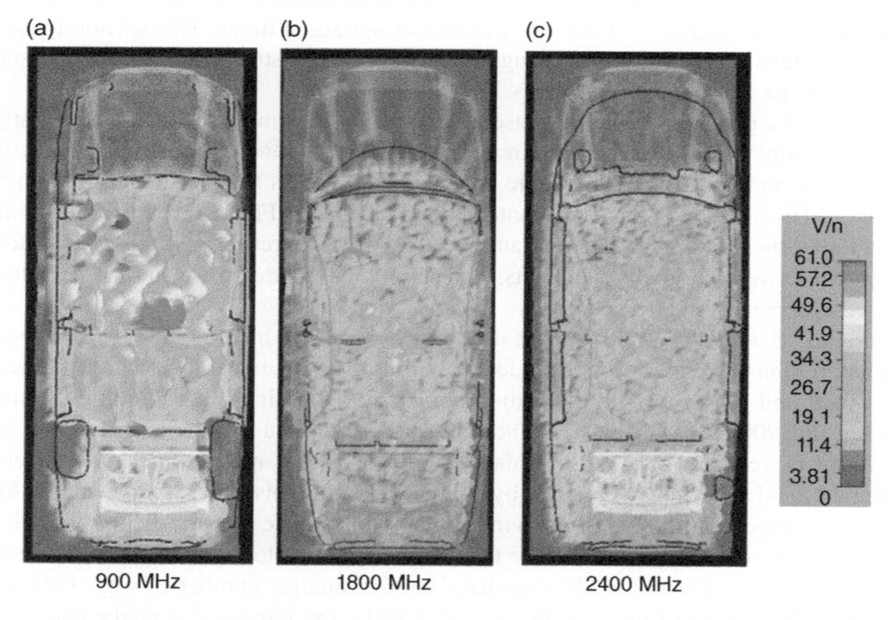

Figure 6.11 Field distribution within vehicle with source antenna at location B.

the 2D field plots shows no significant difference in field levels within the cabin for the different source antenna locations. In such cases, field statistics and a knowledge of the average field level within the cabin can be helpful in determining the optimal location to place the source antenna.

The average field within the cabin is shown in Table 6.3, which indicates higher values for all frequencies when the source antenna is placed at the storage compartment. Histograms summarizing the field levels in the cabin at the three frequencies of interest are shown in Figure 6.12. These plots indicate that a slightly larger proportion of the sampled points are at higher field levels when the source antenna is located on the driver's storage compartment (between the front seats) than when it is mounted at the front of the roof.

The implication of the results obtained is that a larger proportion of energy may be directed outside the vehicle with the antenna placed above the windscreen. The 3D field distribution

Table 6.3 Average electric field within vehicle cabin at various frequencies.

Antenna location	Average electric field (V/m) within cabin		
	900 MHz	1800 MHz	2400 MHz
Storage compartment	19.03	18.89	17.87
Front of roof	13.79	14.53	15.07

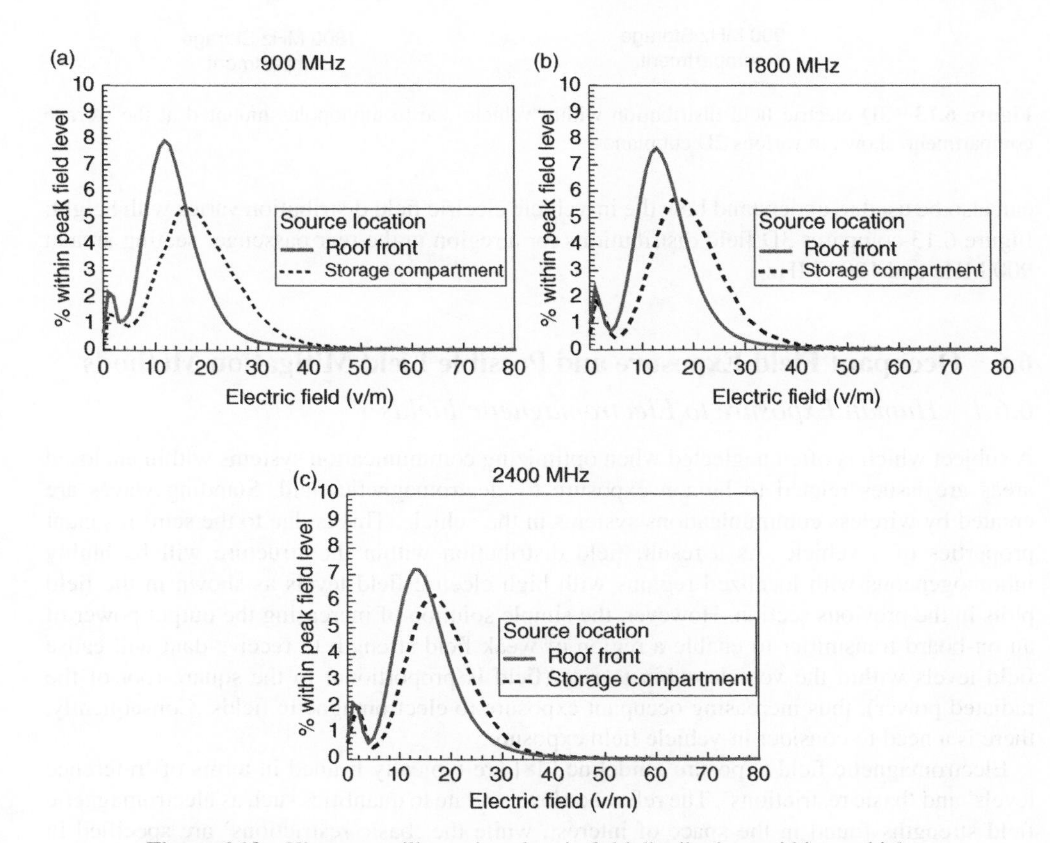

Figure 6.12 Histograms illustrating electric field distributions within a vehicle.

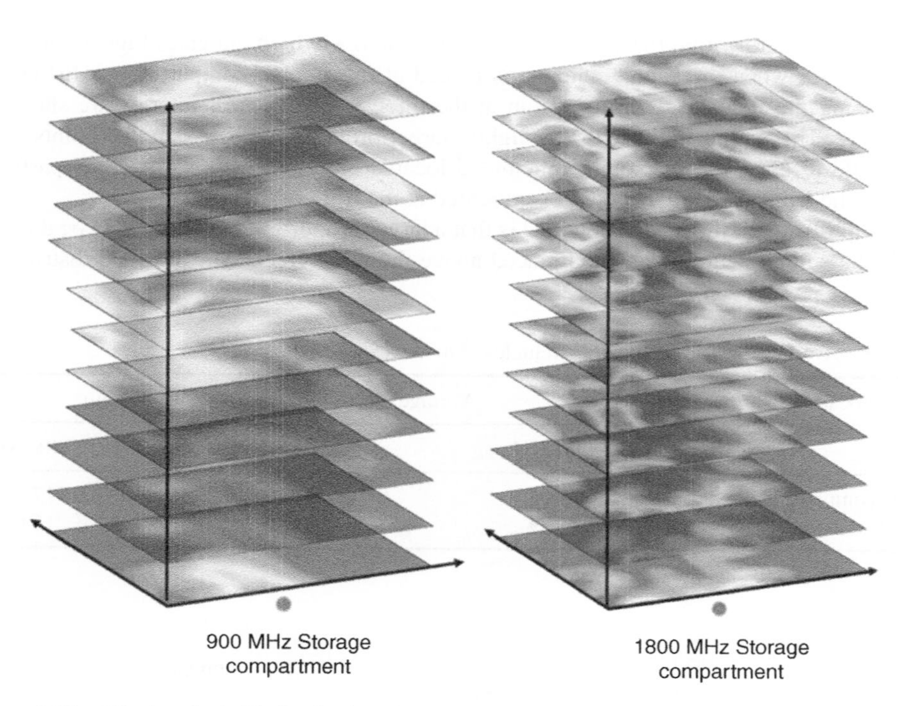

Figure 6.13 3D electric field distribution within vehicle due to monopoles mounted at the storage compartment, shown in various 2D cut planes.

can also be used to understand how the in-vehicle electric field distribution varies with height: Figure 6.13 compares 3D field distributions for a region in the rear passenger seating area at 900 MHz and 1800 MHz.

6.6 Occupant Field Exposure and Possible Field Mitigation Methods

6.6.1 Human Exposure to Electromagnetic Fields

A subject which is often neglected when optimizing communication systems within enclosed areas are issues related to human exposure to electromagnetic field. Standing waves are created by wireless communications systems in the vehicle. This is due to the semi-resonant properties of a vehicle. As a result, field distribution within the structure will be highly inhomogeneous with localized regions with high electric field levels as shown in the field plots in the previous section. However, the simple solution of increasing the output power of an on-board transmitter to enable a region of weak field strength to receive data will cause field levels within the vehicle cabin to rise (field is proportional to the square-root of the radiated power), thus increasing occupant exposure to electromagnetic fields. Consequently, there is a need to consider in-vehicle field exposure.

Electromagnetic field exposure guideline [48] are typically framed in terms of 'reference levels' and 'basic restrictions'. The reference levels relate to quantities such as electromagnetic field strengths found in the space of interest, while the 'basic restrictions' are specified in terms of quantities such as current density and specific absorption rate (SAR) levels induced in the body tissues by local electromagnetic field. The 'reference levels' are derived from basic restrictions by mathematical modelling and extrapolation from laboratory measurements. Safety margins are included to allow for variations between individuals, and for local field enhancements arising from reflections in the environment. It is noted in [48] that the field reference levels 'are intended to be spatially averaged values over the entire body of the exposed individual, but with the important proviso that the basic restrictions on localized exposure are not exceeded'. The use of field reference levels is therefore not considered to be appropriate for highly localized exposures that result from transmitters that are very close to the body. However, for RF sources that are located in a vehicle but not in contact with the body, it would be desirable to use field reference levels for an initial exposure threat assessment. Figure 6.14 shows a brief overview of the steps recommended to assess vehicle occupant exposure to EM fields.

An example of electric field within the vehicle cabin which was used for evaluation against the field reference levels is shown in Figure 6.15, for the configuration with source antenna placed on the storage compartment. The 2D electric field distribution cut-planes approximately represent the occupant locations. The maximum field level was set to 41.3 V/m, which is the electric field reference level specified in [48] for general public exposure at 900 MHz [46].

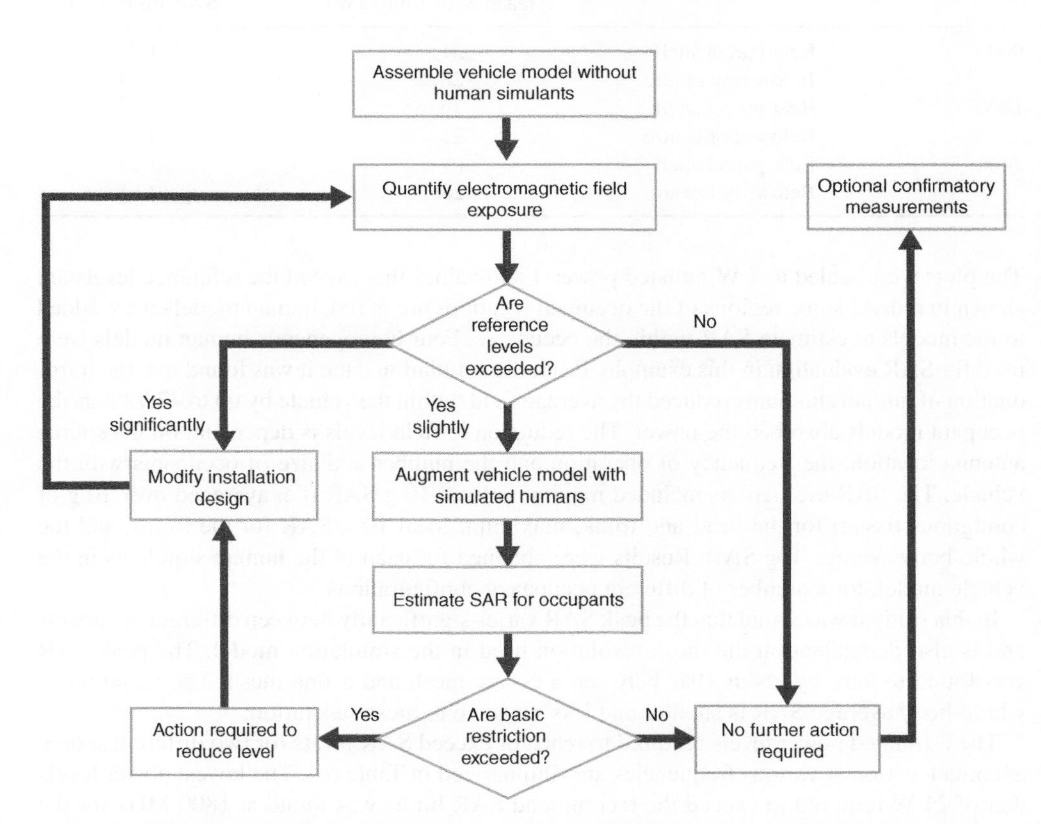

Figure 6.14 Overview of steps involved in assessing vehicle occupant exposure to EM field.

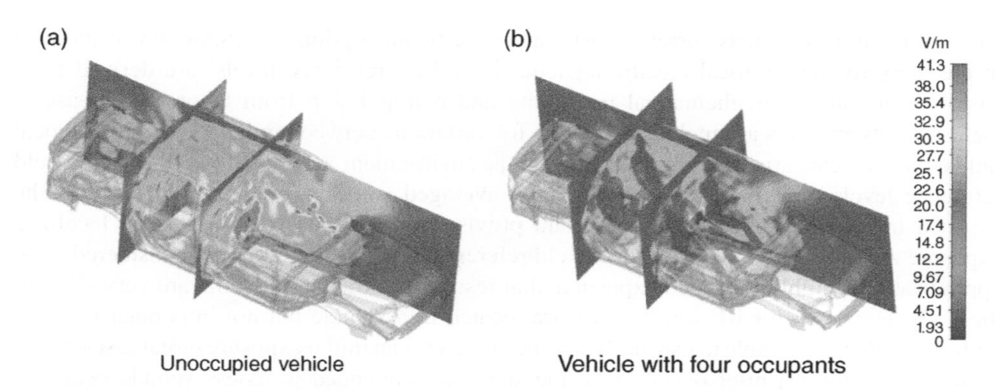

Figure 6.15 Comparison of 900 MHz electric field levels in vehicle with and without human occupants.

Table 6.4 Power levels required to reach or exceed SAR limits.

Frequency (MHz)	Source location	Lowest input power levels required to reach SAR limits (W)	Highest percentage of SAR relative to SAR limits (%)
900	Rear parcel shelf	51	1.94
	Below roof centre	28	3.84
1800	Rear parcel shelf	76	1.31
	Below roof centre	21	4.63
2400	Rear parcel shelf	69	1.21
	Below roof centre	28	3.61

The plots were scaled to 1 W radiated power. Field values that exceed the reference levels are shown in red. As some regions of the occupant locations are in red, human models were added to the models to estimate SAR within the occupants. Four homogeneous human models were used for SAR evaluation in this example. From the simulation data, it was found that the introduction of human simulants reduced the average field within the vehicle by up to 47.92% as the occupant models absorbed the power. The reduction in field levels is dependent on the source antenna location, the frequency of operation and the number and size of occupants with the vehicle. The SAR evaluations included maximum local 10 g SAR (i.e. averaged over 10 g of contiguous tissue) for the head and trunk, maximum local 10 g SAR for the limbs, and the whole-body average 10g SAR. Results were obtained for each of the human simulants in the vehicle model, for a number of different occupancy configurations.

In this study it was found that the peak SAR varies significantly between different occupants and is also dependent on the mesh resolution used in the simulation model. The peak SAR was found to vary by about 10% between a coarse mesh and a fine mesh. The variation of whole-body average SAR is smaller and less sensitive to mesh resolution.

The estimated power levels required to reach or exceed SAR limits for two different source antenna location at various frequencies are summarized in Table 6.4. The lowest power levels that of 21 W required to exceed the recommend SAR limits was found at 1800 MHz for the source antenna placed at the centre of the roof. This power level is unlikely to be transmitted in a vehicle used by the general public.

6.6.2 Field Mitigation Methods

From Table 6.4 it can be seen that an input power of at least 21 W would be required for the SAR limits on human exposure to be exceeded for the antenna configurations considered here. Nevertheless, there could be some situations where field damping may be required to improve the communication channel characteristics by reducing, or moving, the in-vehicle resonances. Typically, specialized field absorbing materials may be used to reduce the effects of cavity resonance and radiation threats. However, many of the materials used are expensive or large in size and may not always be suitable for use in enclosed environments.

The chosen material must have the ability to alter field patterns and levels within semi-resonant cavities and have sufficient bandwidth to cover the frequency bands of interest. In addition, the methods and materials must be cost efficient. Some suggested implementations which could be used as a single solution or in combination to alter electromagnetic fields patterns and levels include:

- use of absorbing materials in strategic locations;
- use of detuning conductors;
- use of frequency selective surfaces or electromagnetic band-gap materials;
- strategic placement of thin resistive sheets;
- changing materials used in the existing structures (reflective or absorbing depending on requirements).

Reflection, transmission and absorption properties of impedance sheets and lossy materials that could be used for field damping were measured at frequencies at 2 GHz (see Table 6.5). Relative permittivity, loss tangent and surface impedance values can be interpolated from the measured data for use in simulations to assess their field damping effects.

Field mitigation within the vehicle was assessed from measurements at 900 MHz and 1800 MHz, for configurations with and without sheets of lossy material placed on the vehicle seats. The equipment described in Section 6.4 was used to automate the electric field measurements. The locations of the sheets of lossy material in the vehicle are shown in Figure 6.16 (coloured in pink). For the field mitigation evaluation, LS-24 [50] was used at 900 MHz and RFLS-5066 [51] was used at 1800 MHz. The source dipole was placed on the parcel shelf.

The changes in average and maximum field levels over the volume measured within the vehicle are summarized in Table 6.6, for LS-24 at 900 MHz and for RFLS 5066 at 1800 MHz. The general observation is a lowering of average field level. Although there is a reduction average field levels, it was also observed that average field levels in some regions not in close proximity to the sheets of lossy foam were enhanced and the maximum field values recorded for some measurement points were higher. This could be due to changes in the field distribution

Table 6.5 RF properties of materials investigated for field mitigation (2 GHz).

Material	Reflection	Transmission	Absorption
Quilt wave [49]	48.16%	9.18%	42.66%
Lossy foam 5066 (10 mm thick)	59.02%	0.23%	40.75%
Rubber (10 mm thick)	50.42%	43.16%	6.42%

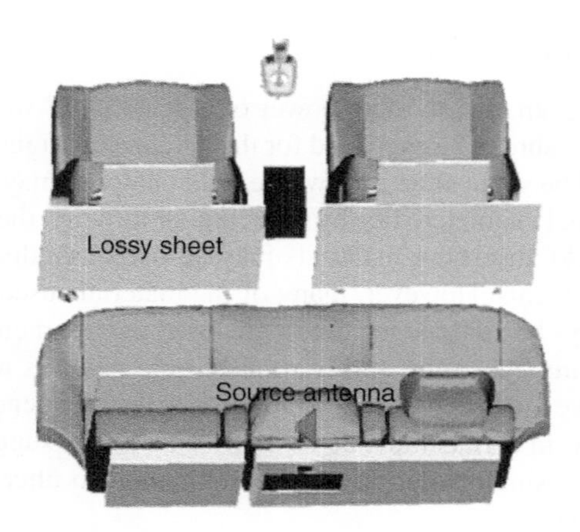

Figure 6.16 Locations of sheets of lossy foam material within vehicle.

Table 6.6 Change in in-vehicle field levels due to sheets of lossy foam placed on front seats.

Test case	In-vehicle field measure	Material configuration	Electric field value (V/m)	Change (%)
LS-24, 900 MHz	Average	With material	6.12	−17.5
		No material	7.42	
	Maximum	With material	17.12	−12.25
		No material	19.51	
RFLS 5066, 1800 MHz	Average	With material	4.47	−20.9
		No material	5.65	
	Maximum	With material	15.56	6.8
		No material	14.51	

resulting from reflections from these materials (see Table 6.5). In regions in close proximity to the lossy foam, field levels were significantly reduced. This demonstrates that the introduction of sheets of lossy foam can be used to provide highly localized field reduction within the vehicle.

The impact of such mitigation material on the far-field radiation pattern of an on-board antenna has also been investigated, using RFLS 5066 at both 900 MHz and 1800 MHz. From Figure 6.17 it can be seen that the introduction of field mitigation material within the vehicle does not results in any significant changes to the far-field radiation patterns. Consequently, such measures are not expected to have significant detrimental effects on the performance of any on-board communications antennas that are already installed.

The above findings indicate that, if field mitigation is required within a vehicle, it could be useful to replace normal seat foam with a material with similar losses to LS-24. Replacing the seat foam would be ideal for field exposure mitigation as the humans are always in close contact to the seats and the foam material is of sufficient thickness to introduce the required losses. If highly localized field reduction is required, then lossy foams can be used. They can enable highly selective reduction in field levels, producing damping where low field levels are

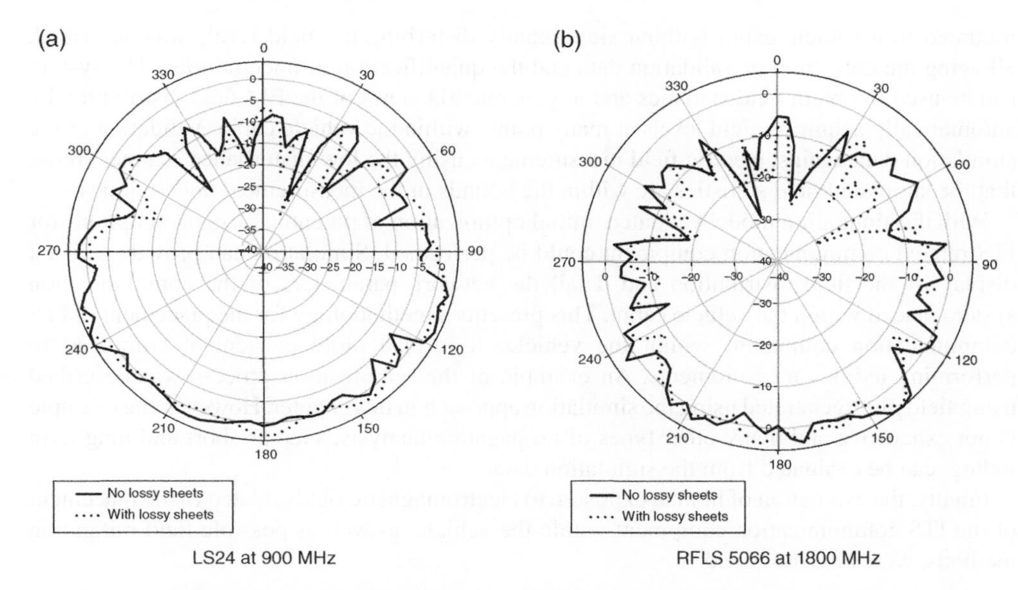

Figure 6.17 Effects of lossy sheet on far-field radiation pattern (Theta = 90°).

required (e.g. locations of critical electronic components and occupant seating locations) whilst still allowing communications devices required for ITS applications to work in regions with higher field levels.

6.7 Conclusions

An overview of issues relating to intra-vehicular communication systems and on-board wireless network nodes is presented in this chapter. The deployment of such technologies to enable emerging ADAS and ITS applications is set to grow in the future. Many of these systems will require reliable wireless communications to communicate with systems installed both within and outside the vehicle. The vehicle body and its complex electronics present a harsh environment for wireless communications. Standing waves within the semi-resonant vehicle interior can cause large variations in electromagnetic field levels within the cabin and could impact on the reliability of intra-vehicle communications.

A survey of the state-of-the-art in channel characterization and intra-vehicle electromagnetic field investigation highlighted the requirement for more detailed studies into the characteristics of propagation channels and the behaviour of electromagnetic field in order to improve the reliability of wireless communication systems. These investigations could be carried out either by simulation or by measurement. As the use of whole vehicle simulation is seldom described in the available literature, this chapter has given a short overview of applicable electromagnetic techniques (e.g. FIT, TLM, MLFMM, GO) that could be used to evaluate the impact of the vehicle body on electromagnetic field in the cabin of a vehicle, as well as some of the practical aspects of constructing such models (e.g. CAD, model content, material properties).

There is also a need to measure the field distributions in order to establish confidence in the simulation data. An automated measurement system which can measure electric field when mounted in a vehicle cabin without significantly disturbing the field levels was described, allowing the collection of validation data and the quantification of uncertainties. The system can be used to obtain field statistics and to generate 3D views of the EM field distribution by automatically acquiring field levels at many points within the vehicle cabin. Validation of the simulation data against electric field measurements using the low-disturbance scanner shows that the simulated field statistics are within the bounds of the measurement uncertainties.

With the simulation model validated, initial optimization of potential placement options for ITS related communication equipment could be performed. Simulation can provide a visual display of the field distribution and detail the network parameters of the communication system placed within the vehicle cabin. This presents a methodology for the placement of ITS communication equipment within the vehicle. It is also more convenient compared to performing tedious measurements. An example of the optimization process was described using field plots generated using the simulation approach in this chapter. However, the example is not exhaustive and many other types of propagation analysis, such as short and long term fading, can be evaluated from the simulation data.

Finally, the evaluation of human exposure to electromagnetic fields related to the placement of the ITS communication equipment within the vehicle, as well as possible field mitigation methods, were also described.

Acknowledgements

The authors acknowledge the support of Prof. Richard Langley (University of Sheffield, UK), Dr Jonathan Rigelsford (University of Sheffield, UK) and Dr Hui Zhang (Cohbam Technologies, USA) as part of the material of this chapter was created during the first author's employment at the University of Sheffield, UK.

Some of the work outlined above was carried out as part of the SEFERE project, a collaborative research project supported by the UK Technology Strategy Board (Technology Programme reference TP/3/DSM/6/1/15266). The project consortium included MIRA Ltd (Project Co-ordinator), ARUP Communications, BAE Systems Ltd, Harada Industries Europe Ltd, Jaguar Cars, UK National Police Improvement Agency, Sheffield University and Volvo Car Corporation (Sweden).

References

1. S. Ezel (2010) *Explaining International IT Application Leadership: Intelligent Transportation Systems, The Information Technology and Innovation* Foundation, January.
2. E.M. Davenport (2007) SEFERE: Simulation of electromagnetic field exposure in resonant environments, *COMIT Community Day, Ilkley, UK, 27 September 2007*, pp. 1–24.
3. N. Lu, N. Cheng and N. Zhang (2014) Connected vehicles: solutions and challenges, *IEEE Internet Things Journal* **1**(4): 289–99.
4. M.J. Cronin (2014) Ford finds its connection. In M.J. Cronin, *Top Down Innovation*. London: Springer International Publishing, pp. 13–24.
5. B. Fleming (2013) Smarter cars: incredible infotainment, wireless device charging, satellite-based road taxes, and better EV batteries, *IEEE Vehicular Technology Magazine* **8**(2): 5–13.
6. OnStar [Online]. Available: https://www.onstar.com (last accessed 27 April 2015).
7. SBD (2012) 2025 every car connected: forecasting the growth and opportunity. SBD & GSMA, Feb. 2012.

8. *Accenture-Connected-Vehicle-Survey-Global*. Available online at: http://www.accenture.com/us-en/landing-pages/products/Documents/ivi/Accenture-Connected-Vehicle-Survey-Global.pdf (last accessed 27 April 2015).
9. M. Pinelis (2013) Automotive sensors and electronics: trends and developments in 2013, *Automotive Sensors and Electronics Expo, Detroit, MI, USA*, **2013**.
10. G. Leen and D. Heffernan (2001) Vehicles without wires, *Computing and Control Engineering Journal* **12**(5): 205–11.
11. M. Kent (1998) Volvo S80 electrical system of the future. Available online at: http://www.artes.uu.se/mobility/industri/volvo04/elsystem.pdf (last accessed 8 May 2015).
12. P.J. Van Rensburg and H.C. Ferreira (2003) Automotive power-line communications: favourable topology for future automotive electronic trends, *Proceedings of the 7th International Symposium on Power-Line Communications and its Applications (ISPLC'03), Kyoto, Japan, March 2003*, pp. 103–8.
13. R.M. Ishtiaq Roufa, H. Mustafaa, S.O. Travis Taylora, *et al.* (2010) Security and privacy vulnerabilities of in-car wireless networks: a tire pressure monitoring system case study, *Proceedings of 19th USENIX Security Symposium*, Washington DC, USA, *August 2010*, pp. 11–13.
14. C.U. Bas and S.C. Ergen (2013) Ultra-wideband channel model for intra-vehicular wireless sensor networks beneath the chassis: from statistical model to simulations, *IEEE Transactions on Vehicular Technology*, **62**(1): 14–25.
15. A.R. Moghimi, H.-M. Tsai, C.U. Saraydar and O.K. Tonguz (2009) Characterizing intra-car wireless channels, *IEEE Transactions on Vehicular Technology* **58**(9): 5299–5305.
16. C.F. Mecklenbrauker, A.F. Molisch, J. Karedal, *et al.* (2011) Vehicular channel characterization and its implications for wireless system design and performance, *Proceedings of IEEE* **99**(7): 1189–1212.
17. D. Balachander and T.R. Rao (2013) In-vehicle RF propagation measurements for electronic infotainment applications at 433/868/915/2400 MHz, *Proceedings of the 2013 International Conference on Advances in Computing and Informatics (ICACCI)*, Mysore, India, August 2013, pp. 1408–13.
18. P. Wertz, V. Cvijic, R. Hoppe, *et al.* (2002) Wave propagation modeling inside vehicle using a ray tracing approach, *Proceedings of IEEE 55th Vehicular Technology Conference, Birmingham, Alabama, USA, May 2002, Vol.* **3**, pp. 1264–8.
19. E.M. Davenport and M.E. Hachemi (2010) Simulation approaches for resonant environments, *Proceedings of the 4th European Conference on Antennas and Propagation*, Barcelona, Spain, April 2010, pp. 1–4.
20. A. Valcarce, D. Lopez-Perez, G. de la Roche, and J. Zhang (2009) Predicting small-scale fading distributions with finite-difference methods in indoor-to-outdoor scenarios, *Proceedings of IEEE 69th Vehicular Technology Conference*, Barcelona, Spain, *April 2009*, pp. 1–5.
21. T.K. Sarkar, J. Zhong, K. Kyungjung, *et al.* (2003) A survey of various propagation models for mobile communication, *IEEE Antennas and Propagation Magazine* **45**(3): 51–82.
22. H. Nagatomo, Y. Yamada, K. Tabira, *et al.* (2002) Radiation from multiple reflected waves emitted by a cabin antenna in a car, *IEICE Transactions on Fundamentals of Electronics, Communications and Computer Sciences*, Vol. **E85-A**, No. 7: 1585–93.
23. F. Harrysson (2003) A simple directional path loss model for a terminal inside a car, *Proceedings of IEEE 58th Vehicular Technology Conference*, Vol. **1**, pp. 119–22.
24. O.K. Tonguz, H.-M. Tsai, T. Talty, *et al.* (2006) RFID technology for intra-car communications: A new paradigm, *Proceedings of IEEE 64th Vehicular Technology Conference*, *Montreal*, Quebec, Canada, September 2006, pp. 1–6.
25. H.-M. Tsai, O.K. Tonguz, C. Saraydar, *et al.* (2007) Zigbee-based intra-car wireless sensor networks: a case study, *IEEE Wireless Communications* **14**(6): 67–77.
26. R. De Francisco, L. Huang, G. Dolmans, and H. De Groot (2009) Coexistence of ZigBee wireless sensor networks and Bluetooth inside a vehicle, *Proceedings of IEEE 20th International Symposium on Personal*, Indoor and Mobile Radio Communications, Tokyo, Japan, September 2009, pp. 2700–4.
27. S. Herbert, I. Wassell, T.H. Loh and J. Rigelsford (2014) Characterizing the spectral properties and time variation of the in-vehicle wireless communication channel, *IEEE Transactions on Communications* **62**(7): 2390–9.
28. S. Herbert, T.H. Loh and I. Wassell (2013) An impulse response model and Q factor estimation for vehicle cavities, *IEEE Transactions on Vehicular Technology* **62**(9): 4240–5.
29. I.G. Zuazola, J.M.H. Elmirghani, and J.C. Batchelor (2011) A telematics system using in-vehicle UWB communications. In B. Lembrikov (ed.), *Novel Applications of the UWB Technologies*, InTech, pp. 195–208. Available online at: http://cdn.intechopen.com/pdfs-wm/17463.pdf (last accessed 8 May 2015).
30. I.G. Zuazola, J.M. Elmirghani and J.C. Batchelor (2009) High-speed ultra-wide band in-car wireless channel measurements, *IET Communications* **3**(7): 1115–23.

31. W. Niu, J. Li and T. Talty (2009) Ultra-wideband channel modeling for intravehicle environment, *EURASIP Journal of Wireless Communications Networks*, Vol. **2009**, Art. 806209.
32. L. Low, H. Zhang, J. Rigelsford and R.J. Langley (2010) Measured and computed in-vehicle field distributions, *Proceedings of the 4th European Conference on Antennas and Propagation*, Barcelona, Spain, *April 2010*, pp. 1–3.
33. A.R. Ruddle (2007) Validation of predicted 3D electromagnetic field distributions due to vehicle mounted antennas against measured 2D external electric field mapping, *IET Science, Measurement and Technology* **1**(1): 71–5.
34. L. Low, H. Zhang, J. Riglesford and R.J. Langley (2009) Computed field distribution within a passenger vehicle at 2.4 GHz, *Proceedings of the 2009 Loughborough Antennas and Propagation Conference*, Loughborough, UK, *November 2009*, pp. 221–4.
35. S. Horiuhci, K. Yamada, S. Tanaka, *et al.* (2007) Comparisons of simulated and measured electric field distributions in a cabin of a simplified scaled car mode, *IEICE Transactions on Communications* **E90**(9): 2408–15.
36. M. Klingler and A. Lecca (2006) Comparison between simulations and measurements of fields created by mounted GSM antenna using a car body, *Proceedings of 6th International Electromagnetic Compatibility Symposium, Barcelona, Spain, September 2006*, pp. 732–42.
37. A.R. Ruddle (2008) Influence of dielectric materials on in-vehicle electromagnetic fields, *IET Seminar on Electromagnetic Propagation in Structures and Buildings*, London, UK, *4 December 2008*, pp. 1–6.
38. Y. Tarusawa, S. Nishiki and T. Nojima (2007) Fine positioning three-dimensional electric-field measurements in automotive environments, *IEEE Transactions on Vehicular Technology* **56**(3): 1295–1306.
39. CST Studio Suite. Available online at: http://www.cst.com (last accessed 28 April 2015).
40. FEKO. Available online at: http://www.feko.info (last accessed 28 April 2015).
41. HFSS. Available online at: http://www.ansys.com/Products/Simulation+Technology/Electronics/Signal+Integrity/ANSYS+HFSS (last accessed 28 April 2015).
42. A.R. Ruddle (2008) Validation of simple estimates for average field strengths in complex cavities against detailed results obtained from a 3D numerical model of a car, *IET Science, Measurement and Technology* **2**(6): 455–66.
43. HP Product Note 8510-3 – *Measuring Dielectric Constant with the HP 8510 Network Analyzer*, Hewlett Packard, January 1983.
44. N. Damaskos (2003) *General Cavity Material Measurement System Manual*, September.
45. H. Zhang, L. Low, J. Rigelsford and R.J. Langley (2008) Field distributions within a rectangular cavity with vehicle-like features, *IET Science, Measurement and Technology* **2**(6): 477–84.
46. L. Low, H. Zhang, J. Rigelsford, *et al.* (2013) An automated system for measuring electric field distributions within a vehicle, *IEEE Transactions on Electromagnetic Compatibility* **55**(1): 3–12.
47. UKAS (2002) *The Expression of Uncertainty in EMC Testing*, 1st edition, Middlesex, UK: UKAS.
48. ICNIRP (1998) Guidelines for limiting exposure to time-varying electric, magnetic and electromagnetic fields (up to 300 GHz), *Health Physics* **74**(4): 494–522.
49. S.M. Tsontzidis, L.C. Lai and N. Zeng (2005) Microwave packaging with indentation patterns, *US Patent* **7319213**.
50. Eccosorb LS-24, Laird Technologies. Available online at: http://www.lairdtech.com/products/eccosorb-ls (last accessed 9 May 2015).
51. RFLS-5066, Laird Technologies. Available online at: http://www.digikey.com/product-detail/en/5066/5066L-ND/4423228 (last accessed 9 May 2015).

7

Novel In-car Integrated and Roof-mounted Antennas

Rus Leelaratne†

7.1 Introduction

The initial automotive antenna for radio reception was a simple quarter wave length monopole, designed specifically for FM frequencies. The rod was typically 75–80 cm in length and still to date remains one of the best performing antennas. The antenna was mounted on a vehicle typically next to the A pillar or on the fender.

As the infotainment system of the car evolved, many additional functions were introduced. These were typically in the form of television, mobile telephone and satellite navigation frequencies in the late 20th century. The early 21st century saw a rapid expansion of frequencies used in the car, with the uptake of additional mobile telephone frequencies, digital radio, satellite radio, Wi-Fi® and more recently emergency call and car to car communications.

The styling requirements of vehicles in the late 20th and early 21st century have also had important implications on how the automotive antenna has evolved. Visible antennas are not welcomed, especially by the European consumer. This resulted in the growth of the hidden antenna, typically printed on a vehicle windshield or hidden underneath a composite body panel.

Enhancements in semiconductor technology and digital signal processing saw the growth of switched diversity and phase diversity as a technique for significantly improving the performance of these hidden antennas. Today, almost all luxury cars utilize phase diversity radio tuners for FM functionality, with the technology slowly moving down to mid range and low end cars.

†Deceased

Intelligent Transport Systems: Technologies and Applications, First Edition. Asier Perallos, Unai Hernandez-Jayo, Enrique Onieva and Ignacio Julio García-Zuazola.

7.2 Antennas for Broadcast Radio

Today, broadcast radio consists of multiple frequency bands. The Long Wave (LW) spectrum occupies frequencies from 148.5 KHz to 283.5 KHz. The Medium Wave (MW) spectrum occupies frequencies from 526.5 KHz to 1705 KHz.

The Frequency Modulation (FM) spectrum is typically 87.5 MHz to 108 MHz. However, in Japan the FM band is 76 MHz to 90 MHz.

Terrestrial Digital Audio Broadcasting (DAB) primarily uses the frequency spectrum from 174 MHz to 240 MHz, with some transmissions in L band between 1452 MHz to 1492 MHz. Digital radio in the USA is provided through a satellite constellation managed by SiriusXM® Radio. The 2320 MHz to 2345 MHz spectrum is utilized.

7.2.1 Roof-mounted Radio Antennas

As vehicle aesthetics became important, the 80 cm quarter wave monopole became shorter in length. The length reduction was typically achieved by replacing the monopole antenna element with a helical coil. The length of the helical windings along with the number of turns determined the operational frequencies.

A rigid spring is attached to the bottom of the helical windings to allow for flexibility in the mast. The mast typically connects to an antenna base, which may or may not contain an amplification circuit, depending on the length of the mast. Figure 7.1 shows typical roof-mounted rod antennas with an amplification circuit.

The roof-mounted rod antennas shown in Figure 7.1 typically are a low-cost solution. AM and FM functionality is a standard requirement with DAB provided as an option.

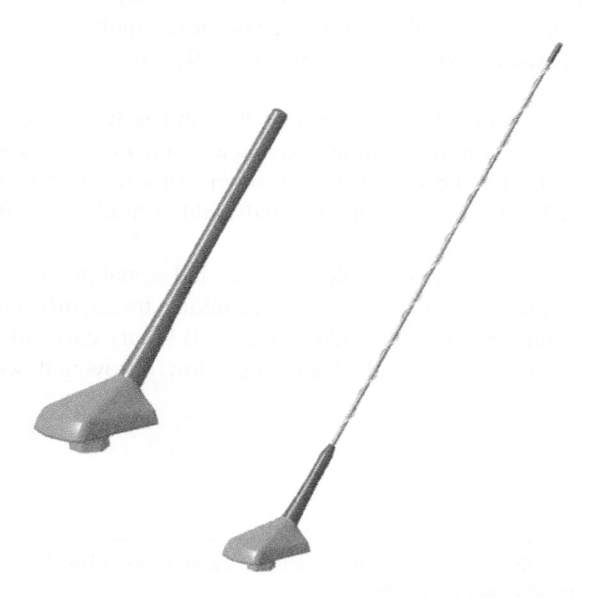

Figure 7.1 Typical roof mounted rod antennas with an amplification circuit.

The inclusion of SiriusXM® Radio requires an antenna with a beam shape that is directional towards the satellites rather than a terrestrial transmitter. Therefore, a typical roof-mounted multiband antenna in the USA will contain a base with a larger footprint as shown in Figure 7.2.

Recent years have also seen the popularity of Sharks fin type antennas for broadcast radio. These antennas are height limited to 70 mm, by EU legislations and practicability for garage-ability and car-wash requirements. Such antennas are challenged by its performance, but have proven to be acceptable by the automotive market. These antennas are common place especially among Japanese car manufacturers. Figure 7.3 presents a typical design configuration used to achieve AM and FM radio functionality.

Figure 7.2 Typical roof mounted multi-band antenna in the USA; contains a base with a larger footprint.

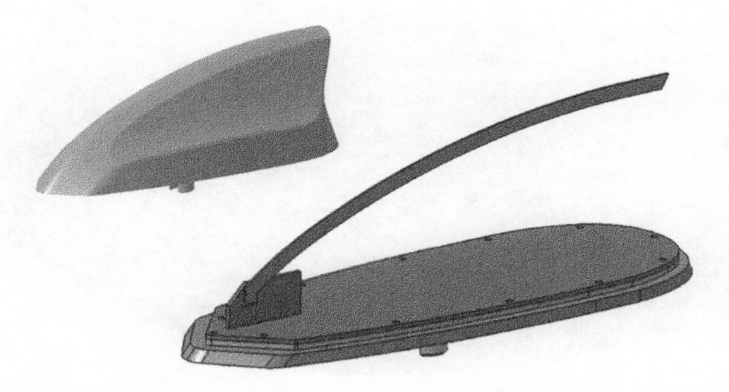

Figure 7.3 Typical design configuration used to achieve AM and FM radio functionality.

7.2.2 *Hidden Glass Antennas*

For premium vehicles visible external antennas are not acceptable. Today, even mid-range vehicles and some of the low-end vehicles find it is unacceptable to have visible antennas. For taller vehicles garage-ability is an issue. Also broken and stolen masts are a problem. This has resulted in the rapid growth of hidden antennas, with the radiating element printed on glass being the most popular technique.

Another reason for the popularity of such hidden antennas is the growth of antenna diversity [1], which is supported by most radio tuners today. Diversity relies on multiple antennas, typically able to receive similar signal strength from different directions. It is not practical to have multiple rod antennas on the roof as this would significantly degrade the vehicle aesthetics. Hidden designs, especially on glass, can support multiple antennas with ease.

The design of such glass antennas remain a challenge for the designer as initial antenna patterns should be developed before prototype vehicles are available. As the vehicle body has a significant impact on the antenna performance, computer modelling is used extensively. One of the most popular techniques used is Method of Moments (MoM) [2].

The most common glass panel used for hidden antennas is the rear screen. With the growth of phase diversity, which ideally requires the antennas to be spaced as far apart as possible, the use of side screens at the rear of the vehicle has become popular.

A typical rear screen antenna configuration is shown in Figure 7.4. The antenna configuration shown supports AM, FM, DAB and Keyless Entry functionality. Two FM antennas are implemented to support phase diversity requirements.

A typical rear side screen configuration as used for phase diversity is shown in Figure 7.5. The presented antenna configuration supports AM, FM, DAB, DVB and Keyless Entry functionality.

Figure 7.6 presents a typical front screen antenna, which integrates AM, FM, DAB, DVB and Keyless Entry functionality. The use of the front screen is popular in sports cars where the engine is located to the back of the vehicle. Locating the antenna as further away as possible from the engine reduces electromagnetic interference, especially at AM frequencies.

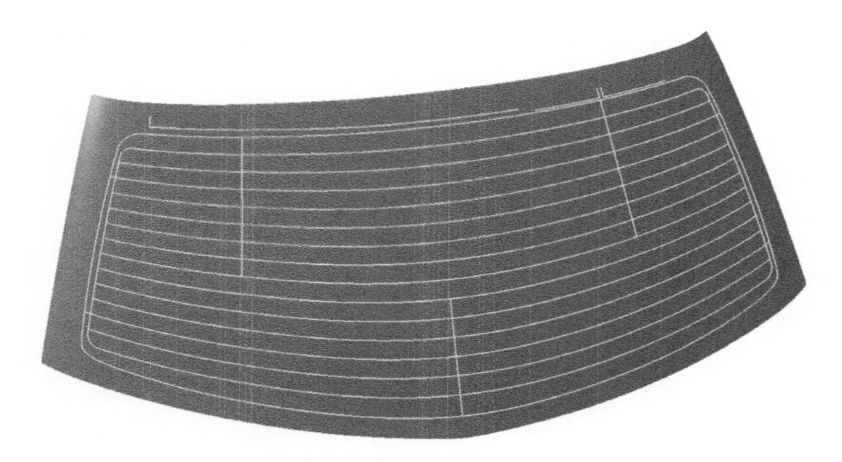

Figure 7.4 Typical rear screen antenna configuration.

Figure 7.5 Typical rear side screen configuration as used for phase diversity.

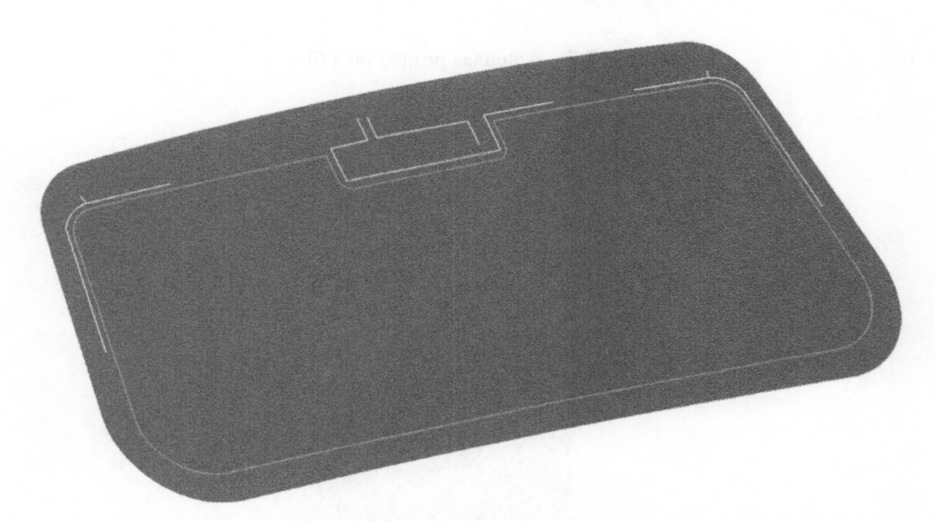

Figure 7.6 Typical front screen antenna, which integrates AM, FM, DAB, DVB and Keyless Entry functionality.

Certain vehicles such as commercial vans may not have any fixed glass panels available, other than the front windscreen. Such vehicles also utilize the front screen for printing the antennas. Any interference from noise sources should be given special consideration during a front screen antenna design process.

A typical screen antenna is not very efficient. The efficiencies for FM frequencies are typically in the ranges of 20– 40%. Therefore, a front-end amplifier is required. Such amplifiers are installed as close to the antenna as possible for optimum performance.

7.2.3 Hidden and Integrated Antennas

Some vehicles designers and stylists prefer to avoid the antennas printed on glass panels. In such a scenario, the antennas may be integrated on to a composite body panel of the vehicle. Such integrations on to composite panels are typically in the form of antennas printed on a film as shown in Figure 7.7. Such films are integrated on to a composite boot lid or roof

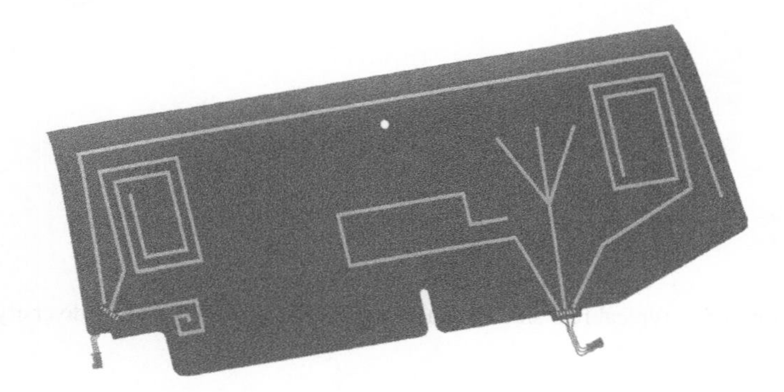

Figure 7.7 Antennas printed on a film.

Figure 7.8 PCB type antenna.

panels. These antennas are extremely challenged for good phase diversity performance as the elements are spaced closer than the ideal requirements.

The use of wing mirrors is popular for integrating hidden antennas. While the antenna proximity is good for phase diversity performance, the designer is challenged by the small size of the typical wing mirrors, lack of a vehicle body ground close by, proximity to nonideal antenna materials and the requirements for water sealing. Wing mirror antennas are common for commercial vehicles where the size of the wing mirror is much larger.

Various technologies are used to obtain reduced size antenna designs. Figure 7.8 shows a PCB type antenna design which can be installed inside vehicles with a composite body, except for carbon fibre. Composite bodies are typical in sports cars. The use of inductive loading loops and fractal type elements are commonly used techniques for designing small printed antennas for integration within the vehicle body [3]. A nonmetallic vehicle body is relatively transparent for electromagnetic signals.

7.3 Antennas for Telematics

The definition of a Telematics antenna typically covers mobile telephone frequencies and satellite navigation frequencies.

In automotive, telephone frequency requirements are typically 3G with the car manufacturers slowly migrating to 4G requirements. It is also not unusual to find a minority of car manufacturers remain with 2G technology. The 2G telephone system is based on two frequency bands of operation, 880–960 MHz and 1710–1880 MHz in Europe and 824–894 MHz and 1850–1990 MHz in the USA. The requirements of 3G introduced a new frequency band at 1920–2170 MHz.

Navigation for many years has relied mainly on GPS only systems. With the introduction of the Russian GLONASS system and the soon to be introduced European Galileo system and the second generation of the Chinese BeiDou system, coverage at multiple frequency bands is required. All the above systems are collectively referred to as the Global Navigation Satellite System (GNSS).

The GNSS systems contain a selection of frequencies operational at and around the frequencies of 1561.098 MHz, 1575.42 MHz and 1598.0625 to 1609.3125 MHz. All systems operate with electromagnetic waves that are Right Hand Circularly Polarized (RHCP).

Typically two types of Telematics solutions exist in automotives. The most common solution is a roof-mounted solution. For vehicles where aesthetics are critical it is usual to install hidden antennas.

Antenna diversity is not required for 2G or 3G telephony or GNSS. However, 4G requires the support of multiple antennas. Today for 4G the solution is based on two antennas typically located in two separate locations of the vehicle to achieve the required isolation between elements.

7.3.1 *Roof-mounted Telematics Antennas*

The roof-mounted antenna is the most popular solution for Telematics applications. Even high-end vehicles utilize roof-mounted solutions due to their good performance. An additional benefit of the roof-mount solution is the ability to carry over the product from one vehicle platform to another, thereby reducing development costs. An antenna integrated into the vehicle body requires optimization for each vehicle platform and therefore increased development costs.

Telematics antennas mounted on the roof are typically shorter than 70 mm to meet European legislations. Figure 7.9 shows a roof-mounted Telematics antenna with a height of 40mm. The design consists of a 3G telephone element and a commonly used ceramic antenna for GPS functionality.

The antenna technology used for the 3G telephone element is known as a Printed Planar Inverted F Antenna [4]. A conventional Planar Inverted F antenna is shown in Figure 7.10. The element in Figure 7.9 is a printed version of the design to achieve a lower cost.

The typical GPS solution for automotive has relied on a microstrip antenna element printed on a ceramic substrate. Depending on the dielectric constant of the substrate material a selection of antenna sizes are available in the automotive marketplace. Typical sizes vary from 25 mm down to 12 mm square solutions.

As GNSS become more and more common meeting the circular polarized bandwidth requirements with ceramic antenna designs has become extremely difficult. This has resulted in the growth of antennas that use two feed locations to excite the element as shown in Figure 7.11. The two feed points are excited with a phase difference of 90 degrees and therefore providing a good axial ratio.

Figure 7.9 Roof-mounted telematics antenna.

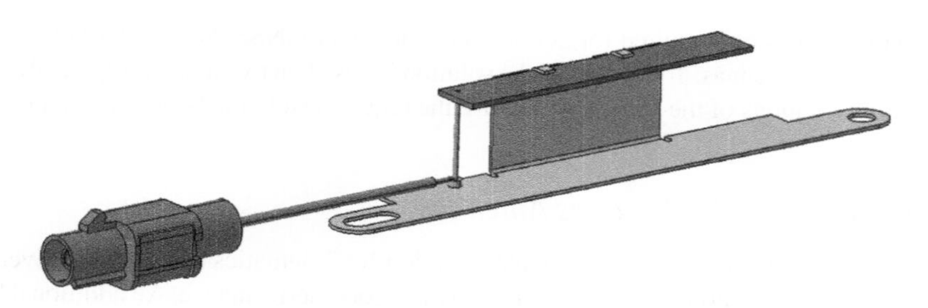

Figure 7.10 Conventional Planar Inverted F antenna.

Figure 7.11 Antennas using two feed locations to excite the element.

Figure 7.12 A unique element design that provides operation for Satellite Radio and GNSS.

Another type of antenna that has found some popularity is solutions with a low dielectric constant. Such antennas are however much larger with sizes typically in excess of 40 mm square.

Some designs have also combined broadcast radio solutions with Telematics solutions. Figure 7.12 shows a solution that has a unique element design that provides operation for Satellite Radio and GNSS. The design also uses a specialist mouldable engineering plastic for a three-dimensional antenna solution.

7.3.2 Hidden Telematics Antennas

Some vehicle manufacturers prefer to hide the Telematics antenna for better aesthetics. Cabriolet vehicles and sports cars are some of the common users of such hidden antennas. Although hidden, the antenna technology is similar to those found in roof-mounted solutions. In most occasions the antenna element is made slightly larger to overcome the loss of performance due to the effects of the vehicle body.

Common packaging locations are behind composite body panels such as boot lids or vehicle roofs, mountings inside the dashboard and behind front or rear windscreens. Some automotive applications also use antennas that are mounted on bumpers, inside wing mirrors and spoilers. Figure 7.13 shows a 3G telephone antenna that is mounted inside a spoiler again utilizing the Printed Planar Inverted F Antenna technology as used in the roof-mounted solution.

Mountings behind windscreens may suffer from dielectric losses present in the glass. Ideally the antenna should be located few millimetres away from the glass, but for most applications the dielectric loss can be tolerated. Figure 7.14 shows a GNSS antenna solution

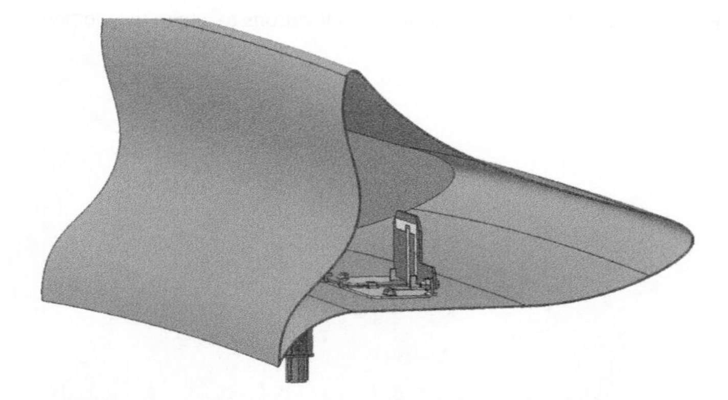

Figure 7.13 3G telephone Printed Planar Inverted F antenna mounted inside a spoiler.

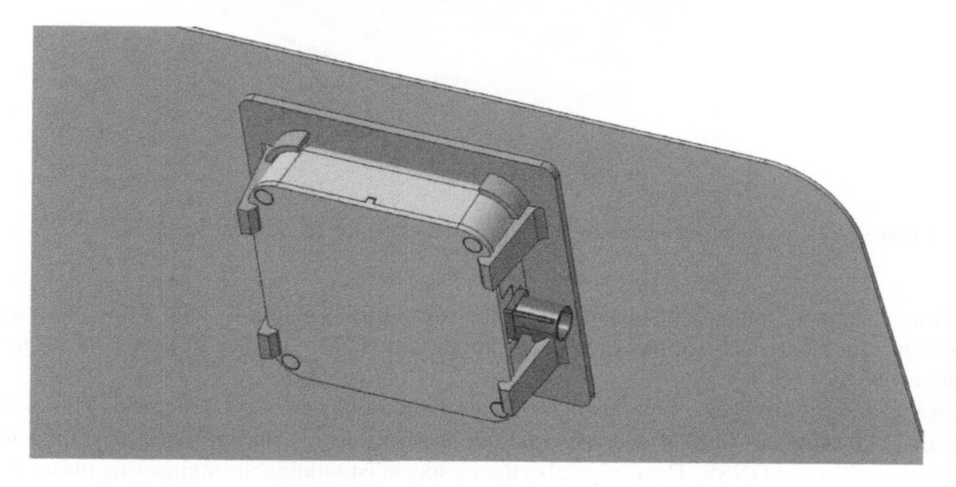

Figure 7.14 GNSS antenna solution mounted on the top of a front windscreen.

mounted on the top of a front windscreen. On this occasion the losses of the glass helped achieve the return loss bandwidth specifications.

7.3.3 Future Trend of Telematics Antennas

Today, the Telematics antenna is an important component of vehicle safety systems. Emergency Call, often referred to as E-Call, is soon to become mandatory in vehicles in the European Union. E-Call relied on GNSS signals for vehicle position information and 3G cellular coverage to communicate with the emergency services.

The system provides the emergency services with the precise location of the accident and can allow an operator to speak to you to provide assistance.

Car 2 Car communications is another application where Telematics information is necessary. As the vehicles become more and more connected and accessing the Internet from the car becomes common place, Automotive Telematics is certain to grow. Wi-Fi® and Bluetooth® are a very important part of the vehicle infrastructure. With the development of Bluetooth® Smart the possibilities are endless. Use of Bluetooth® Smart for Tyre Pressure Monitoring Systems (TPMS) is an example of a market with huge growth potential.

7.4 Antennas for Intelligent Transportation Systems

Today, Intelligent Transportation Systems (ITS) is a significant part of the automotive market. Many technologies encompass the field of ITS systems. Some of the most common technologies discussed in this section are Car2Car communications, E-Call, 4G telephony, Wi-Fi® and Bluetooth®.

Car2Car communications utilize the 5.9 GHz frequency spectrum for the purpose of providing communications between vehicles and vehicle to infrastructure, there by creating a vehicular network including the relevant infrastructures. Such a system aims to provide a new level of safety on roads and an efficient traffic management system by sharing real-time information. It is anticipated that early systems will be introduced in late 2015.

E-call is part of an ITS system which focuses on managing accidents in an efficient manner. In the European Union, the system is expected to be mandatory from 2017 onwards, with the member states responsible for providing the necessary infrastructure. When a collision is detected, a voice call is setup automatically using the mobile phone network and the 112 service. Additional information in the form of data, such as the position of the crash, using the GNSS system will be provided to the relevant authorities. The time of the crash, direction of travel and the vehicle registration will also be provided.

7.4.1 Car2Car Communication Antennas

Car2Car communications are commonly referred to with various names. Car2Car, Car2x, Car2Infrastructure, V2V, V2I, Intellidrive℠ and VANET are some common terms. The technology is based on the IEEE 802.11p standard.

The antenna solution relies on a minimum of two antennas operating at the 5.9 GHz band and a single GNSS antenna. The 5.9 GHz antennas form a MIMO system to provide advantages over noise, fading and multipath characteristics. One of the biggest challenges faced by the antenna designer of a Car2Car system is the requirement for higher gain at the lower elevation angles, which is critical for good performance.

The gain at low elevation angles is affected significantly by the vehicle roof. Additional problems are present for high-sided vehicles, which would require communications with vehicles that have their Car2Car antenna at a lower elevation angle. So, in reality performance is required few degrees below the vehicle roof level.

An antenna design utilized as a four element MIMO architecture is presented in [5]. The paper presents a circular patch antenna element with a diameter of approximately 10 mm, and a shorting post at the centre of the antenna and is operational in a higher order mode and therefore providing coverage at the low elevation angle. The performance of four antennas placed in a uniform spacing of half a wavelength is discussed. The paper also presents the performance improvements that can be achieving by placing parasitic elements in strategic locations.

A disc antenna design with a feed located in the centre is shown in Figure 7.15. Due to the 5.9 GHz frequency, the elements are small in size allowing for multiple antennas to be packaged within a typical automotive Telematics antenna space envelope. An ideal antenna provides a focussed omni-directional beam as shown in Figure 7.16. In addition some

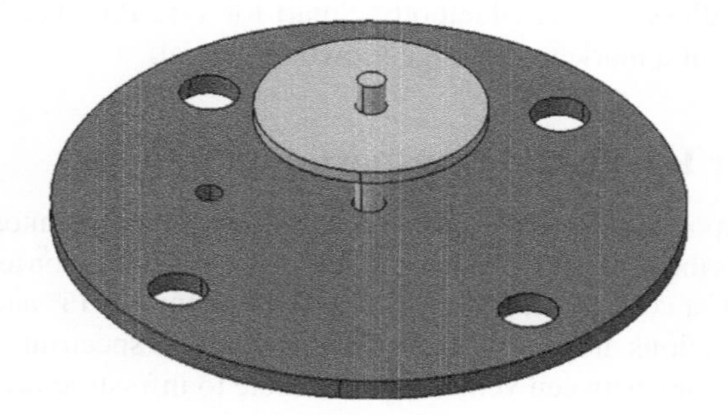

Figure 7.15 Disc antenna design with a feed located in the centre.

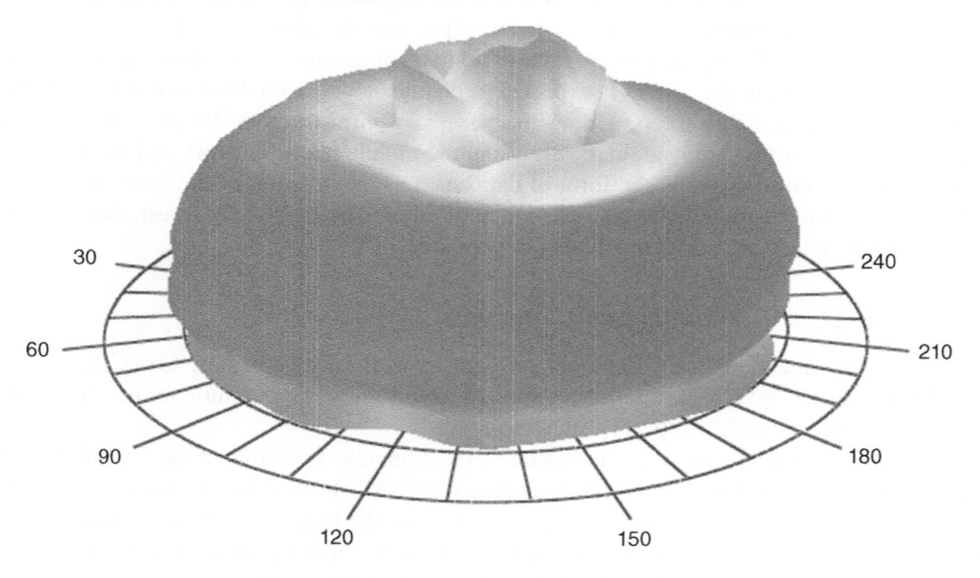

Figure 7.16 Focussed omni-directional beam.

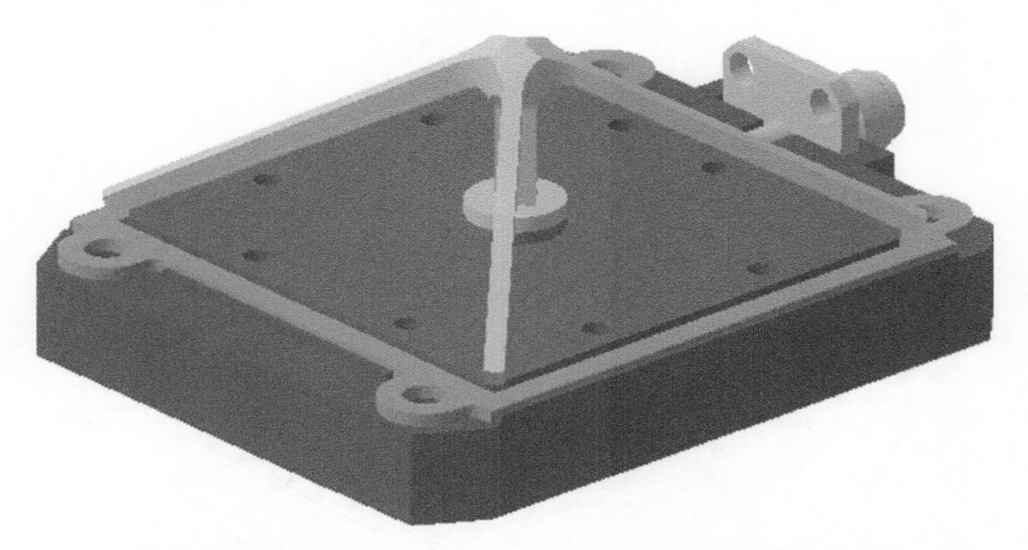

Figure 7.17 Higher order modes of a monopole antenna with a number of arms.

performance is required above the low elevation angles for communication with the infrastructure beacons. A slightly lower gain is acceptable for infrastructure communications due to the efficiency of the infrastructure antennas.

Although packaging multiple antennas for Car2Car communications in a typical Telematics antenna is practical, packaging addition frequency bands such as 4G telephone and GNSS remains a challenge. Such challenges have successfully been overcome by in [6]. In the paper the authors utilize three monopole antennas slightly folded. In addition the authors integrate a multimode radio platform into the antenna assembly thereby avoiding any cable losses of a passive solution.

A more traditional approach without the integrated electronics is presented by [7] whereby a higher-order mode of a patch antenna along with a printed monopole antenna is combined to provide improved Car2Car performance. The roof-mounted antenna solution also integrates 3G telephone functionality.

Various other approaches have also been taken. The design on Figure 7.17 is an approach taken whereby higher order modes of a monopole antenna are used in combination with a number of arms, which are strategically folded to yield appropriate directionality and gain requirements for Car2Car [8]. Figure 7.18 presents the radiation pattern of the antenna. The biggest challenge of the design is the size requirements compared to currently available designs.

A multiband antenna solution with performance at 3G, Wi-Fi® and Car2Car, with a single element is presented in [9]. The paper also presents the use of artificial magnetic conductors to achieve the design objective.

7.4.2 Emergency Call (E-Call) Antennas

It is expected that in 2017 the system will be mandatory in countries that are members of the European Union. The antenna challenge for the system is twofold. The first challenge is to determine whether an antenna solution that is dedicated for E-Call should be available. The second is to determine the ideal locations for the antenna.

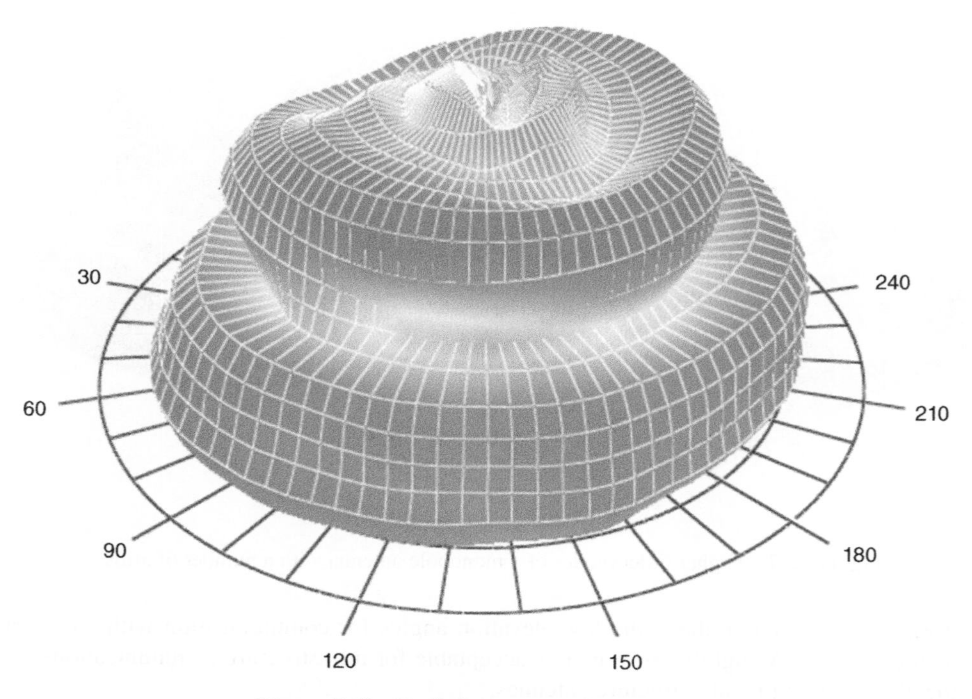

Figure 7.18 Radiation pattern of the antenna.

A multiple antenna solution is based on the argument that in a collision one of the antennas may get damaged with loss of functionality. The classic scenario is an overturned vehicle causing damage to its roof-mounted antenna. Therefore, the tendency is towards equipping the vehicle with two antennas, at least for telephone functionality.

The GNSS functionality may be achieved with a single antenna that uses the CAN bus to output its position data. However, the possible damage to this antenna is also an ongoing argument.

The location of the antenna requires investigation of the mounting environment which is least prone to damage during a collision. This rules out most of the external surfaces such as the roof panel which would be ideal for antenna performance. Locations such as wing mirrors and bumpers carry a high risk of damage from collisions. A general consensus towards the vehicle dashboard is seen, but the location does suffer from a reduction in antenna performance. Vehicles with heated front windscreens are also finding challenges from a dashboard location.

The antenna structures are similar to conventional designs discussed in Section 7.3 and therefore is not discussed in detail in this section.

7.4.3 Other ITS Antennas

ITS systems are fundamentally based on the use of antennas for Telematics and positioning. Therefore the requirements for most systems are currently met with a GNSS antenna and telephone antenna, in most cases with 3G functionality.

As a single antenna is used for multiple systems, it becomes necessary to supply signals to these multiple systems. Conventionally, RF splitters were used to achieve this objective.

However, as the RF signal is divided between the systems, the available RF power becomes less and less. This is one of the key reasons for integrating the relevant electronics with the antenna and outputting the digital data into a vehicle network.

When the digital data is output to a vehicle network, there is no signal attenuation to contend with. The data on the bus is available for all systems on the vehicle that is connected to the vehicle network. Such architecture allows for future upgradeability of the system, without having to consider the installation of another set of antennas.

One of the biggest advantages of locating the receiver and transmitter electronics next to the antenna is system performance. The losses from the coaxial cables connecting the antenna to the electronics are eliminated and noise performance is improved. Availability of electronics next to the antenna allows intelligent antenna configurations, optimizing the antenna based on receive or transmit signal conditions. The architecture also provides for an overall system cost improvement. In Section 7.5, we shall discuss such antenna configurations in more detail.

7.5 Intelligent and Smart Antennas

The growth of vehicle Telematics has seen the growth of in vehicle networking. Typical systems such as LIN and CAN are bandwidth limited for the requirements of today's multimedia systems. CAN is however acceptable for transporting typical data requirements for GNSS systems.

Requirements for bigger bandwidth have resulted in systems such as Media Oriented Systems Transport (MOST®) and Flexray™. However, for today's growing requirements even such systems remain limited. Recent years has seen the rapid growth of automotive Ethernet mainly in the form of BroadR-Reach®, which has been recently incorporated as an IEEE standard 802.3bp.

It is likely that the future will see such wired systems being replaced by wireless systems. For audio and simple control applications, classic Bluetooth® systems have played a significant part. It is anticipated in the near future Bluetooth® Low Energy may also play an integral part in systems like TPMS and Remote Keyless Entry (RKE). In the longer term as faster data rates are required, Wi-Fi® is likely to be a dominant technology.

The growth of in-vehicle networks have allowed for the integration of electronics next to the antenna with digital data being output from the module.

7.5.1 Intelligent Antenna for Broadcast Radio

The integration of electronics next to an antenna has its own challenges. Noise generated from electronics that may be picked up by the antenna is a challenge to be overcome by the designer. For optical systems such as MOST, the switching noise created by the electrical to optical conversion process remains one of the biggest challenges. For broadcast systems such as AM, such noise could result in the loss of even the strongest broadcast stations.

Typically antenna modules are small in size. A classic example is an antenna mounted on a vehicle roof. For aesthetic pleasing, it is required that the electronics are integrated to the antenna base with minimal size increase. Such requirement results in the need for compacting electronic circuitry close together, which could increase the noise problem further.

Intelligent systems require a current consumption much larger than a conventional antenna amplifier, the requirement for a heat sink is a strong design consideration. A vehicle roof does help with this challenge.

Figure 7.19 Electronic circuit of a vehicle roof antenna developed for broadcast radio reception.

Figure 7.19 shows an electronic circuit of a vehicle roof antenna developed for broadcast radio reception. The antenna provides functionality for AM and FM radio. The antenna outputs digital audio in optical form to a vehicle network. The design contains a multilayer printed circuit board for optimum noise performance. For optimum thermal conductivity, the voltage regulators are mounted directly on to the metallic parts of the antenna unit, which in turn is mounted directly to the vehicle roof.

7.5.2 Intelligent Antenna for GNSS

As previous discussed many systems on the vehicle today demands for position data from the vehicle. Conventionally the requirement was limited to the vehicle navigation system. Therefore it was logical to locate the GNSS receiver within the navigation electronics, which typically consisted of the user interface and a display. Today systems such as E-Call, vehicle tracking and systems that utilize apps located in the vehicle demand for position data.

Such systems would typically require multiple splitters to distribute the GNSS signals or multiple antennas to be located on or in the vehicle.

Figure 7.20 presents an electronic circuit board from a module that overcomes this challenge. The module contains a GNSS antenna, which connects to a low-noise amplifier and then a GNSS receiver. The output from the module is connected to the vehicle CAN bus [10].

The CAN bus serves two purposes. The first is to provide acquired data from vehicle sensors to the antenna module for dead reckoning. Dead reckoning is a calculation process in the GNSS receiver that computes the vehicle position based on data received from the vehicle gyro, wheel sensors, steering, vehicle speed, direction and other relevant sensors. In the absence of GNSS signals or when the signals are weak or distorted, the GNSS receiver utilizes the dead reckoning process to output a computed vehicle position.

The second purpose is to transport the vehicle position data output by the antenna module. Such data is available for all systems on the CAN bus. As processed data is output to the vehicle bus multiple systems can receive this data without any degradation in quality.

Figure 7.20 Electronic circuit board from a module that overcomes this challenge.

7.6 Conclusions

The automotive antenna has evolved from a simple design to a complex solution over the last few decades. The 20th century saw the traditional telescopic antenna become a motorized antenna and then on to a short multiband antenna. The high demand for vehicle aesthetics saw the growth of the truly hidden antenna, typically in the form of a printed pattern on glass or a fully hidden film antenna.

As the antenna became hidden possibilities for diversity saw its growth providing superior performance than a single antenna. Today the antenna is growing at its fastest pace ever. The in-vehicle networking and the growth of Intelligent Transportation Systems has introduced many new frequencies for the antenna designer to contend with. The wider bandwidth and the requirement for smaller and smaller antennas and in some cases multiple antennas meant novel solutions were required.

A large amount of such challenges were overcome by the introduction of electronics next to the antenna. Availability of electronics next to the antenna resulted in intelligent solutions which could be optimized for performance based on propagation conditions and characteristics of the required signals for the receiver or from the transmitter at that time.

It is expected that the next few years will see the antenna becoming more and more intelligent as the available semiconductor devices next to the antenna becomes smarter and smarter.

References

1. H. Lindenmeier, J. Hopf and L. Reiter (1992) Antenna and diversity techniques for broadcast reception in vehicles, *Antennas and Propagation Society International Symposium, AP-S. 1992 Digest*, **2**: 1097–1100, July.
2. U. Jakobus, N. Berger and F.M. Landstorfer, Efficient techniques for modelling integrated windscreen antennas within the method of moments, *Proceedings of Millennium Conference on Antennas and Propagation, Davos, Switzerland, April 2000*, pp. 102–5 (last accessed 9 May 2015).
3. R. Breden and R.J. Langley (1999) Printed fractal antenna, *Proceedings of the IEE National Conference on Antennas and Propagation, 1999*, pp. 1–4.

4. R. Leelaratne and R. Langley (2005) Multiband PIFA vehicle Telematics antennas, *IEEE Transactions on Vehicular Technology* **54**(2): 477–85.
5. A. Thiel, O. Klemp, A. Paiera, *et al.* (2010) In-situ vehicular antenna integration and design aspects for vehicle-to-vehicle communications, *Proceedings of the 4th EUCAP, April 2010*, pp. 1–5.
6. T. Smits, S. Suckrow, J. Christ and M. Geissler (2013) Active intelligent antenna system for Car2Car, *International Workshop on Antenna Technology, iWAT 2013*, pp. 67–70 (last accessed 9 May 2015).
7. M. Gallo, S. Bruni and D. Zamberlan (2012) Design and measurement of automotive antennas for C2C applications, *6th European Conference on Antennas and Propagation, EUCAP 2012*, pp. 1799–1803.
8. D.R.V. Leelaratne, Harada Industry Co., Ltd, *Radiation Antenna for Wireless Communication*, US 20130033409, August 2011.
9. N.A. Abbasi and R. Langley (2010) Vehicle antenna on AMC, *Antennas and Propagation Society International Symposium (APSURSI), 2010*, pp. 1–4.
10. M. Merrick, D.R.V. Leelaratne, A. Nogoy, D. Simmonett, Harada Industry Co., Ltd., *Multiband Antenna with GPS Digital Output*, US 20110109522, May 2011.

Part 3

Sensors Networks and Surveillance at ITS

8

Middleware Solution to Support ITS Services in IoT-based Visual Sensor Networks

Matteo Petracca[1], Claudio Salvadori[2], Andrea Azzarà[2], Daniele Alessandrelli[2], Stefano Bocchino[2], Luca Maggiani[2] and Paolo Pagano[1]

[1] *National Laboratory of Photonic Networks, National Inter-University Consortium for Telecommunications, Pisa, Italy*

[2] *Institute of Communication, Information and Perception Technologies, Pisa, Italy*

8.1 Introduction*

In the last several years a large number of projects have been realized with the aim of creating effective Intelligent Transport Systems (ITSs) in which useful and cost-effective services can be provided to the final users. In this direction the European Union (EU) has progressively increased its contribution in funding research activities aiming at: (1) making optimal use of road, traffic and travel data; (2) guaranteeing continuity of traffic and freight management in ITS services; and (3) creating ITS road safety and security applications. Such requirements have been specified by the EU in the Directive 2010/40/EU [1].

In order to create effective ITS solutions able to meet the EU requirements it is necessary to develop layered architectures [2] where a data collection infrastructure is able to collect data pervasively in the monitored area. Although the pervasiveness is a key point for ITS collection layers, such a requirement can be reached in reality only by using low-cost hardware solutions able to communicate among them wirelessly through standard protocols. Nowadays the

* The research leading to these results has received funding from the European Union's Seventh framework Programme (FP7/2007-2013) under grant agreement n° 317671 (ICSI – Intelligent Cooperative Sensing for Improved traffic efficiency).

Intelligent Transport Systems: Technologies and Applications, First Edition. Asier Perallos, Unai Hernandez-Jayo, Enrique Onieva and Ignacio Julio García-Zuazola.

low-cost condition can be reached by using embedded devices based on low-complexity microcontrollers, while providing them advanced communication capabilities able to support well-known Internet-based protocols. In this direction the most suitable technology solution is based on Visual Sensor Networks (VSNs) based on communication standards recently developed for supporting the Internet of Things (IoT) paradigm in resource constrained devices used in Wireless Sensor Networks (WSNs). The use of low-complexity devices coming from IoT-based WSNs and equipped with vision capabilities (i.e. IoT-based VSNs) on the one hand requires the development of advanced low-complexity computer vision algorithms able to extract mobility-related data (e.g. traffic flow, parking lots occupancy levels), while on the other hand strongly reduces installation costs in the ITS scenario. Visual Sensors (VSs) can be installed on existing poles, thus avoiding expensive infrastructure works as required by state-of-the-art solutions based on invasive installations (e.g. inductive loops). Moreover, the use of standard protocols derived from the Internet world, and adapted to the VSNs scenario, enables the pervasiveness of the system, as well as the possibility of creating open and interoperable ITS systems acting as part of a bigger IoT network. By merging together advanced solutions in VSN and IoT research fields, innovative cost-effective and optimized collection layers for ITSs can be created. As a matter of example, each node of the data collection infrastructure can extract mobility related data performing an on-board processing of captured images with the aim of sharing them as resources to be used by other nodes or high level entities to generate complex events (e.g. traffic queue detection by merging features of several nodes) or to feed running applications (e.g. traffic prediction services by sending raw resources to a control centre acting in the cloud).

According to the above-described vision, the intelligence of the whole system is distributed among all the nodes of a global network: inside the VSN acting with IoT protocols, and outside, in remote high-end devices able to elaborate the resources provided by the VSs. Inside the IoT-based VSN the resources exposed by the nodes can be both scalar and vector features obtained by an on-board processing of the captured images. In this way each visual sensor provides features that reflect its environmental understanding based on its field of view. Whenever an in-network processing of such features is necessary by performing their composition, elaboration, and aggregation, this operation can be delegated to a middleware solution able to: (1) manage network transactions among nodes; (2) manage node functionality by interacting with the exposed operating system interfaces; and (3) provide enough flexibility to change at runtime the internal Resource Processing Engine (RPE), thus without changing the node firmware, and enabling a reconfigurable in-network processing in visual sensor networks. A pictorial view of an IoT-based VSN exposing mobility data as resources is depicted in Figure 8.1. In the picture three nodes are monitoring the same parking spaces. While two nodes are considering the highlighted parking space full with a probability equal to 50% and 40% because of a temporarily occlusion of their field of view due to the passing car, the other node is exposing an occupancy level resource suggesting that the monitored parking space is empty. All three resources can be aggregated in any node of the network where a RPE gathering all the occupancy levels, and other possible information, can take the right decision.

In this chapter we first introduce VSNs, by focusing on hardware devices able to support IoT-based communications and developed computer vision applications in the ITS scenario, then we describe IoT enabling protocol solutions and the REpresentational State Transfer (REST) paradigm on which IoT-based WSNs and VSNs are designed. The architecture of the proposed

Figure 8.1 Visual sensor network exposing parking space occupancy levels as resources.

middleware, able to support reconfigurable in-network processing in IoT-based VSNs, is presented in the third part of the chapter, while in the last section the middleware working principles are detailed for the 'parking lot monitoring' use case. The feasibility of the proposed solution in low-cost VSN devices based on microcontroller is proven through a real implementation of the middleware on top of the Contiki OS [3] running in the Seed-Eye board [4]. In the following sections the terms IoT-based WSNs and IoT-based VSNs are used with the same meaning by considering a VSN as a WSN in which nodes embed vision capabilities.

8.2 Visual Sensor Networks and IoT Protocols

In the first part of this section we introduce VSNs by presenting state-of-the-art hardware solutions based on embedded platforms able to support IoT-based communications, as well as already developed applications targeted to the ITS scenario. In the second part the IoT enabling protocols are presented and the use of the REST paradigm in IoT-based VSN is discussed.

8.2.1 Visual Sensor Networks

The idea behind VSNs (aka Smart Camera Networks, or SCN) is the creation of a network of nodes able to locally process the visual information with the aim of extracting a set of features to be sent through the network, thus avoiding the full transmission of raw images. Every VS [5] combines visual sensing, image processing, and network communication in a single embedded platform, permitting to transform traditional cameras into smart sensors. In simple terms, the main components of a VSNs are the embedded VSs and the computer vision algorithms performing the feature extraction. In the following we first present VS hardware solutions, then we introduce VSN applications targeted to the ITS scenario.

8.2.1.1 Visual Sensor Node Solutions

In recent years several research initiatives have produced real prototypes of visual sensors able to perform image processing as well as to communicate wirelessly through the IEEE 802.15.4 [6] standard, the first main requirement to enable IoT-based communications. All the solutions presented in the following embed IEEE 802.15.4 compliant transceivers, thus they are presented by detailing only vision capabilities and main application scenarios.

Among the presented devices the first to be cited is the WiCa [7] platform developed by NXP Semiconductors Research. Such a device is equipped with a NXP Xetal IC3D processor based on an SIMD architecture with 320 processing elements and can host up to two CMOS cameras at VGA resolution (640 × 480). It has been mainly used for image-based local processing and collaborative reasoning applications. The Mesheye [8] project is another research initiative aimed at developing an energy-efficient visual sensor based on an ARM7 processor, and able to run intelligent surveillance applications. Regarding the vision capabilities, the MeshEye mote has an interesting special vision system based on a stereo configuration of two low-resolution, low-power cameras, coupled with a high-resolution VGA color camera. In particular, the stereo vision system continuously determines position, range, and size of moving objects entering its fields of view. This information triggers the color camera to acquire the high-resolution image subwindow containing the object of interest, which can then be efficiently processed.

Another interesting example of low-cost embedded vision system is represented by the CITRIC [9] platform. Such a device integrates a 1.3-megapixel camera sensor, a XScale PXA270 CPU (with frequency scalable up to 624 MHz), a 16 MB FLASH memory, and a 64 MB RAM. CITRIC capabilities have been illustrated by three sample applications: image compression, object tracking by means of background subtraction, and self-localization of the camera nodes in the network. More recent device projects are the Vision Mesh [10] and the Seed-Eye [4] boards. The Vision Mesh integrates an Atmel 9261 ARM9 CPU, 128 MB NandFlash, 64MB SDRAM, and a CMOS camera at VGA resolution. The high computational capabilities of the mounted CPU permit to perform advanced computer vision technique targeted to water conservancy engineering applications. The Seed-Eye board has been specifically designed for image-based applications in sensor networks. Indeed, its design is very similar to that of tiny sensor network devices, with the addition of a more powerful microcontroller, the PIC32MX795F512L manufactured by Microchip, and of a low-cost CMOS camera able to reach up to 1.3-megapixel of resolution. The board has been used for ITS applications by performing an on-board image processing aiming at counting the number of passing vehicles and monitoring the occupancy level of parking spaces.

In all the aforementioned visual sensor devices the image processing tasks are performed through software applications. A new possible hardware-based solution aiming at speeding up processing capabilities in VSN has been recently proposed in [11]. In such a prototype solution an FPGA-based VS is designed with a reconfigurable architecture able to connect optimized hardware processing blocks with the final goal of performing advanced vision tasks. Then, a SoftCore is in charge of the system management by controlling the internal data flow configurations and processing block parameters. Communication capabilities can be demanded to an external microcontroller or to another SoftCore embedded in the FPGA. All the presented visual sensors are depicted in Figure 8.2.

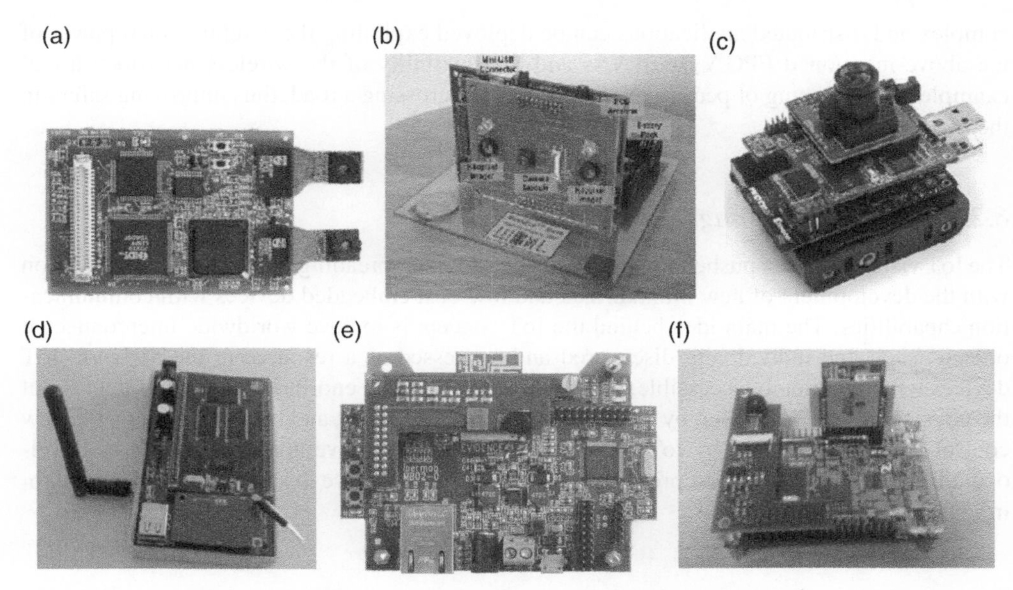

Figure 8.2 From top left to bottom right: WiCa, Mesheye, CITRIC, Vision Mesh, Seed-Eye and FPGA-based VS.

8.2.1.2 Visual Sensor Network Applications in the ITS Scenario

Although WSNs have already proved to be an effective tool for supporting next generation ITS, VSNs capabilities in such a scenario have not been fully exploited. Indeed, even if several real deployments of sensor networks based on scalar sensors (e.g. acoustic sensors, ultrasonic sensors, magnetometers, etc.) can be found in literature (e.g. VTrack [12], ParkNet [13], WITS [14] projects), VSNs testbeds are less common. A recent example in this direction is the testbed developed during the IPERMOB project [15], where a real VSN has been installed at the Pisa International Airport landside to monitor and control urban mobility in real-time. The main difficulty in developing real VSNs targeted to ITS is in adapting already developed computer vision applications to the new resource-constrained scenario. Indeed, the use of low-cost hardware, having limited computational capabilities and on-board memory, makes it unfeasible to use of state-of-the-art computer vision applications, which has to be redesigned or modified for the new application scenario, as proposed by [16].

In recent years, several ITS-related applications based on embedded VSs have been proposed. As a matter of example we can cite [17,18], where two embedded low-complexity computer vision applications have been developed with the aim of detecting the status of parking spaces, and [19,20], where VSs are used to count passing vehicles as well as to measure car speed. Looking forward at future applications, the use of pervasive VSNs will allow the realization of a set of interoperable systems capable to solve several open issues in the ITS field. While on one hand a pervasive deployment of VSs able to extract mobility related parameters will be able to generate open data to be consumed by municipalities for traffic planning purposes, on the other hand they can retrofit old road illumination systems for creating advanced services in the so-called Smart Cities. Moreover, in the same direction, more complex and distributed applications can be deployed exploiting the computational power of the above-mentioned FPGA based VSs and the flexibility of the wireless networks: a real example is the tracking of pedestrians while they are crossing a road, thus improving safety in the road of the future.

8.2.2 *Internet of Things*

The IoT vision has been pushed forward by the worldwide spreading of Internet in combination with the development of new miniaturized and low-cost embedded devices with communication capabilities. The main idea behind the IoT concept is to have worldwide interconnected objects, each one individually discovered and addressed as a resource in the network. IoT devices will be remotely accessible, thus making available an enormous amount of data about the physical world. Moreover, by using pervasive collected data, and by leveraging on the new control possibility offered by IoT enabling solutions, innovative applications can be developed. In the following we first present the IoT protocols, then we discuss the REST paradigm in IoT-based WSNs.

8.2.2.1 IoT Protocols

IoT-based communication relies on standard protocol solutions covering all the layers of the well-known Internet Protocol Suite. The first standard protocol, specifying both the Physical (PHY) and Medium Access Control (MAC) sublayers of the ISO/OSI communication model, is the IEEE 802.15.4 [6]. The standard has been released in its first version in the 2003 with the aim of enabling energy-efficiency communications in Low-Rate Wireless Personal Area Networks (LR-WPANs). In 2007 Kushalnagar *et al.* [21] proposed the adaptation of the IPv6 over Low-Power Wireless Personal Area Networks (6LoWPANs), thus specifying a Network (NET) layer for Internet-like communication in IEEE 802.15.4-based networks. The 6LoWPAN concept comes from the idea that 'the Internet Protocol could and should be applied even to the smallest devices' [22], and that low-power devices with limited processing capabilities should be able to participate in the envisioned IoT. 6LoWPAN defines the frame format for the transmission of IPv6 packets, as well as mechanisms for header compression, and formalizes how to create IPv6 global addresses on top of IEEE 802.15.4 networks. Along with the definition of IPv6-based communications in standard wireless sensor networks, another major point to consider is related to routing protocols. Routing issues are very challenging for 6LoWPAN due to the low-power nature of such networks, multihop mesh topologies to be managed, and topology changes due to node mobility. Successful solutions should take into account the specific application requirements, along with IPv6 behaviour and 6LoWPAN mechanisms. The Routing Protocol for Low-power and Lossy networks (RPL) [23] is the state-of-the-art routing algorithm developed by the networking community. RPL has been proposed by the IETF Routing over Low-power and Lossy networks Working Group (ROLL) as a standard routing protocol for 6LoWPAN, since existing routing protocols do not satisfy all the requirements for Low power and Lossy Networks (LLNs). The last protocol to be cited for enabling IoT solutions is the Constrained Application Protocol (CoAP) [24], which is a standard solution working at the Application (APP) layer, and currently being defined within the CoRE working group of the IETF. It aims at providing a REST-based

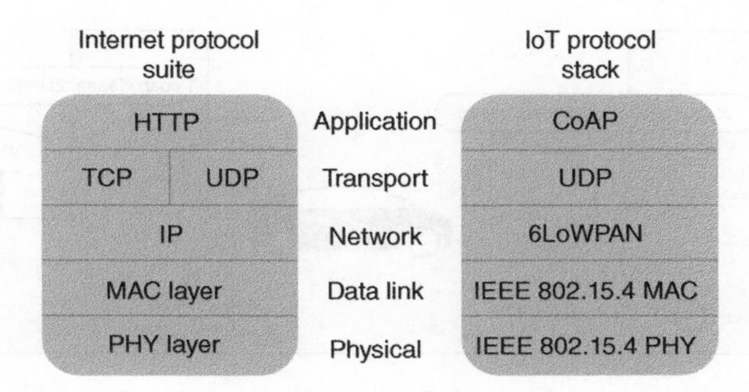

Figure 8.3 IoT protocol stack versus Internet protocol suite.

framework for resource-oriented applications by designing a protocol stack able to cope with limited packet size, low-energy devices and unreliable channels. CoAP is designed for having an easy stateless mapping with HTTP, and for providing Machine-2-Machine interaction. HTTP compatibility is obtained by maintaining the same interaction model, but using a subset of the HTTP methods. Any HTTP client or server can interoperate with CoAP-ready endpoints by simply installing a translation proxy between the two devices. The REST paradigm in IoT-based WSNs will be detailed in the following, while in Figure 8.3 the protocol stack for the IoT is shown and compared with the classical Internet protocol suite.

8.2.2.2 REST Paradigm in IoT-based WSNs

REpresentational State Transfer (REST) [25] is an architectural style for distributed systems introduced and defined in 2000 by Fielding in his doctoral dissertation. RESTful architectures basically consist of clients and servers. Clients send requests to servers, which reply with appropriate responses. Requests and responses are built around the transfer of representations of resources, where a resource can be essentially any coherent and meaningful concept that may be addressed. A representation of a resource is typically a document that captures the current or intended state of the resource. The most relevant example of a system conforming to the REST architectural style is the World Wide Web, in which resources are manipulated using the HTTP protocol.

The REST paradigm can be successfully used in IoT-based WSNs, where resources usually represent sensors, actuators or other possible information. However, as previously introduced, the CoAP protocol is used instead of HTTP, thus allowing sensor nodes to run embedded web services through which their resources can be manipulated. Specifically, CoAP provides four methods for manipulating resources: (1) PUT, which requests to update or create, with the transmitted representation, the resource identified by the URI specified in the request; (2) POST, which requests the process of the representation transmitted in the request; (3) GET, which retrieves a representation of the resource identified by the URI specified in the request; and (4) DELETE, which requests the deletion of the resource identified by the URI specified in the request. CoAP also provides a resource observation mechanism [26] (OBSERVE) which allows a node to receive notifications about changes in resources it has previously subscribed to.

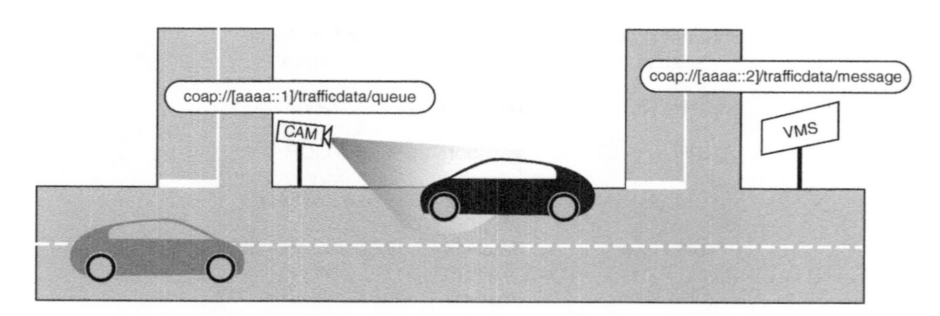

Figure 8.4 Example of an IoT-based WSN supporting ITS services.

Figure 8.4 depicts a simple example of an IoT-based WSN with vision capabilities able to act as an ITS collection layer. In the picture the network is composed by two nodes: one is a vision sensor able to count vehicles and detect possible traffic congestion by analyzing the speed of vehicles (on the left), while the other one is an actuator able to control the messages of a Variable Message Sign (VMS). According to the REST working principles each resource is identified by the node IPv6 address and a symbolic name, and is managed through PUT, POST, GET and DELETE methods. For example, the CAM node can expose a 'queue' resource that can be retrieved by issuing a GET request on the URI *coap://[aaaa::1]/trafficdata/queue*. Similarly the VMS exposes a 'message' resource that can be controlled by sending a PUT request to the URI *coap://[aaaa::2]/trafficdata/message* containing the message to be displayed in its payload. In such a way it is possible to create an application that collects traffic-congestion notifications (sending GET requests to the 'queue' resource) and uses them for suggesting alternative routes to drivers (sending PUT requests to the 'message' resource).

The use of standard protocols (i.e. 6LoWPAN and CoAP) allows nodes to be used for many different applications. However, such protocols alone do not provide a way for changing the devices application logic once nodes are installed and configured for a specific task. As previously stated, such a feature must be demanded to a middleware solution running in visual sensors and able to provide enough flexibility in changing at runtime the internal resource processing engine, as well as to manage network transactions and resources (i.e. the middleware must be able to get data from running computer vision algorithms in order to update the exposed resources).

8.3 Proposed Middleware Architecture for IoT-based VSNs

A middleware is basically a software solution able to interact with both high-level applications and Operating Systems (OSs). Its main purpose is to uniform heterogeneous systems by hiding their specific complexity and by providing a common unified software abstraction. Moreover, a middleware usually provides common configuration and maintenance services, thus enabling an easier management of complex systems. In a WSN scenario a middleware system is usually designed as a tool to bridge the gap between the high-level requirements of the applications and the low-level hardware complexity. WSN middleware are supposed to help programmers by providing an adequate system abstraction, and allowing them to focus on the high-level application logic without caring about low-level implementation details.

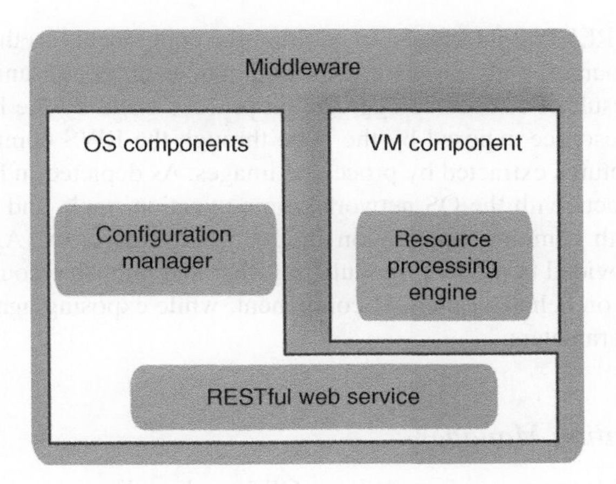

Figure 8.5 Middleware components.

In IoT-based VSNs in which low-complexity computer vision algorithms are executed on VSs with the aim of providing both scalar and vector resources (i.e. scalar and vector features extracted by processed images) to other network nodes, a middleware must be able to: (1) manage network transactions among nodes by leveraging on IoT protocols while using the REST paradigm; (2) manage VS functionality permitting to configure on-board computer vision algorithms through exposed operating system interfaces, and (3) provide enough flexibility to change at runtime the VS RPE. The last requirement is a key point to consider to permit the composition, elaboration, and aggregation of image-related resources, and to enable reconfigurable in-network processing in VSNs. Such middleware high-level requirements can easily be mapped onto middleware components able to interact among them through common interfaces. At a system design level three main middleware components can be identified:

1. RESTful Web Service (RWS);
2. Configuration Manager (CM);
3. Resource Processing Engine (RPE).

The three components are graphically reported in Figure 8.5, while their functionality are detailed in the rest of the section. The first two are identified as OS components, because they mainly interact with OS interfaces to configure the embedded vision logic and to use network services for enabling the REST paradigm. The last component is identified as a Virtual Machine component, because, as it will be better clarified in the following, it provides a platform-independent run-time environment running on top of the operating system. According to the classification presented in [27] this approach is considered a Virtual Machine based design.

8.3.1 RESTful Web Service

The RESTful Web Service component handles all network data inputs and outputs through CoAP transactions; moreover it acts as a resource directory service. In fact, each resource provided by a certain node to other entities of the network is internally registered by such component through RESTful interfaces. In addition the component has the knowledge of the internal exposed resources, both simple (e.g. the output of an on-board running algorithm) and complex (e.g. the result of a resource aggregation process made by the RPE). In IoT-based VSNs the internal resource exposed by the node through the RWS component can be both scalar and vector features extracted by processed images. As depicted in Figure 8.5 the component mainly interacts with the OS network communication stack, and also with the other two components with common Application Programming Interfaces (APIs). In fact, along with the services provided to the RPE module to gather and publish resources, it can manage network transaction on behalf of the CM component, while exposing again as resources the CM configuration parameters.

8.3.2 *Configuration Manager*

The Configuration Manager component is an OS-based application running on top of the OS. The component is responsible for the configuration of some part of the platform. It is in charge of the RPE configuration by changing the RPE logic for the resource processing. In IoT-based VSNs a main feature requested to such a component is the possibility of changing configuration parameters of the running computer vision applications. In VSs based on embedded processors (e.g. WiCa, MeshEye, Seed-Eye, etc.) simple parameters can be easily configured at runtime, such as the resolution of the acquired image, the frame rate, the region of interests to be used by the on-board computer vision applications. In such a category of VSs the CM can be even used to change the running computer vision algorithms, even if this would require firmware updating policies. In VS based on FPGA and internal programmable logic (e.g. FPGA-based VS presented in the previous section) the CM can be used to configure and compose the whole computer vision pipeline, thus changing at runtime the whole on-board processing application. Moreover, because of the interaction with the RWS component the output of any single selected step of a computer vision pipeline can be abstracted as a resource, thus enabling the coexistence of multiple distributed applications running inside the network (i.e. the output of a step can be used by the RPE of a node for a certain application, while another node can use the same value for further elaborations).

8.3.3 *Resource Processing Engine*

The RPE component is the core technical module of the middleware. The component is requested to: (1) monitor one or more input resources; (2) execute some data processing on their values; and (3) send the result to other resources. In order to gather and publish resource values RPE strongly interacts with WBS, while interacts with CM for changing the RPE processing logic. The RPE of the proposed middleware architecture for IoT-based VSNs is mainly based on T-Res [28], a task resource abstraction mechanism presented in 2013 by Alessandrelli *et al.* The main idea behind T-Res is to represent resource processing tasks as CoAP resources. In such a way a processing task resource can be manipulated using CoAP methods, like any other classic resource: it can be created, deleted, modified, or even retrieved (duplicated). In T-Res a processing task is completely defined by its input sources, the processing it performs, and the destination of its output. T-Res keeps separated those three elements by defining a subresource for each of them. A 'Task' resource contains the following subresources:

1. the 'input sources' resource (/is);
2. the 'processing function' resource (/pf);
3. the 'output destinations' resource (/od).

Both the input-source resource (/is) and the output-destination resource (/od) contain URIs (i.e. references to other resources). T-Res monitors input resources for new values by using the CoAP OBSERVE mechanism. The processing-function resource (/pf) contains the code performing the desired processing. Such code is simply a function that is called every time a new input value is available and that may produce an output value which is sent to the output resource. As previously mentioned, the processing function of a task resource contains the code to be executed when a new input is received. In order to fully decouple task processing functions from the sensing infrastructure, as well as to permit to change them through another entity (e.g. CM component) without requiring firmware updates, the processing functions must be platform independent. Therefore, they cannot be written in languages that must be compiled into native code (e.g. C, C++, etc.). The alternative is to adopt languages that can be compiled into bytecode or directly executed by an interpreter. To this end, T-Res uses Python for defining processing functions. Since Python bytecode runs on a Python interpreter hosted by the operating system, T-Res is considered a virtual machine component, and the same classification can be applied to the RPE component of the proposed middleware.

8.4 Middleware Instantiation for the Parking Lot Monitoring Use Case

In this section the working principles of the middleware are detailed for a 'parking lot monitoring' use case by considering VS resources for each middleware component, and by detailing the Python code performing resource processing in RPEs. Moreover, the feasibility of the proposed middleware solution in IoT-based VSNs is proven through a real implementation on the Seed-Eye board.

8.4.1 Use Case Scenario, Exposed Resources and Their Interaction

The considered use case is that of an ITS collection layer based on an IoT-based VSN in which VSs are deployed on the field to evaluate the status of several parking spaces. By referring to Figure 8.1 we see that we have three nodes monitoring (for the sake of simplicity) a total amount of four parking spaces whose identification numbers are 23, 25, 26 and 27. Because of their deployment in the field, all the three nodes have in their field of view the parking space 25. Moreover, each VS runs an on-board computer vision algorithm giving as output the occupancy level for each parking space as a value ranging from 0 to 255. Such a value intrinsically reflects the uncertainty of the decision regarding the state of the parking space. Indeed, considering values in the range (0, 128) associated to an empty state, values close to 128 can indicate a bigger uncertainty. The same consideration is valid for a full state in which the occupancy level is in the range (128, 255). During the installation phase each VS can be remotely configured through a possible gateway entity to set up, for each parking space, a *region of interest (ROI)* to be considered by the algorithm, as well as a *fixed weight* which is a value used by the algorithm to evaluate the occupancy level. For instance, during the deployment, the field of view of a camera can frame only partially a parking space: in this case the algorithm can run on the configured ROI by using a low weight, corresponding to a bigger uncertainty associated to the occupancy level. Considering of having the middleware previously described running in each VS, the resources exposed for each component are:

(a) Parking resources through RWS component
  - /space_xx_level;

(b) CM component
  - /space_xx_roi;
  - /space_xx_weight;

(c) RPE component
  - /tasks/parking/is;
  - /tasks/parking/pf;
  - /tasks/parking/od;
  - /tasks/parking/lo.

The exposed resources are detailed for the three nodes in Figure 8.6, along with a possible interaction model performing an in-network processing to take a global decision on the parking space 25.

As shown in the picture, Node 1 is exposing Parking resources for parking spaces 25 and 27, as well as CM resources related to them. In the example such a node is not exposing RPE resources because it does not provide any processing on them. The same is for Node 3 which only exposes Parking and CM resources for the parking spaces 25 and 26. Node 2, instead, exposes, along with Parking and CM resources, also RPE related resources. Indeed, Node 2 is performing a processing task on resources related to the parking space 25. After the configuration phase, the RPE on Node 2 starts monitoring all the */space_25_level* resources. Every time a VS sends a notification to Node 2 the processing task is activated (*/tasks/parking/pf*) and the decision on the occupancy level of parking space 25 (e.g. through an

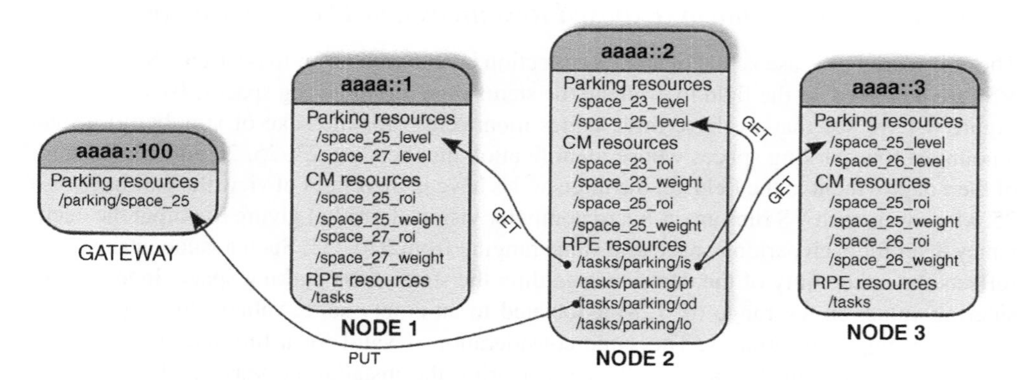

Figure 8.6 Exposed resources and interaction model.

average evaluation of acquired values) is sent to the Gateway and possibly communicated to a parking control center. To perform the processing the RPE */tasks/parking/is* resource contains the following URIs:

```
coap://[aaaa::1]/space_25_level
coap://[aaaa::2]/space_25_level
coap://[aaaa::3]/space_25_level
```

The /tasks/parking/od resource is set as:

```
coap://[aaaa::100]/parking/space_25
```

The */tasks/parking/lo* is a read-only resource, and it can be queried to retrieve the last output produced by the processing function. The */tasks/parking/pf* resource, instead, contains the bytecode generated compiling the following Python script:

```
from rpe import *
isInputs = rpe_getInputs()
sum = 0
for input in isInputs:
      sum += input
avg = sum / len(isInputs)
if avg < 128 :
      rpe_setOutputs("empty")
else:
      rpe_setOutputs("full")
```

Where the *rpe_getInputs()* is a function returning the representation of the input resources, and *rpe_setOutput()* performs a PUT on the output resources. In a real scenario the bytecode of the Python script must be transferred to the VS through a programming module running on an external device and able to interact with the CM component of the middleware. In the presented use case the Programming module can easily run on the Gateway and generates the bytecode to be transferred to the node once a new Python script is available. Moreover, envisaging a more complex scenario, it can even generate scripts to distribute the complexity of the processing among several nodes.

8.4.2 *Middleware Implementation*

To prove the feasibility of the proposed middleware solution we evaluated its requirements both in terms of code size (flash memory occupancy) and memory requirements (RAM occupancy) by implementing its components on Contiki OS, and by considering the Seed-Eye

Table 8.1 Code size and RAM requirements on the Seed-Eye device.

Device	RAM [bytes]	Flash [bytes]
Seed-Eye	16 120 (13%)	171 832 (34%)

board as target device. As first step we ported T-Res and PyMite [29] on the Seed-Eye. PyMite is a reduced Python virtual machine suitable for embedded systems and used by T-Res to run processing functions. In a second stage we implemented a basic CM component with reduced functionality, while RWS is intrinsically provided by Contiki OS thanks to its internal 6LoWPAN and CoAP support.

We tested the resulting implementation on the already presented Seed-Eye platform, which is equipped with a PIC32MX795F512L microcontroller having 128kB of RAM and 512kB of flash memory. Table 8.1 shows the RAM and flash memory occupancy by reporting in percentage their impact on the total microcontroller resources. The proposed middleware is a real feasible solution for low-complexity and low-cost VSs. In the Seed-Eye board it requires only 13% of the available RAM and 34% of the available flash memory. Because of the limited amount of occupied RAM, more processing tasks (Python bytecode) can be instantiated in the RPE component, thus enabling much more complex in-network processing functionality. As a matter of example a node may host several resource composition logics to evaluate possible better composition strategies. Moreover, because of the reduced amount of required flash memory, more complex CM components with respect to the one implemented can be developed, thus fully supporting more flexible configuration policies required by next generation FPGA-based VSs.

8.5 Conclusions

IoT-based VSNs are a suitable solution for the development of low-cost, pervasive and interoperable ITS collection layers able to fully support cost-effective services for to final users. According to such a visionVSs run on-board computer vision algorithms to extract mobility related features that can be represented as network resources thanks to the use of IoT protocols and the REST paradigm. In IoT-based VSNs whenever an in-network processing of exposed resources is necessary, a middleware system can be instantiated, hiding all the complexity and abstracting network functionality. In this chapter we presented a middleware solution suitable for IoT-based VSNs and able to: (1) manage network transactions among nodes using IoT protocols and the REST paradigm; (2) manage VS functionality permitting to configure on-board computer vision algorithms through exposed operating system interfaces, and (3) provide enough flexibility to change at runtime the internal resource processing engine by using a solution based on virtual machine. In the chapter the main middleware components have been first presented before detailing its working principles for a 'parking lot monitoring' use case where several VSs monitor overlapped parking spaces. The feasibility of the proposed middleware is proven by reporting objective data obtained through a real implementation based on the Contiki OS, and running on the Seed-Eye board.

References

1. European Parliament, Directive 2010/40/EU on the framework for the deployment of intelligent transport systems in the field of road transport and for interfaces with other modes of transport, *Official Journal of European Union*, 2010.
2. P. Pagano, M. Petracca, D. Alessandrelli and C. Salvadori (2013) Is ICT technology mature for an EU-wide intelligent transport system?, *Proceedings of IET Intelligent Transport Systems* **7**(1): 151–9.
3. A. Dunkels, B. Gronvall, and T. Voigt (2004) Contiki – a lightweight and flexible operating system for tiny networked sensors, *Proceedings of IEEE International Conference on Local Computer Networks*, pp. 455–62.
4. Seed-Eye: an advanced multimedia Wireless Sensor Network node for ITS applications, http://noes.sssup.it/index.php/hardware/seed-eye (last accessed 28 April 2015).
5. B. Rinner and W. Wolf (2008) An introduction to distributed smart cameras, *Proceedings of the IEEE* **96**(10): 1565–75.
6. IEEE Computer Society (2003) *Wireless Medium Access Control (MAC) and Physical Layer (PHY) Specifications for Low-Rate Wireless Personal Area Networks (LR-WPAN)*, The Institute of Electrical and Electronics Engineers.
7. A.A. Abbo and R.P. Kleihorst (2002) A programmable smart-camera architecture, *Proceedings of Advanced Concepts for Intelligent Vision Systems*.
8. S. Hengstler, D. Prashanth, S. Fong, and H. Aghajan (2007) Mesheye: a hybrid-resolution smart camera mote for applications in distributed intelligent surveillance, *Proceedings of International Symposium on Information Processing in Sensor Networks*, pp. 360–9.
9. P. Chen, P. Ahammad, C. Boyer, *et al.* (2008) Citric: A low-bandwidth wireless camera network platform, *Proceedings of International Conference on Distributed Smart Cameras*, pp. 1–10.
10. M. Zhang and W. Cai (2010) Vision mesh: A novel video sensor networks platform for water conservancy engineering, *Proceedings of International Conference on Computer Science and Information Technology*, pp. 106–9.
11. L. Maggiani, C. Salvadori, M. Petracca, *et al.* (2013) Reconfigurable FPGA architecture for computer vision applications in smart camera networks, *Proceedings of ACM/IEEE International Conference on Distributed Smart Cameras*, pp. 1–6.
12. A. Thiagarajan, L. Ravindranath, K. LaCurts, *et al.* (2009) VTrack: accurate, energy-aware road traffic delay estimation using mobile phones, *Proceedings of ACM Conference on Embedded Networked Sensor Systems*, pp. 85–98.
13. S. Mathur, T. Jin, N. Kasturirangan, *et al.* (2010) Parknet: drive-by sensing of road-side parking statistics, *Proceedings of International Conference on Mobile Systems, Applications, and Services*, pp. 123–36.
14. W. Chen, L. Chen, Z. Chen and S. Tu (2006) WITS: A wireless sensor network for intelligent transportation system, *Proceedings of International Multi-Symposiums on Computer and Computational Sciences*, pp. 635–41.
15. The IPERMOB Project (2009) A Pervasive and Heterogeneous Infrastructure to control Urban Mobility in Real-Time. http://www.iit.cnr.it/en/node/2534 (last accessed 9 May 2015).
16. C. Salvadori, M. Petracca, J.M. Rincon, *et al.* (2014) An optimisation of Gaussian mixture models for integer processing units, *Journal of Real-Time Image Processing*, pp. 1–17.
17. M. Magrini, D. Moroni, C. Nastasi, *et al.* (2011) Visual sensor networks for infomobility, *Pattern Recognition and Image Analysis* **21**: 20–9.
18. C. Salvadori, M. Petracca, M. Ghibaudi and P. Pagano (2012) On-board image processing in wireless multimedia sensor networks: a parking space monitoring solution for intelligent transportation systems. In *Intelligent Sensor Networks: Across Sensing, Signal Processing, and Machine Learning*, London: CRC Press.
19. M. Chitnis, C. Salvadori, M. Petracca, *et al.* (2012) Distributed visual surveillance with resource constrained embedded systems. In *Visual Information Processing in Wireless Sensor Networks: Technology, Trends and Applications*. IGI Global Press, Pennsylvania: IGI Global.
20. C. Salvadori, M. Petracca, S. Bocchino, *et al.* (2014) A low cost vehicles-counter for next generation ITS, *Journal of Real-Time Image Processing*, pp. 1–17.
21. N. Kushalnagar, G. Montenegro and C. Schumacher (2007*)* IPv6 over Low-power Wireless Personal Area Networks (6LoWPANs): Overview, Assumptions, Problem Statement, and Goals, RFC 4919.
22. G. Mulligan (2007) The 6lowpan architecture. In *Workshop on Embedded Networked Sensors*, New York: ACM, pp. 78–82.
23. T. Winter, P. Thubert, A. Brandt, *et al.* (2012) RPL: IPv6 Routing Protocol for Low-Power and Lossy Networks, RFC 6550.

24. Z. Shelby, K. Hartke, C. Bormann, and B.Frank (2013) Constrained Application Protocol (CoAP), IETF Draft Version 18.
25. R.T. Fielding (2000) Architectural styles and the design of network-based software architectures, PhD dissertation, University of California, Irvine.
26. K. Hartke (2012) *Observing Resources in CoAP*, IETF Internet-Draft.
27. S. Hadim and N. Mohamed (2006) Middleware: middleware challenges and approaches for wireless sensor networks, *IEEE Distributed Systems Online* **7**(3): 1–54.
28. D. Alessandrelli, M. Petracca and P. Pagano (2013) T-Res: enabling reconfigurable in-network processing in IoT-based WSNs, *Proceedings of IEEE International Conference on Distributed Computing in Sensor Systems and Workshops*, pp. 1–54.
29. PyMite, http://code.google.com/p/python-on-a-chip (last accessed 28 April 2015).

9

Smart Cameras for ITS in Urban Environment

Massimo Magrini, Davide Moroni, Gabriele Pieri and Ovidio Salvetti
Institute of Information Science and Technologies (ISTI), National Research Council of Italy (CNR), Pisa, Italy

9.1 Introduction

Fully automatic video and image analysis from traffic monitoring cameras is a fast-emerging field based on computer vision techniques with a growing impact on Intelligent Transport Systems (ITS).

Indeed the decreasing hardware cost and, therefore, the increasing deployment of cameras and embedded systems have opened a wide application field for video analytics both in urban and highway scenarios. It can be envisaged that several monitoring objectives such as congestion, traffic rule violation, and vehicle interaction can be targeted using cameras that were typically originally installed for human operators [1].

On highways, systems for the detection and classification of vehicles have successfully been using classical visual surveillance techniques such as background estimation and motion tracking for some time. Nowadays existing methodologies have good performance also in case of inclement weather and are operational 24/7. On the converse, the urban domain is less explored and more challenging with respect to traffic density, lower camera angles that lead to a high degree of occlusion and the greater variety of street users. Methods from object categorization and 3-D modelling have inspired more advanced techniques to tackle these challenges. In addition, due to scalability issues and cost-effectiveness, urban traffic monitoring cannot be constantly based on high-end acquisition and computing platforms; the emerging of embedded technologies and pervasive computing may alleviate this issue: it is indeed challenging yet definitely important to deploy pervasive and untethered technologies such as Wireless Sensor Networks (WSN) for addressing urban traffic monitoring.

Intelligent Transport Systems: Technologies and Applications, First Edition. Asier Perallos, Unai Hernandez-Jayo, Enrique Onieva and Ignacio Julio García-Zuazola.

On the basis of these considerations, the aim of this chapter is to introduce scalable technologies for supporting ITS-related problems in urban scenarios; in particular we survey embedded solutions for the realization of smart cameras that can be used to detect, understand and analyse traffic-related situation and events thanks to an on-board vision logics. Indeed, to suitably tackle scalability issues in the urban environment, we propose the use of a distributed, pervasive system consisting in a Smart Camera Network (SCN), a special kind of WSN in which each node is equipped with an image sensing device. For this reason, SCN are also known as Visual Sensor Networks (VSN). Clearly, gathering information from a network of scattered cameras, possibly covering a large area, is a common feature of many video surveillance and ambient intelligence systems. However, most classical solutions are based on a centralized approach: only sensing is distributed while the actual video processing is accomplished in a single unit. In those configurations, the video streams from multiple cameras are encoded and conveyed (sometimes thanks to multiplexing technologies) to a central processing unit which decodes the streams and perform processing on each of them. With respect to those configurations, the need to introduce distributed intelligent system is motivated by several requirements, namely [2]:

- Speed: in-network distributed processing is inherently parallel; in addition, the specialization of modules permits to reduce the computational burden in the higher level of the network: in this way the role of the central server is relieved and it might be actually omitted in a fully distributed architecture.
- Bandwidth: in-node processing permits to reduce the amount of transmitted data, by transferring only information-rich parameters about the observed scene and not the redundant video data stream.
- Redundancy: a distributed system may be re-configured in case of failure of some of it components, still keeping the overall functionalities.
- Autonomy: each of the nodes may process the images asynchronously and may react autonomously to the perceived changes in the scene.

In particular, these issues suggest moving a part of intelligence towards the camera nodes. In these nodes, artificial intelligence and computer vision algorithms are able to provide autonomy and adaptation to internal conditions (e.g. hardware and software failure) as well as to external conditions (e.g. changes in weather and lighting conditions). It can be stated that in a VSN the nodes are not merely collectors of information from the sensors, but they have to blend significant and compact descriptors of the scene from the bulky raw data contained in a video stream.

This naturally requires the solution of computer vision problems such as change detection in image sequences, object detection, object recognition, tracking, and image fusion for multiview analysis. Indeed, no understanding of a scene may be accomplished without dealing with some of the above tasks. As it is well known, for each of such problems there is an extensive corpus of already implemented methods provided by the computer vision and the video surveillance communities. However, most of the techniques currently available are not suitable to be used in VSN, due to the high computational complexity of algorithms or to excessively demanding memory requirements. Therefore, ad hoc algorithms should be designed for VSN, as we will explore in the next sections.

In this chapter, we first envisage applications of smart cameras and VSN to urban scenarios, highlighting specific challenges and peculiarities. Embedded vision nodes are introduced and a brief survey of existing hardware solutions is provided; the implementation of general computer vision algorithms on smart cameras and VSN is then addressed. We move further describing two sample ITS applications, namely analysis of traffic status and parking lot monitoring. In the first sample application, the estimation of vehicular flows on a lane is performed by using a lightweight computer vision pipeline that is somewhat dissimilar form the conventional one used on standard architecture. In the second sample application, an approach to parking lot monitoring is presented; here the vision nodes can collaborate with each other in producing more accurate and robust results, e.g. by resorting to the middleware for VSN presented in Chapter 8 by Petracca *et al.* A smart camera prototype designed with ITS application in mind is presented in Section 9.5, while in Section 9.6 we present its envisaged application scenarios and experimental results.

9.2 Applications to Urban Scenarios

According to [1], there has been an increased scope for the automatic analysis of urban traffic activity. This is partially due to the additional numbers of cameras and other sensors, enhanced infrastructure and consequent accessibility of data. In addition, the advances in analytical techniques for processing video streams together with increased computing power have enabled new applications in ITS. Indeed, video cameras have been deployed for a long time for traffic and other monitoring purposes, because they provide a rich information source for human understanding. Video analytics may now provide added value to cameras by automatically extracting relevant information. This way, computer vision and video analytics become increasingly important for ITS.

In highway traffic scenarios, the use of cameras is now widespread and existing commercial systems have excellent performance. Cameras are used tethered to ad hoc infrastructures, sometimes together with Variable Message Signs (VMS), RSU and other devices typical of the ITS domain. Traffic analysis is often performed remotely by using special broadband connection, encoding, multiplexing and transmission protocols to send the data to a central control room where dedicated powerful hardware technologies are used to process multiple incoming video streams [3]. The usual monitoring scenario consists in the estimation of traffic flows distinguished among lanes and vehicles typologies together with more advanced analysis such as detection of stopped vehicles, accidents and other anomalous events for safety, security and law enforcement purposes.

Conversely, traffic analysis in the urban environment appears to be much more challenging than on highways. In addition in urban environments several extra monitoring objectives can be supported in principle by the application of computer vision and pattern recognition techniques, including the detection of complex traffic violations (e.g. illegal turns, one-way streets, restricted lanes) [4,5], identification of road users (e.g. vehicles, motorbikes and pedestrians) [6] and of their interactions understood as spatiotemporal relationships between people and vehicle or vehicle to vehicle [7]. For these reasons, it is worthwhile to apply the wireless sensor network approach to the urban scenario.

Generally, we may identify four different scopes that can be targeted thanks to video-surveillance based systems, namely (1) safety and security, (2) law enforcement, (3) billing and (4) traffic monitoring and management. Although in this chapter we focus mostly on the latter, we give a brief overview of each of them.

Safety and security relate to the prevention and prompt notification both of proper traffic events and of roadside events typical of urban environment. From one side, detection of events like car accidents, stopped vehicles, general obstacles, tunnel accidents, floods and landslides is of fundamental importance: real time detections allows for immediate response that might be life-saving. In most cases, the information obtainable thanks to visual nodes most be usefully complemented with other detectors. For example, smoke detectors play a more crucial role than video sensors for dangerous tunnel accidents involving fire. In general, visual information turns out to be essential when complex scenes with nontrivial semantics should be understood. For instance, in case of landslides and obstacle detection, technologies based on radar might provide extended reliability and be fully operational also in case of adverse meteorological conditions (e.g. rain and snow) and low visibility situation (e.g. foggy weather). However, also in this case, integration of video information might be useful in reducing false positives by using object recognition methods thus improving the overall performance. Safety in urban environment regards also the detection of roadside events like crimes and vandalisms. For instance the commercial available solution [8] includes methods for detecting car park surfing, that is the act of a pedestrian getting out as passenger of one car and moving to another. This is indeed the usual hunting behaviour of car thieves.

Law enforcement is based on the detection of unlawful acts and to their documentation for allowing the emission of a fine. Besides well-known and established technologies, e.g. for streetlight violations, vision-based systems might allow for identification of more complex behaviour, e.g. illegal turns or trespassing on a High Occupancy Vehicle (HOV) lane. For instance, Xerox has recently produced a vehicle passenger detection system that uses geometric algorithms to detect whether a seat is vacant or occupied without using facial recognition [9]. Documentation of unlawful acts is usually performed by acquiring a number of images sufficient for representing the violation, combined with automatic number plate recognition (ANPR) for identifying the offender vehicle.

ANPR is also a common component of video-based billing and tolling. Also in this case there are a number of established technologies provided as commercial solutions by many vendors [10]. A peculiarity of urban billing systems with respect to highways is the nonintrusiveness requirement: it is not possible to alter the normal vehicular flow but a free-flow tolling must be implemented. Technologies satisfying this requirement are already available and used in cities such as London, Stockholm and Singapore but their actual cost prevents their massive deployment in medium-size or low-resource cities. Nevertheless, the availability of such billing technologies at a lower cost may pave the way to the collection of fine-grained data analytics of vehicular flows, road usage and congestions, allowing for the implementation of adaptive Travel Demand Management (TDM) policies aimed at a more sustainable, effective and socially acceptable mobility applied to urban and metropolitan contexts. It is likely that other technologies not based on video but for instance on NFC might become widespread in the near future to fill this gap.

Finally, traffic monitoring and management is related to extraction information from urban observed scenes that might be beneficial in several contexts. For instance, real-time vehicle counting might be used to assess level of service on a road and detecting possible congestions. Such real-time information might then be used for traffic routing; either by providing directly suggestion to user (e.g. by VMS) of by letting a trip planner deploys these data to search for an optimal path. Finally, statistics on vehicular flows may be used to understand mobility patterns and help stakeholders to improve urban mobility. Usually, vehicle count is performed by inductive loops which provide precise measurements and some vehicle classification. The major drawback of inductive loops is that they are very intrusive in the road surface and therefore require a rather long and expensive installation procedure. Furthermore, maintenance also requires intervention on the road pavement and therefore is not sustainable in most urban scenarios. Radar-based sensing systems are also used for vehicle counting and simple analytics but in cases of congestions they generally exhibit deteriorated performance. In the last years there has been interest in video-based counting system based on imaging devices, also embedded. Some solutions, such as Traficam [11], are commercially available and provide vehicle count in several lanes at an intersection. A version of Traficam working in the infrared spectrum is also available. Besides vehicle counting, traffic management can include the extraction of other flow parameters, e.g. discriminating the components of flow generated by different vehicle classes (car, track, buses, bike and motorbikes) and assessing the transit speed of each detected vehicle.

Another interesting topic is the monitoring of parking slots. Indeed, although there are several commercial parking slot monitoring solutions, most of them are only suitable for structured and closed parking lots, often requiring great installation costs to be adapted to already existing parking facilities. Visual nodes, instead, are flexible for application to several scenarios, including roadside parking spaces. The visual nodes can then provide information pertaining to the availability or not of a single parking space. This might be useful, for example, in the monitoring of special spaces, such as disabled space or spaces featuring electric vehicle charging station.

From this brief survey of urban scenario applications, we might argue that pervasive technologies based on vision turn out to be of interest when (1) there are some semantics to be understood that cannot be acquired solely on the basis of scalar sensors; (2) there is no possibility or no sufficient revenue in actuating installation of tethered technologies, such as intrusive sensor or high-end devices; and (3) there is the need of a scalable architecture, capable of covering a metropolitan area. Since computer vision is not application specific, an additional feature of a VSN is represented by the fact that it can be re-adapted to the changing urban environment and reconfigured even for supporting new scene understanding tasks by just updating the vision logics hosted in each sensor. Conversely, scalar sensors (like inductive loops) and specific sensors like radar have no flexibility in providing information different from the one they were built for.

In summary, with respect to more conventional ITS, that are often limited to close and rich systems, pervasive technologies based on VSN can thus provide a cost-effective collaborative sensing infrastructure which has intrinsic scalability features (since the architecture is made out of logical islands corresponding to VSN segments), can be adapted to several – even unstructured – scenarios and employs advanced yet low-cost technologies. Thus VSN may be exploited at several levels, impacting on transportation systems to be set up in small, mid-size and big cities as well as in unstructured road networks.

9.3 Embedded Vision Nodes

Following the trends in low-power processing, wireless networking and distributed sensing, VSN are experiencing a period of great interest, as shown by the recent scientific production [12]. A VSN consists of tiny visual sensor nodes called camera nodes, which integrate the image sensor, the embedded processor and a wireless RF transceiver. The large number of camera nodes forms a distributed system where the camera nodes are able to process image data locally (in-node processing) and to extract relevant information, to collaborate with other-cameras – even autonomously – on the application specific task, and to provide the system user with information-rich descriptions of the captured scene.

9.3.1 Features of Available Vision Nodes

In the last years, several research projects produced prototypes of embedded vision platforms which may be deployed to build a VSN. Among the first experiences, the Panoptes project [13] aimed at developing a scalable architecture for video sensor networking applications. The key features of the Panoptes sensor are a relatively low power and high-quality video capturing device, a prioritizing buffer management algorithm to save power and a bit-mapping algorithm for the efficient querying and retrieval of video data. Nevertheless the size of the sensor, its power consumption, and its relatively high computational power and storage capabilities makes the Panoptes sensor more akin to smart high-level cameras than to untethered low-power low-fidelity sensors. The Cyclops project [14] provided another representative smart camera for sensor networks. The camera nodes are equipped with a low-performance ATmega128 8-bit RISC microcontroller. From the storage memory point of view the system is very constrained, with 128 KB of Flash program memory and only 4 KB of SRAM data memory. The CMOS sensor supports three image formats of 8-bit monochrome, 24- bit RGB colour, and 16-bit YCbCr colour at CIF resolution (352 × 288). In the Cyclops board, the camera module contains a complete image processing pipeline for performing demosaicing, image size scaling, colour correction, tone correction and colour space conversion.

In the MeshEye project [15] an energy-efficient smart camera mote architecture was designed, mainly with intelligent surveillance as target application. MeshEye mote has an interesting special vision system based on a stereo configuration of two low-resolution low-power cameras, coupled with a high resolution colour camera. In particular, the stereo-vision system continuously determines position, range, and size of moving objects entering its fields of view. This information triggers the colour camera to acquire the high-resolution image subwindow containing the object of interest, which can then be efficiently processed. Another interesting example of low-cost embedded vision system is represented by the CMUcam series [16], developed at the Carnegie Mellon University. More precisely the third generation of the CMUcam series has been specially-designed to provide an open-source, flexible and easy development platform with robotics and surveillance as target applications. The hardware platform is more powerful with respect to its predecessors and may be used to equip low-cost embedded system with simple vision capabilities, so as to obtain smart sensors. The hardware platform is constituted by a CMOS camera, an ARM7 processor and a slot for MMC cards. Standard RF transceiver (e.g. TELOS mote) can be easily integrated. CMUcam4 is now on the market, featuring a Parallax P8X32A and an Arduino compatible shield.

More recently, the CITRIC platform [17] integrates in one device a camera sensor, a CPU (with frequency scalable up to 624 MHz), a 16 MB Flash memory and a 64 MB RAM. Such a device, once equipped with a standard RF transceiver, is suitable for the development of VSN. The design of the CITRIC system allows performing moderate image processing task in-network that is along the nodes of the network. In this way, there are less stringent issues regarding transmission bandwidth than with respect to centralized solutions. Such results have been illustrated by three sample applications, namely (1) image compression, (2) object tracking by means of background subtraction and (3) self-localization of the camera nodes in the network. The aforementioned electronics projects are examples of existing devices that can be turned into sensor nodes of a visual wireless sensor network. In Section 9.5 we will present an alternative smart camera prototype.

9.3.2 *Computer Vision on Embedded Nodes*

Embedded nodes equipped with an image sensor need special computer vision algorithms to be turned into actual smart cameras. The versatility of computer vision offers the possibility to tackle a great range of problems, by drawing on the extensive literature on the subject. Indeed for most of computer vision tasks, such as change detection, object detection, object recognition, tracking, and image fusion for multiview analysis, there exists an arsenal of already implemented methods (see [18] for a survey of change detection algorithms); however, most of the techniques currently available are not suitable to be used in VSN. Indeed, as shown by the examples reported in previous sections, embedded nodes usually have very constrained memory and computational power. Sometimes microcontroller architectures are also used, where floating point operation are not natively supported. In addition, power consumption is often limited in self-powered or battery-powered sensors: intensive operations might reduce autonomy below acceptable levels. For these reasons, conventional computer vision pipeline used in standard centralized infrastructure cannot be used on VSN, but a redesign of the employed algorithms is necessary. The redesign may range from an optimization for the embedded architecture (use of lookup tables, approximation in computations and introduction of heuristics) to more drastic changes in the pipeline in order to implement a more lightweight approach.

Some attempts to employ nontrivial image analysis methods over VSN have been done. For example [19] presents a VSN able to support the query of a set of images in order to search for a specific object in the scene. To achieve this goal, the system uses a representation of the object given by the Scale Invariant Feature Transform (SIFT) descriptors [20]. SIFT descriptors are indeed known to support robust identification of objects even among cluttered background and under partial occlusion situations, since the descriptors are invariant to scale, orientation, affine distortion and partially invariant to illumination changes. In particular, using SIFT descriptors allows retrieving the object of interest from the scene, no matter at which scale it is imaged. Interesting computer algorithms are also provided on the CMUcam3 vision system. Besides basic image processing filters (such as convolutions), methods for real-time tracking of blobs on the base either of colour homogeneity or frame differencing are available. A customizable face detector is also included. Such detector is based on a simplified implementation of Viola-Jones detector [21], enhanced with some heuristics to further reduce the computational burden. For example, the detector does not search for faces in the regions of the image exhibiting low variance. Machine-learning classifiers, such as the Viola-Jones detector, are very useful for deployment on embedded nodes, in order to provide a semantic interpretation of the scene. Indeed, the automatic detection of semantic concepts in videos and images represents an attempt to overcome the semantic distance between machines and humans, a distance that can be defined as the lack of coincidence between the information that one can extract from the sensor data and the interpretation that this same data can give a user in a given situation [22]. Bridging this gap with vision logics that can automatically recognize certain semantic concepts (such as *car*, *person*, or *obstacle*) is the real strength that makes VSN unparalleled with respect scalar sensors.

The detection of basic concepts can be performed using supervised learning methods, where a sufficient set of labelled data (annotated so that they contain, or do not contain, the concept to be detected) is used in a training phase to learn a model of the concept. The system learned in a supervised manner (e.g. a Support Vector Machine, SVM [23]) extracts some features of the images and their labels as input and learns a relationship model between these visual features and the concept. We can then classify new images not used in the training process using the learned model. The low-level visual features typically used are the colour histograms depicting the image or parts of the image, histograms of gradients, points of interest [20], edges, motion and depth to cite a few.

In more complex cases, it is required to detect events instead of simple objects; events can be formally represented and recognized as a set of objects (including people) interacting in time and space, e.g. a group of pedestrian crossing the street, a load loss, a car accident and car park surfing for instance. The regions in images of a video sequence are labelled by objects, and the spatial relationships between objects changes between images as a result of their interactions [24]. Machine learning algorithms require a preliminary (and generally computational intensive) learning phase to produce a trained classifiers. Clearly, when the methods are to be deployed on a VSN, the preliminary learning phase may be accomplished off-site, while only the already trained detectors need to be ported to the visual nodes. Among machine learning methods, a common and efficient one is based on the sliding windows approach; namely rectangular subwindows of the image are tested sequentially, by applying a binary classifier able to distinguish whether they contain an instance of the object class or not.

A priori knowledge about the scene or – if available – information already gathered by other nodes in the network may be employed to reduce the search space either by (a) disregarding some region in the image and (b) looking for rectangular regions within a certain scale range (e.g. rectangular regions covering less than 30% of the whole image area). For example, since licence plates have standard sizes, if we know roughly the scale of the image, we could expect to observe a plate only if the size in pixel of the area is compatible with the actual physical size. Regarding the binary classifiers themselves, among various possibilities, the Viola-Jones method is particularly appealing for use on VSN. Indeed, such a classifier is based on the use of the so-called rejection cascade. A window which fails to meet the acceptance criterion in some stage of the cascade is immediately rejected and no further processed. In this way, only detection should go through the entire cascade. The cascade also permits adapting the response of the detector to the particular use of its output in the network, also in a dynamical fashion, in order to properly react to changes in internal and external conditions. First of all, the trade-off between reliability of detections and needed computational time may be controlled by adaptive real-time requirements of the overall network. Indeed, the detector may be interrupted at an earlier stage in the cascade, thus producing a quick, even, though less reliable output, which in any case may be sufficient for solving the current decision-making problem. In the same way, by controlling the threshold in the last stage of the cascade, the VSN may dynamically select the optimal trade-off between false alarm rate and detection rate needed in a particular context.

One advantage of VSN with respect to monocular systems is the fact that they can inherently exploit multiview information. Due to bandwidth and efficiency considerations, however, images cannot be routinely shared on the network, so that no dense computation of 3D properties (like disparity maps and depth) can be made over a VSN.

Nevertheless, the static geometrical entities observed in the scene may be suitably codified during the setup of the acquisition system. In addition, specially designed references may be introduced in the scene for obtaining an initial calibration of the views acquired by each camera, thus permitting us to find geometrical correspondences among regions or points of interest seen by different nodes. To this end, a coordinator node, aware of the results of such a calibration step, may be considered, so as to translate events from image coordinates to physical world coordinates. Such an approach may produce more robust results as well as a richer description of the scene. Such ideas are used for tackling the parking lot monitoring problem by using multiple cameras and a special middleware layer for complex event composition (Section 9.4).

9.4 Implementation of Computer Vision Logics on Embedded Systems for ITS

In this section, two sample ITS applications based on computer vision over VSN are considered. The first one concerns the estimation of vehicular flows and is based on a lightweight computer vision pipeline that is dissimilar from the conventional one used on standard architectures. In the second sample application, an approach to parking lot monitoring is presented; here the vision nodes can cooperate among each other for producing more accurate and robust results, performing an inter-node decision regarding parking slot occupancy status. The inter-node decision logics can be implemented in an Internet of Things (IoT) framework, e.g. by using the middleware for VSN presented in Chapter 8 by Petracca *et al.* The presented applications are based on and extend previous work [25,26].

9.4.1 Traffic Status and Level of Service

The analysis of traffic status and the estimation of level of service are usually obtained by extracting information on the vehicular flows in terms of passed vehicles, their speed and typology. Conventional pipelines start with (1) background subtraction and move forward to (2) vehicle detection, (3) vehicle classification, (4) vehicle tracking and (5) final data extraction. On VSN, instead, it is convenient to adopt a lightweight approach; in particular data only in Region of Interest (RoI) is processed, where the presence of a vehicle is detected. On the basis of these detections, then, flow information is derived without making explicit use of classical tracking algorithms.

In more detail, background subtraction is performed only on small quadrangular RoIs. Such a shape is sufficient for modelling physical rectangles under perspective skew. In this way, when low-vision angles are available (as common in urban scenarios), it is possible to deal with a skewed scene even without performing direct image rectification, which can be computationally intensive on an embedded sensor. The quadrangular RoI can be used to model lines on the image (i.e. a 1 pixel thick line) either.

On such RoI, lightweight detection methods are used to classify a pixel as *changed* (in which case it is assigned to the foreground) or *unchanged* (in which case it is deemed to belong to the background). Such a decision is obtained by modelling the background. Several approaches are feasible. The simplest one is represented by straightforward *frame differencing*. In this approach, the frame previous to the one that is being processed is taken as background. A pixel is considered changed if the frame difference value is bigger than a threshold. Frame differencing is one of the fastest methods but has some cons in ITS applications; for instance, a pixel is considered changed twice: first when a vehicle enters and secondly when it exits from the pixel area. In addition, if a vehicle is homogeneous and it is imaged in more than one frame, it might not be detected in the frames after the first.

Another approach is given by *static background*. In this approach, the background is taken as a fixed image without vehicles, possibly normalized to factor illumination changes. Due to weather, shadow, and light changes, the background should be updated to yield meaningful results in outdoor environments. However, strategies for background update might be complex; indeed it should be guaranteed that the scene is without vehicles when updating. To overcome these issues, algorithms featuring *adaptive background* are used. Indeed this class of algorithms is the most robust for use in uncontrolled outdoor scenes. The background is constantly updated fusing the old background model and the new observed image. There are several ways of obtaining adaptation, with different levels of computational complexity. The simplest is to use an *average image*. In this method, the background is modelled as the average of the frames in a time window. Online computation of the average is performed. Then a pixel is considered changed if it is more different than a threshold from the corresponding pixel in the average image. The threshold is uniform on all the pixels. Instead of modelling just the average, it is possible to include the standard deviation of pixel intensities, thus using a statistic model of the background as a single *Gaussian distribution*. In this case, both the average and standard deviation images are computed by an online method on the basis of the frames already observed. In this way, instead of using a uniform threshold on the difference image, a constant threshold is used on the probability that the observed pixel is a sample drawn from the background distribution which is modelled pixel by pixel as a Gaussian. *Gaussian Mixture Models* (GMM) are a generalization of the previous method. Instead of modelling each pixel in the background image as a Gaussian, a mixture of Gaussians is used. The number k of Gaussians in the mixture is a fixed parameter of the algorithm. When one of the Gaussian has a marginal contribution to the overall probability density function, it is disregarded and a new Gaussian is instantiated. GMM are known to be able to model changing background even in cases where there are phenomena such as trembling shadows and tree foliage [27]. Indeed in those cases pixels clearly exhibit a multimodal distribution. However, GMM are computationally more intensive than a single Gaussian. *Codebooks* [28] are another adaptive background modelling technique presenting computational advantages for real-time background modelling with respect to GMM. In this method, sample background values at each pixel are quantized into codebooks which represent a compressed form of background model for a long image sequence. That allows us to capture even complex structural background variation (e.g. due to shadows and trembling foliage) over a long period of time under limited memory.

Several ad hoc procedures can be envisaged starting with the methods just described. In particular, one important issue concerns the policy by which the background is updated or not. In particular, if a pixel is labelled as foreground in some frame, we might want this pixel not to contribute in updating the background or for it to contribute to a lesser extent. Similarly, if we are dealing with a RoI, we might want to fully update the background only if no change has been detected in the RoI; if a change has been detected instead, we may decide not to update any pixel in the background.

The data extraction procedure starts taking in input one or more RoIs for each lane suitably segmented in foreground/background by the aforementioned methods. When processing the frame acquired at time t, the algorithm decides if the RoI R_k is occupied by a vehicle or not. The decision is based on the ratio of pixels changed with respect to the total number of pixels in R_k, i.e. $a_k(t) = \#(\text{changed pixels in } R_k) / \#(\text{pixels in } R_k)$. Then $a_k(t)$ is compared to a threshold τ in order to evaluate if a vehicle was effectively passing on R_k. If $a_k(t) > \tau$ and at time $t-1$ no vehicle was detected, then a new transit event is generated. If a vehicle was already detected instead at time $t-1$, no new event is generated but the time length of the last created event is incremented by one frame. When finally at a time $t + k$ no vehicle is detected (i.e. $a_k(t) < \tau$), the transit event is declared as accomplished and no further updated. Assuming that the vehicle speed is uniform during the detection time, the number of frames k in which the vehicle has been observed is proportional to the vehicle length and inversely proportional to its speed. In the same way, it is possible to use two RoIs R_1 and R_2, lying on the same lane but translated by a distance Δ, to estimate the vehicle speed (see Figure 9.1).

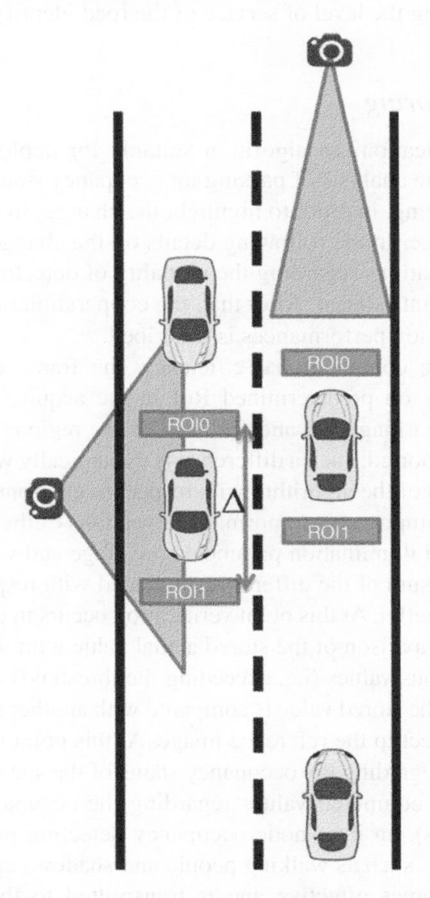

Figure 9.1 RoI configuration for traffic flow analysis.

Indeed, if there is a delay of δ frames, the vehicle speed can be estimated as $v = \Delta/(\delta * \nu)$ where ν is the frame rate. The vehicle length can in turn be estimated as $l = k/v$. Clearly the quality of these estimates varies greatly with respect to several factors, and is in particular due to (a) frame rate and (b) finite length of RoIS. Indeed, the frame rate generates a quantization error which leads to the estimation of the speed range; therefore the approach cannot be used to compute the instantaneous speed. In the case of (b), an ideal detection area is represented by a detection line, having a length equal to zero. Otherwise, a localization error affects any detection, i.e. it is not know exactly where the vehicle is inside the RoI at detection time. The use of a 1-pixel thick RoIs alleviates the problem but it results in less robust detections. This problem introduces some issues both in vehicle length and speed computations, because in both formulas we use the nominal distance Δ and not the precise (and unknown) distance between the detections. This is the drawback in not using a proper tracking algorithm in the pipeline, which would, however, require computational resources not usually available on embedded devices. Nevertheless, it is possible to provide a speed and size class for each vehicle. For each speed and vehicle class a counter is used to accumulate the number of detections. Temporal analysis on the counter is sufficient for estimating traffic typologies, average speed and analysing the level of service of the road identifying possible congestions.

9.4.2 Parking Monitoring

As a second sample application, an algorithm suitable for deployment on VSN has been studied and designed for the analysis of parking lot occupancy status. The approach followed is based on frame differencing, in order to highlight the changes in the RoIs, with respect to a background reference image. In the following details on the change detection algorithm will be given, and then specifications regarding the modality of detecting the occupancy rate of a single parking slot will be introduced. After that, the cooperation of nodes within the VSN in order to improve the detection performances is described.

In order to improve the computational efficiency, the frame differencing for detecting changes is performed only on predetermined RoI in the acquired frame. Each of the RoI corresponds to a specific parking slot, and for each of the regions the absolute accumulated pixel-wise differences is reported; such a difference is dynamically weighted in order to correct and improve the robustness of the algorithm with respect to environmental light changes.

In order to perform this improvement, normalized versions of the images are computed and used, with respect to global illumination parameters (average and variance of both the current and reference image). The sum of the differences is scaled with respect to the size of the RoI, and finally it is stored in a buffer. At this point verification occurs in order to detect the eventual change. In particular, a comparison of the stored actual value with the historical values allows filtering out possible spurious values (i.e. exceeding the threshold) due to e.g. the presence of shadows. In the same way the stored value is compared with another threshold in order to detect possible changes with respect to the reference image. At this point the algorithm yields a first outcome which is a value regarding the occupancy status of the specific parking slot.

Once the algorithm has computed values regarding the occupancy for each parking slot (corresponding to the RoIs), an intra-node occupancy detection process occurs. In order to avoid transitory events (e.g. such as walking people and shadows casted by external objects), the occupancy status becomes effective and is transmitted to the VSN, only after being observed consecutively for a specific number of acquired frames.

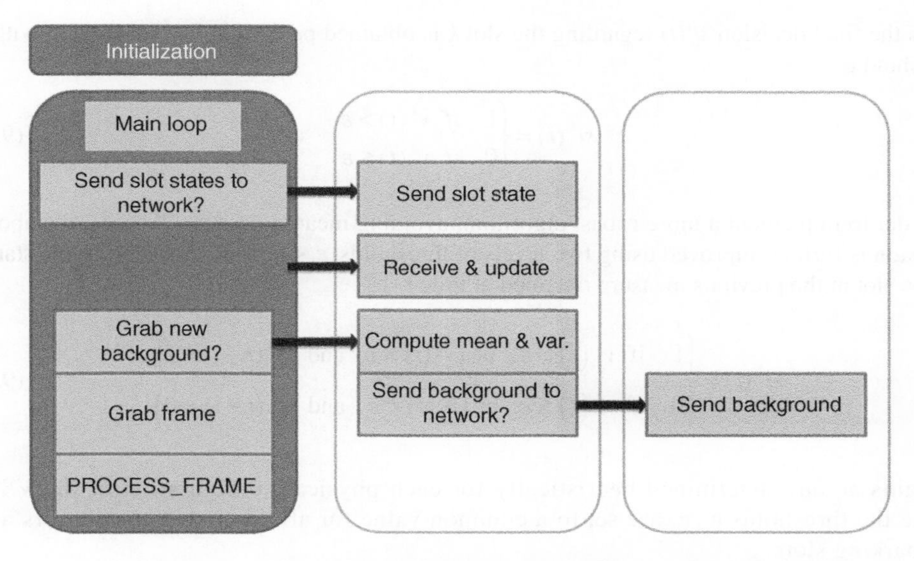

Figure 9.2 Flow chart representation of the parking lot occupancy algorithm.

For each parking slot, the algorithm yields a confidence value in the range [0.255], meaning that values next to 0 represent almost no change detected with respect to the reference value, and thus the slot is likely to be free; higher values, on the other hand, indicate that main changes have occurred in the observed scene, and thus the slot is likely to be occupied. Figure 9.2 shows the flow chart of the algorithm.

At a higher level of the VSN, the confidence values produced by the single nodes as a 256 levels number should be transformed to binary values corresponding either to free or busy parking slots, thus taking a final decision regarding the parking availability.

To this end, local confidence values will be propagated through the VSN thanks to a middleware layer. In particular when a parking space is monitored by more than one sensor node, the final decision regarding its occupancy is obtained at an inter-node level.

In more detail, this final decision is obtained by aggregating all the confidence values produced by the different nodes (which are statically dislocated and have static tables of the monitored parking slots). If a slot k is monitored by $n=n(k)$ sensor nodes, and being $v_1^k(t),\cdots,v_n^k(t)$ the confidence values measurements from each single sensor node at time t, then the aggregated measure is computed as:

$$v^k(t)=\sum_{i=1}^{n}\omega_{i,k}v_i^k(t) \tag{9.1}$$

Where the $\omega_{i,k}$ are the nonnegative weights and:

$$\sum_{t=1}^{n}\omega_{i,k}=1 \tag{9.2}$$

Thus the final decision $st^k(t)$ regarding the slot k is obtained performing a comparison with a threshold ε:

$$st^k(t) = \begin{cases} 1 & if\ v^k(t) > \varepsilon \\ 0 & if\ v^k(t) \leq \varepsilon \end{cases} \tag{9.3}$$

In order to implement a more robust algorithm, avoiding meaningless oscillations, the above decision is further improved using two levels of thresholds $\varepsilon_1 < \varepsilon_2$, and considering the status of the slot at the previous measure obtained at time t–1:

$$st^k(t) = \begin{cases} 1 & \text{if } v^k(t) > \varepsilon_2 \text{ or } \left(v^k(t) > \varepsilon_1 \text{ and } st^k(t-1) = 1\right) \\ 0 & \text{if } v^k(t) \leq \varepsilon_1 \text{ or } \left(v^k(t) < \varepsilon_2 \text{ and } st^k(t-1) = 0\right) \end{cases} \tag{9.4}$$

Weights $\omega_{i,j}$ are determined heuristically for each physical configuration of the VSN, while the thresholds $\varepsilon_1, \varepsilon_2$ are set to a common value for all the nodes, the sensors and the parking slots.

9.5 Sensor Node Prototype

In this section the design and development of a sensor node prototype based on VSN concepts is presented. This prototype is particularly suited for urban application scenarios. In particular, the prototype is a sensor node having enough computational power to accomplish the computer vision task envisaged for urban scenarios as described in the previous section. Along with such computational power, the prototype is completed with a networking board in order for it to be included within the sensor network, and for dispatching and receiving data, through a deployed event-based middleware. Finally an energy harvesting module, implemented to keep the node autonomous, is included and described.

In the following an overview of the prototype implementation is given, starting with the description of the architecture of the implemented prototype, followed by the features of the single hardware components, namely: the vision board, the networking board, characteristics of the acquisition sensor and the energy harvesting module. Thereafter the layout of designed and implemented board will be presented.

For the design of the prototype an important issue to follow has been the use of low-cost technologies. In particular, the node is using sensors and electronic components at low cost, so that once engineered, the device can be manufactured at low cost in large quantities. In the design and planning of the architectural side, an important issue is represented by the ease of installation of the device; thus the protective shield that has been considered for the sensor node is compact but able to accommodate all components of the device.

Going into detail, the single sensor node can be divided into two main parts: the *vision board* equipped with the camera sensor and the logics for image analysis and the *networking board* connected to the wireless communication module (RF Transceiver).

They have respectively the following tasks: (1) acquire and process images and (2) control the device to coordinate the processes of transmission of all the information extracted about the scene.

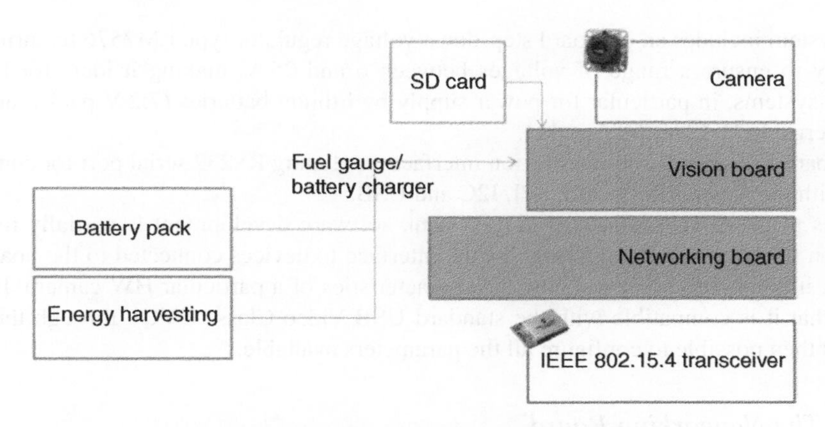

Figure 9.3 Architecture of the sensor node.

Other components of the sensor node are given by the power supply system that controls charging and permits choice of optimal energy savings policies. The power supply system includes the battery pack and a module for harvesting energy, e.g. through a photovoltaic panel. Figure 9.3 shows the last design of the sensor node architecture.

9.5.1 *The Vision Board*

For the realization of the vision board, an embedded Linux architecture has been selected in the design stage for providing enough computational power and ease of programming. A selection of ready-made Linux based prototyping boards had been evaluated with respect to computing power, flexibility/expandability, price/performance ratio and support. For example, the following candidates were considered Raspberry Pi Model B (ARM11, 700 MHz) [29], Phidget SBC (ARM9, 400 MHz) [30] and BeagleBone – TI Sitara AM3359 (Cortex A8, 720 MHz) [31].

All these candidates have as common disadvantages high power consumption and the presence of electronic parts which are not useful for the tasks foreseen here.

It has therefore been decided to design and realize a custom vision component by designing, printing and producing a new PCB. The new PCB was designed in order to have the maximum flexibility of use while maximizing the performance/consumption ratio. A good compromise has been achieved by using a *Freescale* CPU based on the ARM architecture, with support for MMU-like operating systems GNU/Linux.

This architecture has the advantage of integrating a Power Management Unit (PMU), in addition to numerous peripherals interface, thus minimizing the complexity of the board. Also the CPU package of type TQFP128 helped us to minimize the layout complexity, since it was not necessary to use multilayer PCB technologies for routing. Thus, the board can be printed also in a small number of instances. The choice has contributed to the further benefit of reducing development costs, in fact, the CPU only needs an external SDRAM, a 24 MHz quartz oscillator and an inductance for the PMU. It has an average consumption, measured at the highest speed (454 MHz), of less than 500 mW.

The system includes an on-board step-down voltage regulator type LM2576 featuring high efficiency to ensure a range of voltages between 6 and 25 V, making it ideal for battery-powered systems, in particular for power supply by lithium batteries (7.2 V packs) and lead acid batteries (6 V, 12 V, 24 V packs).

The board has several communication interfaces including RS232 serial port for communication with the networking board, SPI, I2C and USB.

Thanks to the GNU/Linux operating system, software development is partially relieved, relying on libraries already available for the interface to devices connected to the board. For example, it is not necessary to know the characteristics of a particular HW camera, but it is enough that it is compatible with the standard USB Video Class (UVC); through the UVC API, it is then possible to configure all the parameters available.

9.5.2 The Networking Board

For the realization of the networking board, it has been decided to use a microcontroller-based device with a 32 bit architecture. For radio communication, a transceiver compliant with IEEE 802.15.4 has been required in line with modern approaches to IoT. With regards to the software, it has been decided to adopt Contiki [32] as the operating system. Contiki provides the uIPv6 stack, which deals with IPv6 networking. The IPv6 stack also contains the 6LoWPAN header compression and adaptation layer for IEEE 802.15.4 links. Therefore the operating system is well capable of supporting an event-based middleware for VSN. An analysis of the boards available on the market has shown that there exist devices satisfying all the above requirements. In particular, the Evidence SEED-EYE board [33] has been selected, which is particularly suited for implementing low-cost multimedia WSN.

9.5.3 The Sensor

For the integration of a camera sensor on the vision board, some specific requirements were defined in the design stage for providing easiness of connection and to the board itself and management through it, and capability to have at least a minimal performance in difficult visibility condition, i.e. night vision. Thus the minimal constraints were to be compliant with USB Video Class device (UVC) and the possibility to remove IR filter or capability of Near-IR acquisition. Moreover, the selection of a low-cost device was an implicit requirement considered for the whole sensor node prototype. Among a very large list of UVC compliant devices [34], an easy-to-buy and cheap camera was selected (TRUST SpotLight Webcam [35]). Moreover the camera is equipped with an IR filter, designed to reduce the noise from IR light sources, which is easily removable for our purposes of acquiring images even in low light conditions.

9.5.4 Energy Harvesting and Housing

The previously described boards and camera are housed into an IP66 shield. Another important component of the node is the power supply and energy harvesting system that controls charging and permits the choice of optimal energy savings policies. The power supply system

Figure 9.4 General setup of the VSN sensor node prototype with energy harvesting system.

includes the lead (Pb) acid battery pack and a module for harvesting energy through photovoltaic panel.

In Figure 9.4, the general setup of a single node with the electric connections for the involved components is shown. Notice that, in order to implement energy savings policies, the vision board has also been used to measure the charging status of the batteries. To this end, an ADC Conditioning module has been used to adapt the voltage level of the power supply system to the voltage range of the vision board ADC input.

9.5.5 *The Board Layout*

After having introduced the main features of the selected hardware, this section presents the layout of the vision board. Indeed the creation of a vision board having the basic features as previously described has required the design of a schematic in which to allocate and organize all modules and components required for its operation. As a result, in Figure 9.5 the layout of the vision board is shown, respectively through a 3D rendering of the board and a printed sample of the board.

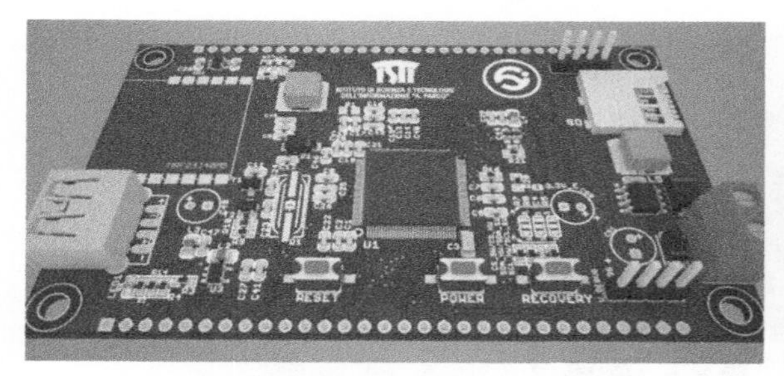

Figure 9.5 The 3D rendering of the vision board and a printed sample of the board.

9.6 Application Scenarios and Experimental Results

Nowadays most of the available sensors for traffic monitoring are usually focused on structured environments and are based on tethered sensors. Moreover, their cost usually prevents their massive use for covering large areas, since the ratio cost/benefit is not favourable. The proposed and developed embedded system and low-cost camera sensor make it possible to conceive sensor-based pervasive intelligent systems centred on image data. Following this concept the application scenarios identified for an urban area case study are capable of covering several different situations that typically occur in this kind of area: restricted traffic area, ecological area, intermodal node services, intermodal parking, car sharing services, electric vehicle charging stations. In particular, different scenarios for the evaluation of both parking lots and traffic flow have been set up. For the parking lot scenario the set-up consists in a set of VSN nodes equipped with cameras having partially overlapping field of views. The goal was to observe and estimate the availability and location of parking spaces. A basic assumption was made on the geometry of the parking: each camera knows the positions of the parking slots under its monitoring. In addition, we assume that a coordinator node knows the full geometry of the parking lot as well as the calibration parameters of the involved cameras, in order to properly aggregate their outcomes.

For the traffic flow, the set-up consists in a smaller set of VSN nodes, which are in charge of observing and estimating dynamic real-time traffic related information, in particular regarding traffic flow and the number and direction of the vehicles, as well as giving a rough estimate about the average speed of the cars in the traffic flow.

Regarding the experimentation results, first data from the parking lot are presented and analysed; in Figure 9.6 an example of parking lot image is reported and then in Figure 9.7 the differences between the sensor values and the ground-truth recorded for a sample parking slot acquisition is presented, showing the good separation obtained between different events.

Regarding the traffic flow monitoring, two versions of the algorithm were implemented on the VSN. In the first, the solutions used three different frames (using frame differencing) to obtain a binary representation of the moving objects in the reference frame. Analysing the connected components, blobs are detected, and then it is verified whether these can be referred to objects moving through (i.e. traffic flow) a predefined RoI [25]. The final was designed to eliminate the analysis of connected components, using the algorithm presented in Section 9.4.

Figure 9.6 Car detection and analysis of parking lot occupancy status.

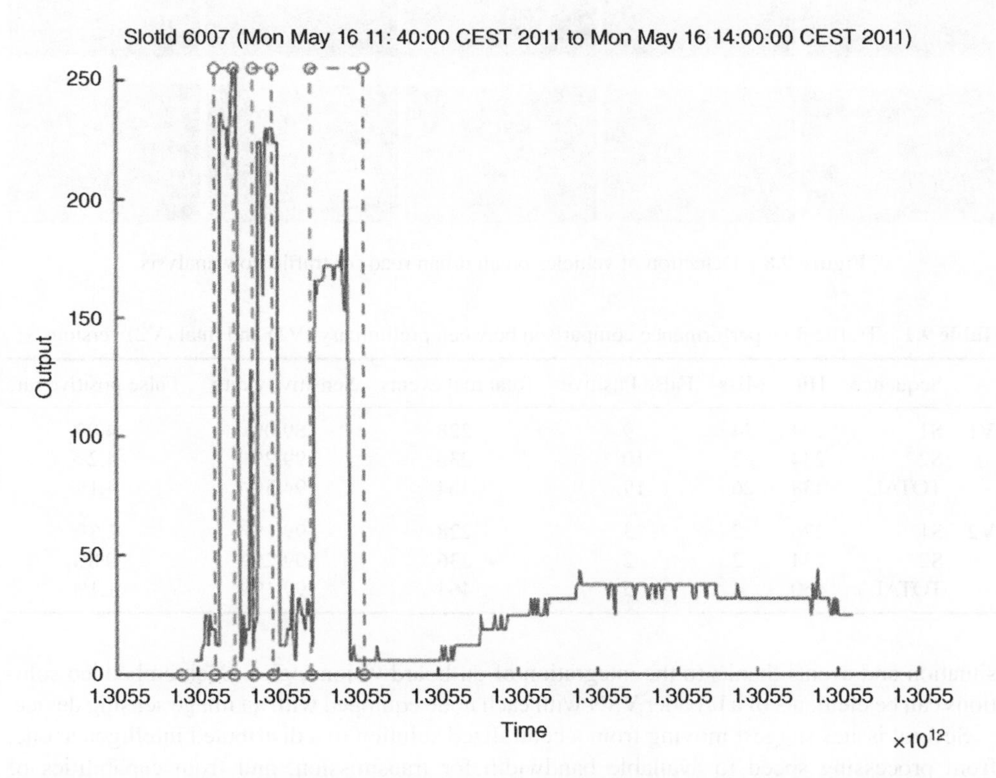

Figure 9.7 Data collected for a parking slot on the case study site. Sensor confidence values are shown in blue, while the ground-truth recorded is shown in red with circles representing change-event.

In Figure 9.8 a sample of the acquired and processed images for traffic flow analysis is reported and in Table 9.1 the results of the traffic flow case study are reported, showing the improvement in performance between the preliminary version and the final implemented solution.

9.7 Conclusions

In this chapter a scalable technological solution has been introduced for supporting ITS-related problems in an urban scenario. The survey mainly addressed embedded solutions for the realization of smart cameras that can be used to detect, understand and analyse traffic-related

Figure 9.8 Detection of vehicles on an urban road for traffic flow analysis.

Table 9.1 Traffic flow performance comparison between preliminary (V1) and final (V2) versions.

	Sequence	Hit	Miss	False Positive	Total real events	Sensitivity rate	False positive rate
V1	S1	204	24	9	228	89.0%	4.0%
	S2	234	2	10	236	99.2%	4.2%
	TOTAL	438	26	19	464	94.3%	4.1%
V2	S1	226	2	3	228	99.1%	1.3%
	S2	234	2	2	236	99.2%	0.8%
	TOTAL	460	4	5	464	99.1%	1.1%

situation and events thanks to the integration of on-board vision logics. Such embedded solutions can be elements of a broader VSN with each node equipped with an image sensing device.

Several issues suggest moving from a centralized solution to a distributed intelligence one, from processing speed to available bandwidth for transmission, and from capabilities of redundancy in a distributed solution to the autonomy granted to each distributed node. In such nodes, artificial intelligence and computer vision algorithms are able to provide autonomy and adaptation both to internal as well as to external conditions.

The smarter part is that these nodes are not just gatherers of information from the cameras, but they can extract significant and compact descriptors of the scene from the data acquired in the video stream. We then show that currently available techniques are not suitable to be used in those networks of vision due to the complexity of the algorithms and to the excessive hardware requirements; thus ad hoc algorithms needed to be designed and implemented.

After introducing the developed embedded vision node, the implementation of computer vision algorithms for smart cameras (in a VSN) has been addressed. In particular, the two sample ITS applications, for the analysis of traffic flow status and parking lot monitoring, were described. The designed and realized smart camera prototype was presented and the envisaged application scenarios described with the promising results obtained.

References

1. N. Buch, S.A. Velastin and J. Orwell (2011) A review of computer vision techniques for the analysis of urban traffic, *IEEE Transactions on Intelligent Transportation System* **12**(3), 920–39.
2. P. Remagnino, A.I. Shihab and G.A. Jones (2004) Distributed intelligence for multi-camera visual surveillance. *Pattern Recognition* **37**(4): 675–89.
3. J. Lopes, J. Bento, E. Huang, *et al.* (2010) Traffic and mobility data collection for real-time applications, *Proceedings of the IEEE Conference. ITSC*, 216–23.
4. H. Guo, Z. Wang, B. Yu, *et al.* (2011) TripVista: triple perspective visual trajectory analytics and its application on microscopic traffic data at a road intersection. *Pacific Visualization Symposium (PacificVis), 2011 IEEE*, 163–70.
5. Z. Wang; M. Lu; X. Yuan, *et al.* (2013) Visual traffic jam analysis based on trajectory data, visualization and computer graphics, *IEEE Transactions on Visualization and Computer Graphics*, **19**(12): 2159–68.
6. N. Buch, J. Orwel and S.A. Velastin (2010) Urban road user detection and classification using 3D wire frame models, *IET Computer Vision* **4**(2): 105–16.
7. J. Candamo, M. Shreve, D.B. Goldgof, *et al.* (2010) Understanding transit scenes: A survey on human behavior-recognition algorithms, *IEEE Transactions on Intelligent Transportation System* **11**(1): 206–24.
8. Ipsotek. Available at: http://www.ipsotek.com (last accessed 29 April 2015).
9. Xerox Technology Showcase at 21st ITS World Congress, Detroit, 2014. http://itsworldcongress.org/events/xerox-vehicle-passenger-detection-system (last accessed 29 April 2015).
10. Digital Recognition. Available at: http://www.digital-recognition.com (last accessed 29 April 2015).
11. Traficam. Available at: http://www.traficam.com (last accessed 29 April 2015).
12. D. Kundur, C.-Y. Lin and C.-S. Lu (2007) Visual sensor networks. *EURASIP Journal on Advances in Signal Processing*, Article ID 21515, 3 pp.
13. W. Feng, B. Code, E.C. Kaiser, *et al.* (2003) Panoptes: scalable low-power video sensor networking technologies, *Proceedings of the Eleventh ACM International Conference on Multimedia, Berkeley, CA, USA, 2–8 November 2003*, pp. 562–71.
14. M.H. Rahimi, R. Baer, O.I. Iroezi, *et al.* (2005) Cyclops: in situ image sensing and interpretation in wireless sensor networks, *Proceedings of the 3rd ACM Conference on Embedded Networked Sensor Systems (SenSys), 2005*, pp. 192–204.
15. S. Hengstler, D. Prashanth, Sufen Fong and H. Aghajan (2007) Mesheye: a hybrid-resolution smart camera mote, *Proceedings of the 6th International Conference on Information Processing in Sensor Networks, ACM, 2007*, pp. 360–9.
16. CMUcam. Available at: http://www.cmucam.org (last accessed 29 April 2015).
17. P. Chen, C. Boyer, S. Huang, *et al.* (2008) Citric: A low-bandwidth wireless camera network platform, *Proceedings of the ACM/IEEE International Conference on Distributed Smart Cameras, September 2008*, pp. 1–10.
18. R.J. Radke, S. Andra, O. Al-Kofahi and B. Roysam (2005) Image change detection algorithms: a systematic survey, *IEEE Transactions on Image Processing* **14**(3): 294–307.
19. T. Yan, D. Ganesan and R. Manmatha (2008) Distributed image search in camera sensor networks, *Proceedings of the 6th International Conference on Embedded Networked Sensor Systems, SenSys 2008, Raleigh, NC, USA, 5–7 November 2008*, ACM, pp. 155–68.
20. Lowe, D.G. (2004) Distinctive image features from scale-invariant keypoints, *International Journal of Computer Vision* **60**(2): 91–110.
21. P.A. Viola and M.J. Jones (2004) Robust real-time face detection. *International Journal of Computer Vision* **57**(2): 137–54.
22. A.W.M. Smeulders, M. Worring, S. Santini, *et al.* (2000) Content-based image retrieval, the end of the early years, *IEEE Transactions on Pattern Analysis and Machine Intelligence*, pp. 1349–80.
23. C. Cortes and V. Vapnik (1995) Support-vector networks, *Machine Learning*, **20**: 273–97.
24. D. Ayers and R. Chellappa (2000) Scenario recognition from video using a hierarchy of dynamic belief networks, *Proceedings of the 15th International Conference on Pattern Recognition*, pp. 835–8.
25. M. Magrini, D. Moroni, G. Pieri and O. Salvetti (2012) Real time image analysis for infomobility, *Lecture Notes in Computer Science* **7252**: 207–18.
26. M. Magrini, D. Moroni, C. Nastasi, *et al.* (2011) Visual sensor networks for infomobility. In *Pattern Recognition and Image Analysis*, vol. **21**. London: Springer, pp. 20–9.
27. C. Stauffer and W.E. Grimson (1999) Adaptive background mixture models for real-time tracking, *Proceedings of the Conference on Computer Vision and Pattern Recognition, Fort Collins, CO, USA, 1999*, **2**: 246–52.

28. K. Kim, T. Chalidabhongse, D. Harwood and L. Davis (2004) Background modeling and subtraction by codebook construction, *IEEE International Conference on Image Processing (ICIP), 2004*, pp. 2–5.
29. Raspberry Pi. Available at: http://www.raspberrypi.org (last accessed 29 April 2015).
30. Phidgets board. Available at: http://www.phidgets.com (last accessed 29 April 2015).
31. Beagleboard. Available at: http://beagleboard.org (last accessed 29 April 2015).
32. Contiki OS. Available at: http://www.contiki-os.org (last accessed 29 April 2015).
33. SEED-EYE Board. Available at: http://www.evidence.eu.com/it/products/seed-eye.html (last accessed 29 April 2015).
34. UVC devices. Available at http://www.ideasonboard.org/uvc/ (last accessed 6 May 2014).
35. Trust Spotlight. Available at: http://www.trust.com/it-it/all-products/16429-spotlight-webcam (last accessed 29 April 2015).

Part 4

Data Processing Techniques at ITS

10

Congestion Prediction by Means of Fuzzy Logic and Genetic Algorithms

Xiao Zhang[1], Enrique Onieva[2], Victor C.S. Lee[1] and Kai Liu[3]
[1] *City University of Hong Kong, Hong Kong*
[2] *University of Deusto, Bilbao, Spain*
[3] *Chongqing University, China*

10.1 Introduction

Traffic congestion causes energy waste, requires investment in public infrastructure, and threatens urban environmental quality. These challenges are even more pressing if the forecast of growth in transport is considered. The business costs caused by congestion will increase by approximately 50% by 2050, according to the Transport White Paper (European Commission, March 2011). Intelligent Transportation Systems (ITS) consider easing highway traffic congestion a significant issue to be investigated. Therefore, the prediction and identification of traffic congestion play a vital part in ITS, which intends to make it more efficient, safer and energetically sustainable. Accurate traffic prediction reporting can be adopted either by drivers to avoid traffic jams, or traffic management systems to take measures in advance to ensure traffic flow.

In the last few decades, the most commonly used techniques for traffic forecasting are on the basis of the Kalman Filter (KF) [1,2] and the Autoregressive Integrated Moving Average (ARIMA) [3,4]. Although these techniques can obtain good results, there are still some weaknesses. For example, KF tends to generate overestimation or underestimation that deteriorates the prediction accuracy when traffic conditions undergo very significant changes; on the other hand, ARIMA mainly targets single variable time-series data, instead of using multiple input variables.

Intelligent Transport Systems: Technologies and Applications, First Edition. Asier Perallos, Unai Hernandez-Jayo, Enrique Onieva and Ignacio Julio García-Zuazola.

With respect to the studies on multiple-variable datasets [5–8], the factors including traffic flow, occupancy and speed have been considered suitable for traffic conditions forecasting. Particularly, the findings in [9] revealed that the prediction accuracy of traffic flow and occupancy was higher than that of the speed. Results in [10] implied some phenomena that predictions based on traffic flow were more reliable by analysing real traffic data; and the use of occupancy indicated the traffic condition more accurate. Nevertheless, there are still several contradictory results, where the speed information is more acceptable than the flow and the occupancy information, because it is more effective and meaningful for the first-hand users [6].

Among all the studies on rich traffic datasets on highway scenarios, four input variables fit for diagnosing and forecasting traffic conditions have occupied most literatures: mainline flow, occupancy, speed, and ramp flow.

In recent literatures, soft computing techniques have been considered as powerful tools for traffic forecasting. These techniques mainly consist of Support Vector Machines (SVM) [11–13], Neural Networks (NN) [14–17], Fuzzy Rule-Based Systems (FRBS) [18,19], and Genetic Algorithms (GAs) [20]. Moreover, hybrid methods have also exported good results, such as the genetic optimization of NN [21] and other combinations [22,23].

In the aspect of shortcomings, due to the appropriate kernel function selection, too many input variables may come up with unreliable prediction by SVMs [12]. NN has obtained better effect on traffic condition prediction. However, the local optimum and generalization ability limit its function [15]. Moreover, the qualities of datasets and parameters have particular influence on its performance.

Fuzzy logic is often used to deal with complex traffic situations [24,25]. It can process uncertain data with simple rules. Traffic condition information, such as speed, occupancy and flow, are classified into finite categories, such as high, medium, and low. Then, rules are formulated to relevant traffic states with congestion detection output. Among the systems of Fuzzy Logic, FRBS is the most representative case.

GAs are well-known heuristic algorithms, which are widely used to explore a large search space for suitable solutions [26,27]. In addition to their ability to find approximate optimal solutions in complex search spaces, the generic code structure and independent performance features of GAs make them suitable for incorporating a priori knowledge. In the case of FRBSs, this a priori knowledge may be in the form of linguistic variables, fuzzy membership function parameters, fuzzy rules, number of rules, etc. These capabilities broaden the function of GAs in the development of learning and tuning FRBSs over the last few years [28], which is commonly called Genetic FRBS (GFRBS) [25,29].

The design of an FRBS is generally a time-consuming and complex process. When a traditional FRBS is faced with a high number of input variables, the number of rules increases exponentially while the obtained FRBS is barely accurate or interpretable. One feasible improvement is to decompose the fuzzy system into a hierarchical structure, which is known as hierarchical FRBS (HFRBS) [30,31]. It shows that the total number of rules grows only linearly with the number of input variables.

With regard to implementation of these systems, there are many proposals [30,32–35]. Some of them identify common parts of the rule set and generate them by producing submodules [32,36]. In other studies, the hierarchical layer of each module deals with an increase in the granularity of the input variables [33]. The researchers in [34] propose a limpid-hierarchical fuzzy system, which aims to overcome the drawback that the outputs of the intermediate layers do not possess physical meaning. The result presented in [35] gives an introduction to hierarchical fuzzy control system by using evolutionary algorithms. The researchers in [30] propose to use multi-objective evolutionary algorithms (MOEAs) to optimize HFRBS, with the objectives of reducing the size of the rules and improving accuracy. However, the curse of dimensionality is still an unsolved and difficult problem in fuzzy logic control theory [37].

This work is motivated by the circumstance that previous methods have their pros and cons, and previous approaches may not be able to deal with traffic congestion prediction with a high number of input variables over the durations of the prediction period. For the purpose of achieving more accurate and robust traffic prediction, we propose a Genetic Hierarchical FRBS (GHFRBS) capable of predicting traffic congestion in multiple prediction horizons.

The main contributions of the chapter are as follows: (1) with a hierarchical structure, this model allows the system to identify and rank input variables among a large number of input variables; (2) GHFRBS significantly reduces the complexity of the fuzzy rules; (3) GHFRBS performs a lateral tuning of the database of the fuzzy systems to improve the accuracy.

The rest of this chapter is organized as follows: Section 10.2 gives preliminary concepts of HFRBS. Section 10.3 introduces the implementation of GHFRBS. Section 10.4 gives the experimentation and evaluation results. Finally, Section 10.5 is the conclusion and future work.

10.2 Hierarchical Fuzzy Rule-based System (HFRBS)

In HFRBS, the number of rules is reduced by decomposing the FRBS into a set of simpler fuzzy subsystems linked hierarchically. In this hierarchical structure, the first level of the FRBS obtains an approximate output, and then it is tuned by the second-level FRBS. This procedure can be iterated in subsequent levels.

In order to predict congestion at a desired point, an HFRBS with serial distribution [30,38] is used for methodology development and discussions throughout this study, in order to predict congestion at a desired point.

Figure 10.1 presents an example of an HFRBS with 4 input variables (right), in comparison with a classical FRBS with 4 input variables (left). Assuming the use of three member functions (MFs) to fuzzify each input variable, and a complete rule base that covers each combination of variables, we observed that:

- The conventional FRBS (Figure 10.1 (left)) must be composed of $3^4 = 81$ rules with 4 antecedents each.
- However, using the HFRBS (Figure 10.1 (right)), each low-dimensional fuzzy system consists of $3^2 = 9$ rules. Therefore, the total number of rules is $3 \cdot 3^2 = 27$.

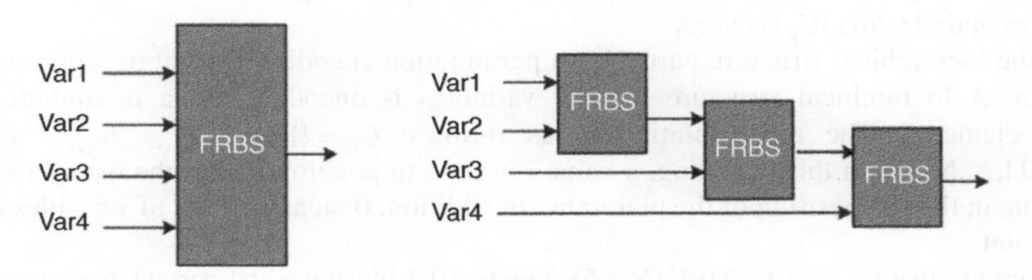

Figure 10.1 Fuzzy system examples: (left) nonhierarchical FRBS; (right) hierarchical FRBS.

This indicates that the system enjoys a significant reduction in the total number of rules due to its hierarchical structure.

In this way, suppose a fuzzy system with N input variables and M MFs for each variable; the rule base would be composed of M^N rules in a conventionally structured FRBS, and $M^2 \cdot (N-1)$ rules in a hierarchically structured FRBS. Thus, given the 26 input variables in this study, $3^{26} (= 2.5419 \cdot 10^{12})$ rules would be necessary in the nonhierarchical FRBS, while only $9 \cdot (26-1)(= 225)$ rules would be used in the HFRBS. In addition, in a hierarchical structure, typically, the most influential input variables are chosen as the system variables in the first level, while the next most important variables are chosen in the second level, and so forth [12,38].

In many traffic congestion studies, different variables, such as traffic flow, occupancy, speed, and their derivatives etc., have been used in a variety of prediction models. However, it is often debated which variables are better suited for this purpose. Therefore, another great advantage of HFRBS for traffic prediction is the ability of obtaining a better understanding of the usability of the available dataset, and determining which of its attributes are the most valuable.

10.3 Genetic Hierarchical Fuzzy Rule-based System (GHFRBS)

This section explains the algorithm proposed for dealing with the problem of defining and tuning an HFRBS. As mentioned in Section 10.2, a serial distribution of modules will be introduced in this chapter, where each FRBS in the hierarchy only uses two input variables. The first FRBS employs two external variables and the remaining ones use an internal and an external variable (Figure 10.1 (right)).

In this proposal, an evolutionary process which evolves the HFRBS incorporates a set of functions that allow us to tackle traffic prediction problem with a large number of variables. These functions are: (1) evolution of the hierarchical structure, including variable selection and ranking; (2) lateral tuning of the MFs used to codify the input variables in each fuzzy system in the HFRBS and (3) optimization of the rule base of each fuzzy system in the hierarchy.

In the following subsections, important aspects of the algorithm are introduced. Then the mechanisms of the algorithm framework and specific characteristics are described.

10.3.1 Triple Coding Scheme

A triple coding scheme for hierarchical structure (C_H), tuning (C_T), and rule-base consequences optimization (C_R) is used.

In the hierarchical structure part (C_H), a permutation encoding is used to represent the system. A hierarchical structure with N variables is encoded into a permutation of N+1 elements. The representation is as follows: $C_H = \{h_1, h_2, \ldots, h_i, \ldots, h_{N+1}\}$, where $h_i \in \{0, 1, \ldots, N\}$. With this encoding, a value j in the i th position means the use of the j th variable in the i th position of the hierarchy. In addition, 0 means no use of variables from this point.

Assuming that $C_H = \{4, 1, 5, 2, 0, 3\} (N = 5)$, Figure 10.2 illustrates the permutation-structural representation of the hierarchical fuzzy model.

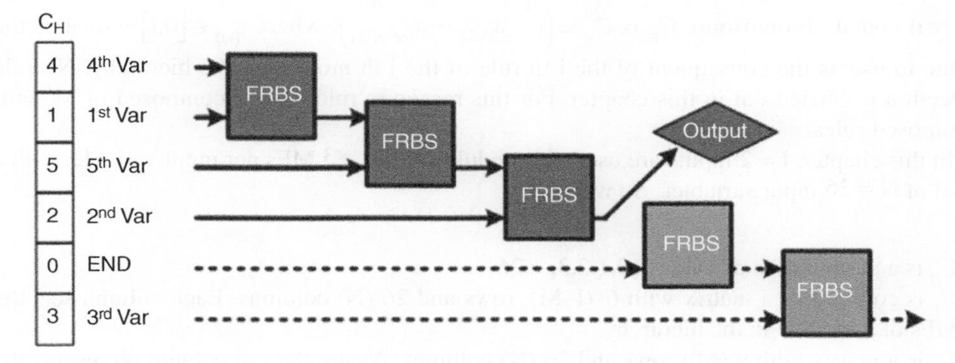

Figure 10.2 Permutation representation of the hierarchical model with {4,1,5,2,0,3}.

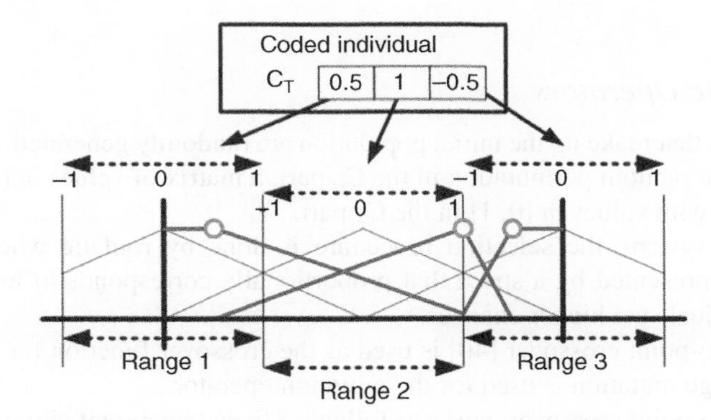

Figure 10.3 Example of lateral tuning codification.

In regard to the lateral tuning part (C_T), a real matrix with dimensions $I \cdot N \cdot M$ is adopted in the proposed model, where I is the number of input variables per module, M is the number of MFs to codify each input variable, and N is the total number of variables in the dataset. It could be represented formally as follows: $C_T = \{t_{(1,1,1)}, \cdots, t_{(1,1,M)}; \cdots, t_{(I,N-1,M)}\}$, where $t_{i,j,k} \in [-1,1]$ is used to modify the k th membership function of the i th input variable of the j th module in the hierarchy.

Instead of using the absolute position of the core of the MFs, the genetic lateral tuning process proposed in [39] is employed. Specifically, for each set of MFs, a real value makes the core of the MF move within a predefined range centered on the position it would occupy in an equally distributed configuration. Figure 10.3 shows an example to illustrate how the codification works. The initial uniformly distributed labels and the ranges where the MF can move are shown in the gray cell. The final distribution obtained for an individual coded as {0.5, 1, –0.5} is presented by the gray lines. Each value represents the displacement along the range in which the MF moves.

A real coded matrix is used for the rule-base consequences part (C_R), where we consider the number of labels per variable M, the total number of variables N, and the number of input variables for each FRBS in level I. The complete consequence base encoded as a part of real coded chromosome C_R is $C_R = (r_{(1,1)}, r_{(1,2)}, \cdots, r_{(M^I,N-1)})$, where $r_{(i,j)} \in [0,1]$ indicates the value to use as the consequent of the i th rule of the j th module in the hierarchy. No rule selection is carried out in this chapter. For this reason, a rule base is composed of M^I and-composed rules.

In this chapter, $I = 2$ inputs are used per module, and $M = 3$ MFs per input variable, with a total of $N = 26$ input variables. So we have:

- C_H is a permutation of values $\{0,1,2,3,\ldots 26\}$.
- C_T is codified as a matrix with 6 $(I \cdot M)$ rows and 26 (N) columns. Each column sets the MFs of a module in the hierarchy.
- C_R is a matrix with 9 (M^I) rows and 26 (N) columns. Again, the i th column represents the rule base to implement by the i th module in the hierarchy.

10.3.2 Genetic Operators

The chromosomes that make up the initial population are randomly generated. Each individual is initialized with a random permutation in the C_H part, a matrix of values in [−1, 1] in the C_T part, and a matrix with values in [0, 1] in the C_R part.

For the whole system, the selection procedure is done by roulette wheel, where each chromosome is represented by a space that proportionally corresponds to its fitness. In this way, better individuals get higher chances.

The ordered two-point crossover [40] is used as the crossover function for the hierarchical part C_H. Interchange mutation is used for the mutation operator.

The ordered two-point crossover works as follows. Given two parent chromosomes, firstly, two random crossover points are selected to cut each parent into a left, a middle, and a right part. One offspring inherits its left and right parts from the first parent, and its middle part is filled by the sequence of genes in the second parent. The second offspring is generated the other way round: using the left and right parts from the second parent and the middle of the first one. This is one of the most popular and effective crossovers for dealing with permutations [41]. After the crossover operation, the implementation of the mutation is done by the interchange operator. It randomly selects two genes within a specific range (a relatively small interval), in order to be further improved by fine-tuning the chromosome.

For the real coded parts of the individual (C_T and C_R), BLX$-\alpha$ crossover [42] and BGA mutation [43] are used. Both are widely employed in real-coded GA [44–46].

Given two parents $X = (x_1 \ldots x_m)$ and $Y = (y_1 \ldots y_m)$, for each i, BLX$-\alpha$ crossover creates two offspring by generating random values in the interval: $[\min(x\_i, y\_i) - \alpha \cdot |x\_i - y\_i|, \max(x\_i, y\_i) + \alpha \cdot |x\_i - y\_i|]$ with $\alpha \in [0,1]$.

For the same assumed $X = (x_1 \ldots x_m)$ as above, the x_i obtained from the BGA mutation operator is: $x_i' = x_i \pm \text{range}_i \cdot \sum_{k=0}^{15} (\alpha_k 2^{-k})$, where range_i defines the mutation range: it is set to $0.5 \cdot (b_i - a_i)$ in this chapter. The sign (+ or −) is chosen with a probability of 0.5 and $\alpha_k \in \{0,1\}$ is randomly generated with $p(\alpha_k = 1) = \frac{1}{16}$.

10.3.3 Chromosome Evaluation

In this chapter, we use the well-known mean absolute error (MAE):

$$\text{MAE} = \frac{1}{N_t}\sum_{i=1}^{N_t}|O_i - E_i| \tag{10.1}$$

where E_i and O_i are the desired and the obtained outputs for the i th sample in the dataset, and N_t represents the number of training data. In order to apply the roulette wheel operator for the crossover selection process, the value $(1 - MAE)$ is used.

10.3.4 Mechanism and Characteristics of the Algorithm Framework

The proposed evolutionary algorithm framework is based on the well-known steady-state GA [47], which is shown in Algorithm 10.1. The steady-state GA benefits from the elite replacement strategy. Specifically, if two new individuals are better adapted than the two worst individuals of the population, the worst ones are replaced by new ones. Accordingly, the use of the steady-state GA allows a faster convergence and smaller number of evaluation.

Algorithm 10.1 Steady-state genetic algorithm used in the present work.

```
Input: N_P (population size), M_E (maximum number of
  evaluations)
Output: Best individual from P

e ← 0
P ← Initial population
Evaluate individuals in P
e ← e + N_P
While (e < M_E)
  Select two individuals P_1 and P_2 from P
  Apply crossover and mutation over P_1 and P_2
  Evaluate P_1 and P_2
  e ← e + 2
  Find the two worst individuals Worst_(1,2) from P
  P_(1,2) are better than Worst_(1,2)
  Replace Worst_(1,2) by P_(1,2)
Return best individual in P
```

10.4 Dataset Configuration and Simplification

The traffic data is collected from Performance Measurement System (PeMS[1]). The traffic system compiles data [48] from over 30.000 miles of highway in the state of California, USA, which is used to assemble a history of traffic measurements for every loop detector station in the site. This system supports multiple publicly accessible web applications, which are available for statistical and academic purposes.

[1] http://pems.dot.ca.gov/

The study was conducted on a segment of the I-5 highway, in San Diego, California, 10.07 miles long and with changes from four to six lanes on the northbound side. Loop detectors were stationed on the highway mainline approximately every one-third of a mile, as well as in each on-ramp and off-ramp segment. To suppress the noise information in traffic data, the traffic measures aggregated by the detector stations over each 5 minutes are the average speed, occupancy, and flow. The data was collected from 30 consecutive days (April 1 06:00 to April 30 22:59 2013).

Data from sensors in the main road are collected at the beginning, the middle and the ending points, in which the middle one is the point where congestion prediction is desired, or the point of interest (PoI). In addition, on-ramps and off-ramps' reported values are aggregated depending on whether the ramp is before or after the point of interest.

It is important to note that loop detectors located in the main road report three different values: *flow* (number of vehicles per time period), *occupancy* (percentage of time that the detector is switched on) and *speed* (average vehicle speed measured). On the other hand, the loop detectors located at the on-ramps and off-ramps only report *flow*. In addition, the derivatives of the variables are involved. These derivatives are calculated as the difference between two consecutive samples, divided by the time step. In this way, 26 variables would be considered eventually. They are represented graphically in Figure 10.4, over a schematic view of the road used in this study.

The variables are summarized as follows:

- $\{F_1, F_2, F_3\}$ represent the average flow at points $1,2$ and 3, respectively.
- The same procedure is followed with the speed and occupancy, obtaining respectively, variables named and $\{S_1, S_2, S_3\}$ and $\{O_1, O_2, O_3\}$.
- As to the on-ramps and off-ramps, flow values from the ramps before and next to PoI are aggregated, yielding $\{IF_1, IF_2, OF_1, OF_2\}$ as the result.
- The derivatives of the previous variables are also calculated and included in the dataset. They are denoted by $\{\Delta F_1, \Delta F_2, \ldots \Delta OF_1, \Delta OF_2\}$.

In summary, the obtained 26 variables are illustrated graphically in Figure 10.4, which can be used to predict the highway congestion at the point of interest at the next time interval.

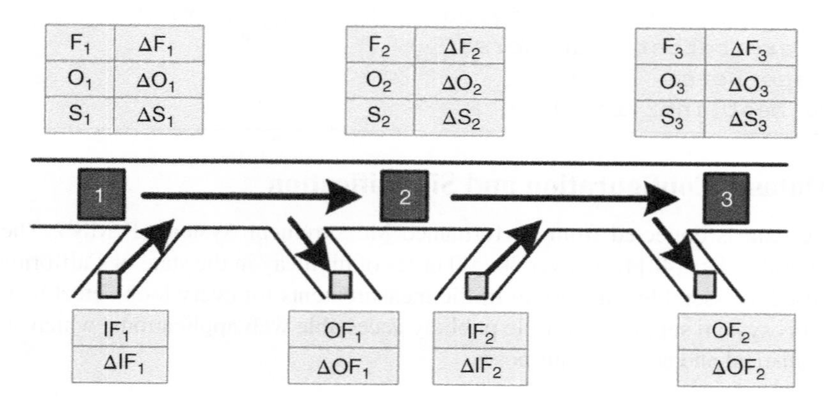

Figure 10.4 Graphical representation of the highway after simplification of the variables.

Thus, the congestion prediction problem is modeled as follows: $C(t+h)=f(F_1(t), F_2(t),\cdots,\Delta OF_2(t))$ where $C(t+h)$ represents the predicted congestion state at time $t+h$, $h=\{5,15,30\}$ minutes.

As reported in many works, such as CoTEC [49], by characterizing highway traffic congestion in several US cities, the traffic state with speed 30–50 mph is categorized as the slight level of congestion. In addition, in order to test our system more effectively, we rationalize the distribution rate of congestion incidents in the pre-processed datasets by analyzing the congestion cases in the collected datasets. In view of the above considerations, a mean speed below 45 mph is considered to be congestion in this study.

10.5 Experimentation

To evaluate the feasibility of the proposed method, namely, GHFRBS, an experimental study was carried out, which was organized as follows: Section 10.5.1 presents the experimental setup, including its datasets, techniques and parameters. Section 10.5.2 elaborates the obtained results. Section 10.5.3 analyzes the obtained results in details, including the accuracy and complexity of the obtained models.

10.5.1 Experimental Setup

This section presents all the datasets, techniques and parameters used during the experimentation, with the purpose of validating the performance of the implemented GHFRBS.

Three datasets are generated in order to test the implemented system for congestion prediction. Each one of them is designed to predict whether congestion is going to occur in the next 5, 15 and 30 minutes, and they are referred to as $TRAFFIC_5$, $TRAFFIC_{15}$ and $TRAFFIC_{30}$, respectively.

In order to study whether the behavior of GHFRBS is comparable to the most used soft computing techniques, its performance will be compared with the following techniques in a variety of aspects: *Fuzzy AdaBoost* [50] is an improved version of the boosting algorithm AdaBoost [41] to deal with fuzzy rules. *Fuzzy LogitBoost* [51] employs a GA to extract fuzzy rules iteratively, which are then combined to decide the output. *Fuzzy Chi-RW* [52,53] identifies the relations between the variables of the problem and makes an association between the feature space and the space of classes, with the aim of generating the rule base. All of the above methods are run by using KEEL (www.keel.es), and the parameter values are configured according to the recommended setting.

The proposed algorithm has been run with the following parameter values: $M_E=100.000$ as the maximum number of evaluations, $N_P=100$ as the population size, $p_c=0.8$ as the crossover probability, $p_m=0.2$ as the mutation probability per chromosome. We set $M=3$ labels per variable for all the problems and methods in order not to make the rule base grow excessively. All the experiments reported in this chapter have been performed by a 5×2 cross-validation model.

10.5.2 Results

In this case, the model produces the final binary output: congestion or no congestion. It can be considered as a binary classification task. For the purpose of evaluating the classification accuracy of GHFRBS and comparing it with other techniques, the MAE (Equation 10.1) measure is chosen.

Figure 10.5 collects the results in terms of average and standard deviation of MAE obtained in both training (top) and test (bottom) datasets. In addition, Table 10.1 presents a comparison about the complexity of the models obtained, where #R and #V stand for the average number of fuzzy rules and the average number of input variables used by the system, respectively; #V/R is the number of variables per rule. In the case of the presented GHFRBS, the number of variables (#V) also gives us the number of modules in the system, which can be calculated as # V −1 (see Figure 10.1 (right)).

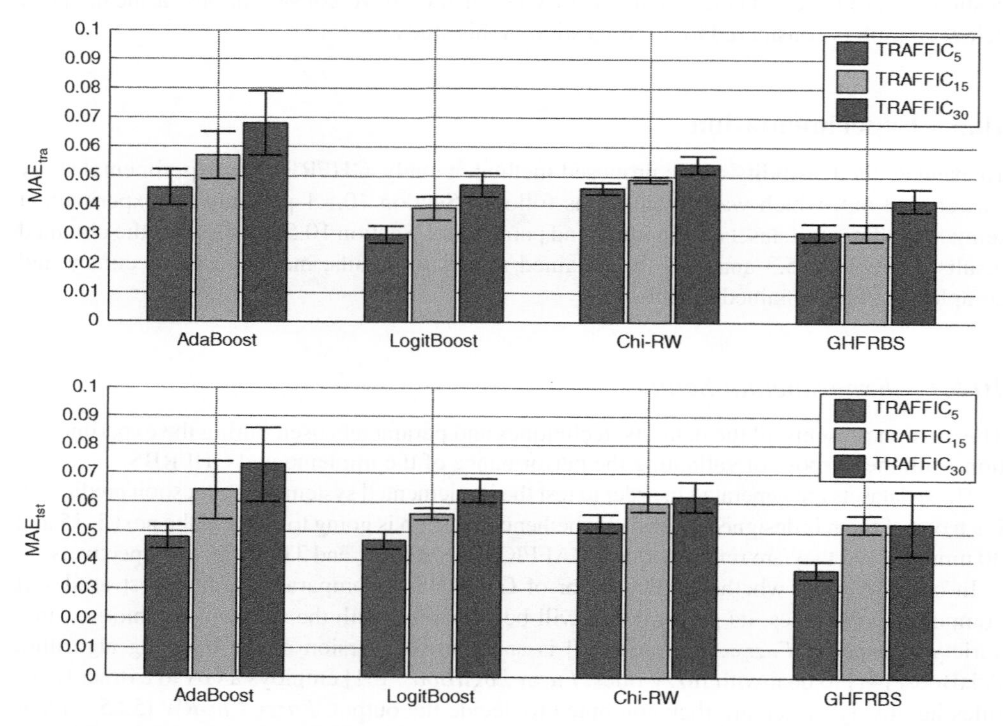

Figure 10.5 Comparison of the performance of Fuzzy AdaBoost, Fuzzy LogitBoost and Fuzzy Chi-RW with GHFRBS in terms of average MAE±σ in the training set (top), and average MAE±σ in the test set (bottom).

Table 10.1 Comparison of the complexity of Fuzzy AdaBoost, Fuzzy LogitBoost and Fuzzy Chi-RW with GHFRBS in terms of #V, #R and #V/R.

Dataset	TRAFFIC$_5$			TRAFFIC$_{15}$			TRAFFIC$_{30}$		
	#V	#R	#V/R	#V	#R	#V/R	#V	#R	#V/R
Fuzzy AdaBoost	26	8	18.93	26	8	19.38	26	8	18.93
Fuzzy LogitBoost	26	25	19.94	26	25	19.46	26	25	19.84
Fuzzy Chi-RW	26	511.4	20	26	506.2	26	26	503.6	26
GHFRBS	**10**	81	**2**	**11.8**	97.2	**2**	**11.4**	93.6	**2**

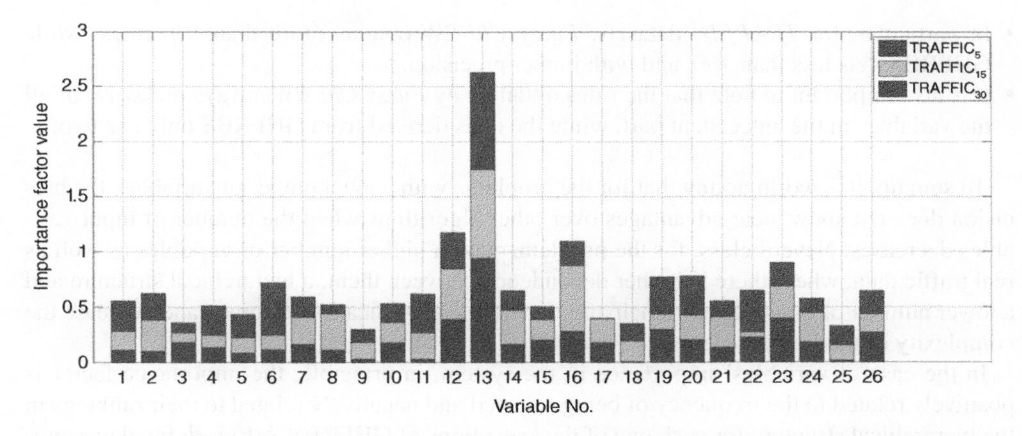

Figure 10.6 The variable importance factor obtained from the three TRAFFIC datasets by GHFRBS.

Moreover, Figure 10.6 shows the importance factor of 26 variables that compound the TRAFFIC datasets. With this, it is expected to be able to identify the most important variables in the dataset.

10.5.3 Analysis of the Results

In this section, a detailed analysis of the obtained results is presented in terms of prediction accuracy, size and complexity of the obtained rule sets. Most importantly, a multiperspective analysis of the results for the TRAFFIC problems is given.

With respect to the accuracy of the algorithms ($1 - MAE_{(traltst)}$ in Figure 10.5), it can be highlighted that with the TRAFFIC datasets (26 variables), the results clearly show that the proposed approach is capable of predicting traffic congestion from 5 minutes to 30 minutes into the future with a high degree of accuracy (94–96%). Meanwhile, it has obtained the highest degree of accuracy in all three TRAFFIC cases when compared with the rest of techniques.

Summarizing, the proposed GHFRBS obtains, in all cases, comparable performance to the most-used techniques, the best results in TRAFFIC problems.

When analyzing the size and complexity of the obtained models, it is worth noting that, given the parameters recommended for their use, Fuzzy AdaBoost and LogitBoost use a fixed number of rules (pre-established), while Fuzzy Chi-RW uses a variable number of rules and GHFRBS uses a variable number of modules (used variables) with a fixed number of rules.

In general, the incorporation of a variable selection mechanism by the GHFRBS allows reducing the complexity of the datasets by using only less than one-half (on average) of the number of variables in the original datasets. In particular, the number of variables is reduced to 40% approximately.

More specifically, as to the complexity of the models, it is worth highlighting that:

- *Fuzzy AdaBoost* and *LogitBoost* always obtain fewer rules than GHFRBS, since the number of rules is pre-established as a parameter of the method.
- Furthermore, the complexity of the obtained rules is much higher than that of the rules obtained by GHFRBS. Rules from Boost methods are formed by about 19 clauses.

- In particular, for *TRAFFIC* datasets, *Fuzzy Chi-RW* obtains more than 500 rules, while GHFRBS gets less than 100, and with higher precision.
- It is also important to note that the rules obtained by *Fuzzy Chi-RW* always make use of all the variables in the antecedent part, while the ones derived from GHFRBS only use two.

To sum up, it is worth noting that for the problems with a low number of variables, the precision does not show clear advantages over other algorithms when the number of input variables decreases. Nevertheless, for the problems with a higher number of variables as well as real traffic data, where there is higher dependence between them, a hierarchical structure and a lower number of input variables help to improve the classification accuracy and decrease the complexity of the fuzzy systems.

In the case of variables' importance in the system, in principle, the importance factor is positively related to the frequency of being selected and negatively related to their rankings in the hierarchical structure for each one of the executions of GHFRBS. δ(k) is defined to represent the importance factor of the *i-th* variables to the classification problem, as shown in Equation 10.2. *Pki* denotes the ranking of *k-th* attribute in hierarchical variables set θi of the *i-th* of n executions result, and $|\theta_i|$ denotes the number of selected variables.

$$\delta(k) = \sum_{i=1}^{n} \frac{Pki}{|\theta i|} \tag{10.2}$$

In order to analyze the importance of each input variable involved (numbered from 1 to 26) in the congestion prediction problem. Figure 10.6 shows variables' importance factor obtained from TRAFFIC datasets by GHFRBS.

By analyzing Figure 10.6, it can be concluded that: O_2, ΔO_2, O_3, ΔF, S_2, ΔS_2, and ΔOF_2 are found to be the variables with the highest correlation with traffic congestion at the target point, while IF_2, OF_2, S_3 and ΔF_3 are minimally correlated with traffic congestion. From the viewpoint of input variables, the results indicate that occupancy's prediction performance is better than that of the speed or the flow. The same level of speed or flow may correspond to two distinct traffic states (free-flow or congested). This is most probably because it is affected by traffic composition and the vehicles' length. From the perspective of location, the input variables at the point of interest are mostly selected, because traffic conditions are mostly associated with their own past values.

10.6 Conclusions

In this chapter, we have proposed a novel approach, based on fuzzy logic and genetic algorithms, to build traffic congestion prediction systems from a high number of input variables.

The proposed GHFRBS has the following novel features: (1) a smaller number of input variables are generated, and the selected input variables are distributed in a set of serial modules; (2) the complexity of the fuzzy rules is reduced significantly because the number of input variables per module is set to two; (3) a Steady-State Genetic Algorithm and a lateral tuning process of the MFs are employed to automatically construct the FRBSs.

In order to test the performance of the new approach, we have applied it to the prediction of traffic congestion with real traffic data collected from PeMS. The obtained results are promising, and have demonstrated the simplicity of the fuzzy rules of the obtained models, and the effectiveness of GHFRBS for predicting traffic congestion with time horizons of 5, 15 and 30 minutes.

Last but not least, benefiting from its automatic ranking and selection of the input variables, GHFRBS may obtain a better understanding of the traffic datasets. This will be regarded as our future research.

Acknowledgment

The authors would like to thank the EU *Intelligent Cooperative Sensing for Improved Traffic Efficiency* (ICSI) project (FP7-ICT-2011-8) for its support in the development of this research.

References

1. Jin, S., Wang, D.-H., Xu, C. and Ma, D.-F. (2013) Short-term traffic safety forecasting using Gaussian mixture model and Kalman filter, *Journal of Zhejiang University Science* **A14**(4): 231–43.
2. Okutani, I. and Stephanedes, Y. (1984) Dynamic prediction of traffic volume through Kalman filtering theory, *Transportation Research Part B: Methodological* **18**(1): 1–11.
3. Ahmed, M. and Cook, A. (1979) Analysis of freeway traffic time-series data by using Box–Jenkins techniques, *Transportation Research Record* **722**: 1–9.
4. Williams, B.M. and Hoel, L.A. (2003) Modeling and forecasting vehicular traffic flow as a seasonal ARIMA process: Theoretical basis and empirical results, *Journal of Transportation Engineering* **129**(6): 664–72.
5. Fung, R.Y., Liu, R. and Jiang, Z. (2013) A memetic algorithm for the open capacitated arc routing problem, *Transportation Research Part E: Logistics and Transportation Review* **50**, 53–67.
6. Samoili, S. and Dumont, A.-G. (2012) Framework for real-time traffic forecasting methodology under exogenous parameters, *Proceedings of the 12th Swiss Transport Research Conference*, pp. 512–22.
7. Stathopoulos, A. and Karlaftis, M.G. (2003) A multivariate state space approach for urban traffic flow modeling and prediction, *Transportation Research Part C: Emerging Technologies* **11**(2), 121–35.
8. Vlahogianni, E.I., Geroliminis, N. and Skabardonis, A. (2008) Empirical and analytical investigation of traffic flow regimes and transitions in signalized arterials, *Journal of Transportation Engineering* **134**(12): 512–22.
9. Dougherty, M.S. and Cobbett, M.R. (1997) Short-term inter-urban traffic forecasts using neural networks, *International Journal of Forecasting* **13**(1): 21–31.
10. Balke, K.N., Chaudhary, N., Chu, C.-L., *et al.* (2005) Dynamic traffic flow modeling for incident detection and short-term congestion prediction: Year 1 progress report. Tech. rep., Texas Transportation Institute, Texas A&M University System.
11. Sapankevych, N. and Sankar, R. (2009) Time series prediction using support vector machines: A survey, *IEEE Computational Intelligence Magazine* **4**(2): 24–38.
12. Wang, J. and Shi, Q. (2013) Short-term traffic speed forecasting hybrid model based on chaos–wavelet analysis-support vector machine theory, *Transportation Research Part C: Emerging Technologies* **27**, 219–32.
13. Wu, C.-H., Ho, J.-M. and Lee, D., 2004. Travel-time prediction with support vector regression, *IEEE Transactions on Intelligent Transportation Systems* **5**(4), 276–81.
14. Chan, K., Dillon, T., Singh, J. and Chang, E. (2012) Neural-network-based models for short-term traffic flow forecasting using a hybrid exponential smoothing and Levenberg–Marquardt algorithm, *IEEE Transactions on Intelligent Transportation Systems* **13**(2): 644–54.
15. Hodge, V., Krishnan, R., Jackson, T., *et al.* (2011) Short-term traffic prediction using a binary neural network. In: *43rd Annual UTSG Conference*, Open University, Milton Keynes, UK.
16. Karlaftis, M. and Vlahogianni, E. (2011) Statistical methods versus neural networks in transportation research: Differences, similarities and some insights, *Transportation Research Part C: Emerging Technologies* **19**(3), 387–99.
17. Smith, B.L. and Demetsky, M.J. (1994. Short-term traffic flow prediction: Neural network approach, *Transportation Research Record* **1453**, 98–104.

18. Dimitriou, L., Tsekeris, T. and Stathopoulos, A. (2008) Adaptive hybrid fuzzy rule-based system approach for modeling and predicting urban traffic flow, *Transportation Research Part C: Emerging Technologies* **16**(5), 554–73.
19. Zhang, Y. and Ye, Z. (2008) Short-term traffic flow forecasting using fuzzy logic system methods, *Journal of Intelligent Transportation Systems* **12**(3), 102–12.
20. Abdulhai, B., Porwal, H. and Recker, W. (2002) Short-term traffic flow prediction using neuro-genetic algorithms, *Journal of Intelligent Transportation Systems: Technology, Planning, and Operations* **7**(1), 3–41.
21. Vlahogianni, E.I., Karlaftis, M.G. and Golias, J.C. (2005) Optimized and meta-optimized neural networks for short-term traffic flow prediction: A genetic approach, *Transportation Research Part C: Emerging Technologies* **13**(3), 211–34.
22. Quek, C., Pasquier, M. and Lim, B. (2006) Pop-traffic: A novel fuzzy neural approach to road traffic analysis and prediction, *IEEE Transactions on Intelligent Transportation Systems* **7**(2): 133–46.
23. Zheng, W., Lee, D.-H. and Shi, Q. (2006) Short-term freeway traffic flow prediction: Bayesian combined neural network approach, *Journal of Transportation Engineering* **132**(2): 114–21.
24. Liu, K. and Fei, X. (2010) A fuzzy-logic-based system for freeway bottleneck severity diagnosis in a sensor network, *Transportation Research Part C: Emerging Technologies* **18**(4): 554–67.
25. Onieva, E., Milanés, V., Villagra, J., *et al.* (2012. Genetic optimization of a vehicle fuzzy decision system for intersections, *Expert Systems with Applications* **39**(18): 13148–13157.
26. Holland, J.H. (1992) *Adaptation in Natural and Artificial Systems*. Cambridge, MA: MIT Press.
27. Konar, A. (2005) *Computational Intelligence: Principles, Techniques and Applications*. Berlin: Springer-Verlag.
28. Badie, A. (2010) Genetic fuzzy self-tuning PID controllers for antilock braking systems, *Engineering Applications of Artificial Intelligence* **23**(7): 1041–52.
29. Cordon, O., Gomide, F., Herrera, F., *et al.* (2004) Ten years of genetic fuzzy systems: Current framework and new trends, *Fuzzy Sets and Systems* **141**(1): 5–31.
30. Benítez, A.D. and Casillas, J. (2013) Multi-objective genetic learning of serial hierarchical fuzzy systems for large-scale problems, *Soft Computing* **17**(1): 165–94.
31. Raju, G., Zhou, J. and Kisner, R. (1991) Hierarchical fuzzy control, *International Journal of Control* **54**(5): 1201–16.
32. Cala, S. and Moreno-Velo, F.J. (2010) XFHL: A tool for the induction of hierarchical fuzzy systems, *IEEE International Conference on Fuzzy Systems*, pp. 1–6.
33. Cordón, O., Herrera, F. and Zwir, I. (2003) A hierarchical knowledge-based environment for linguistic modeling: Models and iterative methodology, *Fuzzy Sets and Systems* **138**(2), 307–41.
34. Lee, M.-L., Chung, H.-Y. and Yu, F.-M. (2003) Modeling of hierarchical fuzzy systems, *Fuzzy Sets and Systems* **138**(2): 343–61.
35. Stonier, R.J. and Mohammadian, M. (2004) Multi-layered and hierarchical fuzzy modelling using evolutionary algorithms, *International Conference on Computational Intelligence for Modelling, Control and Automation, University of Canberra, Canberra, Australia*, pp. 321–44.
36. Torra, V. (2002) A review of the construction of hierarchical fuzzy systems, *International Journal of Intelligent Systems* **17**(5): 531–43.
37. Zajaczkowski, J. and Verma, B. (2010) An evolutionary algorithm based approach for selection of topologies in hierarchical fuzzy systems, *IEEE Congress on Evolutionary Computation*, pp. 1–8.
38. Zajaczkowski, J. and Verma, B. (2012) Selection and impact of different topologies in multi-layered hierarchical fuzzy systems, *Applied Intelligence* **36**(3): 564–84.
39. Alcalá, R., Alcalá-Fernández, J. and Herrera, F. (2007) A proposal for the genetic lateral tuning of linguistic fuzzy systems and its interaction with rule selection, *IEEE Transactions on Fuzzy Systems* **15**(4): 616–35.
40. Goldberg, D.E. (1989) Genetic Algorithms in Search, Optimization and Machine Learning, *Boston, MA: Addison-Wesley*.
41. Freund, Y. and Schapire, R.E. (1996) Experiments with a new boosting algorithm, International Conference on Machine Learning. pp. 148–56.
42. Herrera, F., Lozano, M. and Verdegay, J. (1998) Tackling real-coded genetic algorithms: Operators and tools for behavioural analysis, *Artificial Intelligence Review* **12**(4): 265–319.
43. Schlierkamp-Voosen, D. (1993) Predictive models for the breeder genetic algorithm, *Evolutionary Computation* **1**(1): 25–49.
44. Goldberg, D. and Lingle, R. (1985) Alleles loci, and the traveling salesman problem, *Proceedings of the 1st International Conference on Genetic Algorithms. Lawrence Erlbaum, Hillsdale, NJ, USA*, pp. 154–9.

45. Hadavandi, E., Shavandi, H. and Ghanbari, A. (2010) Integration of genetic fuzzy systems and artificial neural networks for stock price forecasting, *Knowledge-Based Systems* **23**(8): 800–8.
46. Lacroix, B., Molina, D. and Herrera, F. (2012) Region based memetic algorithm with LS chaining, *IEEE Congress on Evolutionary Computation*, pp. 1–6.
47. Whitley, D. and Kauth, J. (1988) *GENITOR: A Different Genetic Algorithm*. Colorado State University, Department of Computer Science.
48. Chen, C. (2003) *Freeway Performance Measurement System (PeMS)*. California PATH Research Report.
49. Bauza, R. and Gozalvez, J. (2013) Traffic congestion detection in large-scale scenarios using vehicle-to-vehicle communications, *Journal of Network and Computer Applications* **36**(5): 1295–1307.
50. del Jesus, M. J., Hoffmann, F., Navascués, L. J., Sánchez, L., 2004. Induction of fuzzy-rule-based classifiers with evolutionary boosting algorithms. *IEEE Transactions on Fuzzy Systems* **12** (3), 296–308.
51. Otero, J. and Sánchez, L. (2006) Induction of descriptive fuzzy classifiers with the LogitBoost algorithm, *Soft Computing* **10**(9): 825–35.
52. Chi, Z., Yan, H. and Tuan, P. (1996) *Fuzzy Algorithms: With Applications to Image Processing and Pattern Recognition*, Vol. **10**. Singapore: World Scientific.
53. Ishibuchi, H. and Yamamoto, T. (2005) Rule weight specification in fuzzy rule-based classification systems, *IEEE Transactions on Fuzzy Systems* **13**(4): 428–35.

11

Vehicle Control in ADAS Applications: State of the Art

Joshué Pérez, David Gonzalez and Vicente Milanés
National Institute for Research in Computer Science and Control (INRIA), Le Chesnay, France

11.1 Introduction

The aim of the Advanced Driver Assistance Systems (ADAS) is mainly linked to aid drivers in safety critical situations rather than to replace them. However, in recent years, many research advances have been done in this field, indicating that fully autonomous driving is closer to becoming a reality. In the literature, the control of autonomous vehicles is divided into lateral and longitudinal actions.

The purpose of this work, dealing with the development of Arbitration and Control strategies and algorithms, is to analyse the existing vehicle control solutions for future autonomous and semi-autonomous vehicles. Both longitudinal and lateral control will be addressed, and these perspectives will consider the driver in the control loop.

This chapter is structured as follows: Section 11.2 describes the vehicle control solutions in ADAS applications. Different control levels are defined in Section 11.3. Some of the most relevant and related previous works (research level) are presented in Section 11.4. A general description of key factors for vehicle control applications (based on ADAS) in the market is presented in Section 11.5. Section 11.6 describes some applications and experimental platforms. Finally, conclusions are given in Section 11.7.

11.2 Vehicle Control in ADAS Application

Some research groups and vehicle manufacturers around the world are technologically ready to provide fully autonomous driving [1]. However, the complexity of traffic scenarios, some legal constraints and drivers' acceptance allow us to forecast a soft transition between manual

Intelligent Transport Systems: Technologies and Applications, First Edition. Asier Perallos, Unai Hernandez-Jayo, Enrique Onieva and Ignacio Julio García-Zuazola.

and fully autonomous driving. In this context, the control capacities in Advanced Driver Assistance Systems (ADAS) have a very important role.

A classification and explanation of the most relevant aspects for the control functions autonomous and semi-autonomous platforms are presented. These are based on the basic control modules presented in interactIVe and DESERVE projects. Three main blocks or platforms are defined: perception, application and Information Warning Intervention (IWI) platform [2].

The perception platform is in charge of analysing all the information from different acquisition modules. These include: external devices (cameras, laser, radar, Global Positioning System (GPS), Inertial measurement unit (IMU), biosensor, among others) and internal buses (odometry and Controller Area Network (CAN) information – speed, angle position and operating signals). Most of these functions are related to different perception sources, object recognition and lane keeping. The information provided by perception software modules is highly important for vehicle control, allowing the definition of use cases of control functions.

The perception modules are detailed in [2]. Some of these modules are: Frontal object perception (FOP), ADASIS Horizon (ADA), Vulnerable road users (VRU), Driver monitoring motorcycle (DMM), Vehicle trajectory calculation (VTC).

To infer the driver intentions is playing an increasing role in the development of vehicle control functions in ADAS applications, because driver behaviour (e.g. distraction or concentration) is one the most sudden causes of fatal accidents [3]. Some researchers are focused in the developing of human-centric intelligent driver assistance systems, which can be based on cognitive knowledge. Usually these techniques use data base information from expert drivers, and then the driver models previously validated are compared with current driver behaviour [4]. The application platform estimates the intention of the driver and risks associated with the current situation based on vehicle performance and vehicle sensors.

Moreover, the arbitration between the fully autonomous driving and the only driver taking the control is considered. The arbitration in the driving process involves the necessity of sharing the control of the vehicle based on the different level described in the next section. The level of assistance provided by the autonomous vehicle to the driver might change depending on the driver's state and on the situation to handle (imminence of danger).

Finally, the IWI platform informs the driver in case of warning conditions, and it also activates the systems related to the longitudinal and/or lateral actions. The functional descriptions of each module are linked to: HMI, lights, lateral and longitudinal actuations. Most of the actions are warning signalizations and control functions.

11.3 Control Levels

Modern ADAS functions help drivers in different tasks or situations in the driving process. Some studies show that nowadays there are many distractions for drivers, causing multiple task overloads (e.g. GPS, panel recognition, security alarms, among others) [5].

One of the most common causes of driver distraction is monotonous driving. It causes mental underload, which decreases vigilance on the route and then generates dangerous situations. Moreover, stress, mental overload and highly difficult situations are important factors to be considered in defining the interaction between driver and fully autonomous functions.

The project HAVEit has established a pragmatic approach for levels of assistance and automation in vehicles. Five levels are defined in this project, considering driver actions and fully autonomous vehicle functions [6]. The stepwise transfer of the driving task forms the basis for optimum task repartition in the fully autonomous driving system. These levels are defined as follows:

- Driver only: the driver has full control of the vehicle without any warning or assistance from the vehicle, e.g. vehicle without ADAS installed.
- Driver assisted: the driver is still in full control, but one co-system supports with lights or acoustic assistances, e.g. during lane change manoeuvres.
- Semi-automated: it can integrate partial control of vehicle in some specific scenarios and/or conditions, e.g. Adaptive Cruise Control (ACC) (only throttle and brake pedals) or parking assist systems (specific configurations).
- Highly automated: the co-system does most of the driving, but the driver is still in the loop and can take over the driving task anytime, e.g. lane departure warning system.
- Fully automated: the vehicle is completely autonomous, even in difficult situations. So far, most of the systems are highly automated; therefore the driver always has the last decision.

11.4 Some Previous Works

In recent years, many safety and comfort improvements have been implemented in commercial vehicles through ADAS, which is one of the systems most studied in the Intelligent Transportation Systems (ITS) field [1,7]. Most of these implementations are focused on constrained scenarios (e.g. Intelligent Park Assist), warning systems (e.g. Blind Spot Warning Systems or Night Vision Systems), and some partial control systems, such as Adaptive Cruise Control (ACC), Pre-crash systems and Low Speed Collision Avoidance Systems (LSCAS), among others. So far, only partial executions on driverless vehicles are available.

The first developments in the ITS field (before the so-called advanced vehicle-control systems – AVCS) occurred in the early 1960s, performed by the General Motors Research Group. This primitive AVCS provided safety warnings or assistance in controlling vehicles, including the extension to full control of vehicle motions. Researchers at General Motors (GM) developed an automatic vehicle (steering – lateral control, speed, and braking – longitudinal control, on test tracks [8]. Later on, other research groups began to improve lateral and longitudinal control of autonomous vehicles. At the end of the 1960s, Ohio State University and the MIT worked on the application of new control techniques to urban transportation problems: longitudinal (spacing) control, steering control, etc. [9].

The first broad-scale investigation of the application of automation technologies to urban transportation problems appeared in MIT's Project METRAN (Metropolitan Transportation) [10]. This project was dedicated to developing an integrated, evolutionary transportation system for Urban Areas with a fully autonomous vehicle.

Early developments in the United States were also paralleled in Europe and Japan. For example in England, the Road Research Laboratory conducted experiments on automatic steering control in the late 1960s, and at the Ministry of International Trade and Industry (MITI) in Japan autonomous driving was performed. These Asian initiatives ended in different Advanced Safety Vehicle (ASV) projects, where several developments in external airbags or the windshield protection system, among others, were achieved [11].

Thanks to advances in the electronics and computer science, the modern history of AVCS development began in 1986 with parallel actions in Europe and the USA. In Europe, the most important project (or programme) was PROMETHEUS [12]. The EUREKA Prometheus Project (PROgraMme for a European Traffic of Highest Efficiency and Unprecedented Safety) was one of the largest R&D projects ever in the field of driverless cars (1987–1995) in Europe. It was coordinated by Bundeswehr Universität München, collaborating with Daimler-Benz.

Other important works in the field of autonomous and assisted driving were carried out by the California Partners for Advanced Transit and Highways (PATH) Program, which was founded in 1986 (still active). Their main motivation is that traffic congestion was becoming an increasingly acute public concern. The initial PATH research topics are: navigation (enhanced information about traffic); automation (impact of ITS); and roadway electrification (roadway-powered electric vehicle) [13].

The last decade has been the most dynamic in the autonomous vehicle field. It is beyond the scope of this chapter to explain all the projects and researches carried out. However, some of the most relevant researches and initiatives developed in Europe are as follows:

- Project VIAC (2007–10): international travel with autonomous vehicles from Parma to Shanghai (Vislab, Parma University);
- Project SPITS (2008–11): communication among intelligent vehicles; more than 100 vehicles involved in the final demonstration (TNO, Helmond, Holland);
- Project HAVEit (2008–11): energy efficiency, driving comfort and control; some of the partners are working in Deserve (7th Framework Program);
- Cybercars-2 and CityMobil (2005–08 and 2008–11): cooperation between autonomous, manual vehicles and Cybercars; other projects have been started in Asia (Inria);
- Initiative e-Safety (2002–13): development of intelligent vehicle safety; this initiative is working for faster development of smart road safety and eco-driving technologies;
- GCDC Competition (2009–11): it was the first European completion among autonomous vehicles; more than 10 team, from different countries.

Other initiatives have been undertaken in the United States, such as the Grand DARPA challenge 2005 and the Urban DARPA challenge 2007, and most recently the Google Driverless Car [14], one of the most disseminated experiments with autonomous vehicles [15]. Another important demonstration was carried out by the Vislab group (University of Parma), in the framework of the PROUD-Car Test 2013 event. An autonomous vehicle was guided in a mixed traffic route (rural, freeway, and urban) open to public traffic [16].

Since today international laws do not allow fully automated driving, some recent projects have started to consider the driver within the control loop. This means that drivers can take over control in order to avoid accidents. The 7th Framework Program (FP7) HAVEit project describes highly automated driving as the next step towards the long-term vision of safe, comfortable and efficient transport for people and goods [6]. The project developed, validated and demonstrated important intermediate steps towards highly automated driving for passenger cars, buses and trucks. The architecture was scalable in terms of safety. This project is one of the pioneers considering mental overload and underload in automated assistance.

However, some nontrivial issues remained open, e.g. managing between manual and fully autonomous driving. For this reason, other concepts and problems, such as longitudinal and lateral control strategies, and their integration in the same platform, as well as an arbitration and control module, have to be included in the control architecture of the DESERVE project.

Research in autonomous driving is also starting to be taken into account from automotive companies (Tier1, OEMs). For example Bosch has developed an autonomous vehicle to solve the door-to-door problem in urban and highways scenarios. The vehicle has the capability of switching between automated and manual driving modes. They use different ADAS in the same platform, such as: Lane Keeping Assistant, Adaptive Cruise Control and Lane Changing Assistant, which were demonstrated in real urban scenarios [17].

11.5 Key Factor for Vehicle Control in the Market

The ADASs are one of the objectives of the ITSs along with intelligent infrastructure and autonomous driving developments. Green, safe and supportive transportation, in relation particularly to accident free mobility scenarios, is the major motivation for the development of ADASs.

Recently, the ADASs are more accepted by consumers. In [18] a study of awareness and interest in the Driver Assistance Systems and Active Safety Features in vehicles is described considering: driver warning, assistance and map enabled systems, HMI preferences and user-friendliness of current safety systems, perceived benefits of integration with chassis and powertrain and other future safety systems and technologies (V2V or V2I communications etc.)

Some of the most demanded ADASs are: Driver Warning and Awareness systems, Driving Assistance and Collision Avoidance Systems, Vehicle Stability Systems and Exterior Lighting Control, among others.

Some other motivations to the development and improvement of these systems are:

- *Legislation*: different initiatives like European New Car Assessment Programme (Euro NCAP) have been developed recently. The most relevant advanced safety technologies are rewarded for the Euro NCAP, such as: Blind Spot Monitoring, Lane Support Systems and Emergency Braking, among others [19].
- *Cost*: studies show that customers are willing to pay more for avoidance systems than other ADAS in their vehicles [18].
- *Market indicator*: some of them show that the ADAS will have a high influence in the market in recent years [18]. For example, in 2016 the overall annual market for Lane Departure Warning systems is expected to reach over 22 million units/year (corresponding to 14.3 billion dollars).
- *Safety*: on board electronics systems have become critical to the functioning of the modern automobile, as mentioned by the National Research Council report [20].

The individual functions of the ADAS are designed from the beginning in such a way that they operate within a common environment. Different ADAS functions will not simply live together but nevertheless coexist and deeply cooperate by providing their assistance to the drivers simultaneously. Some of the available technologies are:

- *Interconnections*: electronics systems are being interconnected with one another and with devices and networks external to the vehicle to provide their desired functions [20].

- *Fusion*: new vehicle capabilities have to be adapted to human behaviour in the driving process, and electronics, perception system information and HMI inputs become relevant in order to handle the best decision in each situation.
- *New infrastructures*: a recent study report from the European Commission explains the need of vehicle and infrastructure systems designed for Automated Driving [21].
- *Sensor technologies*: are becoming more sophisticated and varied, especially to support the functionality of many new convenience-, comfort- and safety-related electronic systems. Some of the sensors technologies used in ADAS developments are:
  - Ultrasound (parking assist ...).
  - Inertial sensors (stability control, air bag deployment ...).
  - Radar and Lidar technologies (ACC, AEB ...).
  - Cameras (Lane Keeping, ACC ...).
  - GPS (Advanced ACC, speed advice ...).
  - HMI: interaction interface between humans and ADAS functionalities. It must be understandable and easy to use, otherwise the advantages can be misunderstood by customers.

Many efforts have been recently done by different manufactures and also ARTEMIS Subprogram related to methods and processes for safety-relevant embedded systems [3]. The aim is to improve functional safety in ADAS embedded systems.

Furthermore, the Human Machine Interface (HMI) evolution requires special attention. Recently, the popularization of smartphones and communication systems has allowed customers to be more familiar with handling touchscreen and interactive HMIs. Graphic capabilities are most demanded with real-time information in new generation vehicles. One of the most remarkable is the HMI solution presented by Mercedes-Benz, which is focused particularly on two topics: safe operation and optimum readability of the displays in order to minimize additional driver stress during top priority driving tasks.

Management of the over-information (board panel, smartphones, GPS, road panel, among others), when the driving process is being executed, is a challenge that all manufactures (Tier1, OEMs) are dealing with. For example, in Continental Automotive are developing the integration of monitoring systems to detect different driving situations. However, an arbitration that covers most ADAS solutions is needed.

The wider diffused ADAS system is the ultrasonic park assist [18]. It is followed by the Adaptive Cruise Control and Stop & Go. In these systems longitudinal control is managed only by the throttle pedal. In other low-speed safety systems such as Low Speed Collision Avoidance systems the brake action is implemented too. So far, the ACC is not considering steering wheel actions.

11.6 ADAS Application From a Control Perspective

In this section the existing control vehicle solutions in different ADAS applications are described.

Ten groups of ADAS are defined. These applications are currently available or will soon be introduced into the automotive market. These are described in each subsection from a control perspective (if it applies, since sometimes these are only warning functions).

11.6.1 Lane Change Assistant Systems

The lane change manoeuvre is one of the most dangerous situations for drivers. In 2009 alone, 13% of all accidents on German freeways (injured persons) were caused by vehicles driving laterally in the same direction, which was during a lane change manoeuvre [22]. Many manufactures have developed Lane Change Decision Aid Systems (LCDAS). This system warns the driver about the dangerous lane change situation when the turn indicator is activated [23] by visual elements in the outer mirror, haptic warning signals at the steering wheel [24] or acoustic signals.

Some of the implementations of LCA systems are the Lane Departure Warning (LDW) and Blind Spot Warning (BSW) systems (Figure 11.1). The LDW helps the driver to stay in the same lane by alerting him or her to any unintended lane departure. The BSW system prevents potential dangerous situations during the lane change process, alerting the driver of unseen vehicles in his blind spot. The Lane Keeping Assistance systems (LKAS) have been developed in laboratory platforms, being the goal of manufacturers in the short and medium term.

From the vehicle control point of view, both lateral and longitudinal controllers are considered. These systems are depending on sensor information (perception of the environment) and the arbitration control based on driver behaviour.

11.6.2 Pedestrian Safety Systems

Pedestrian detection systems are mostly used in the urban environment. Pedestrians are considered vulnerable road users, since they are not protected and are even unaware of dangerous situations.

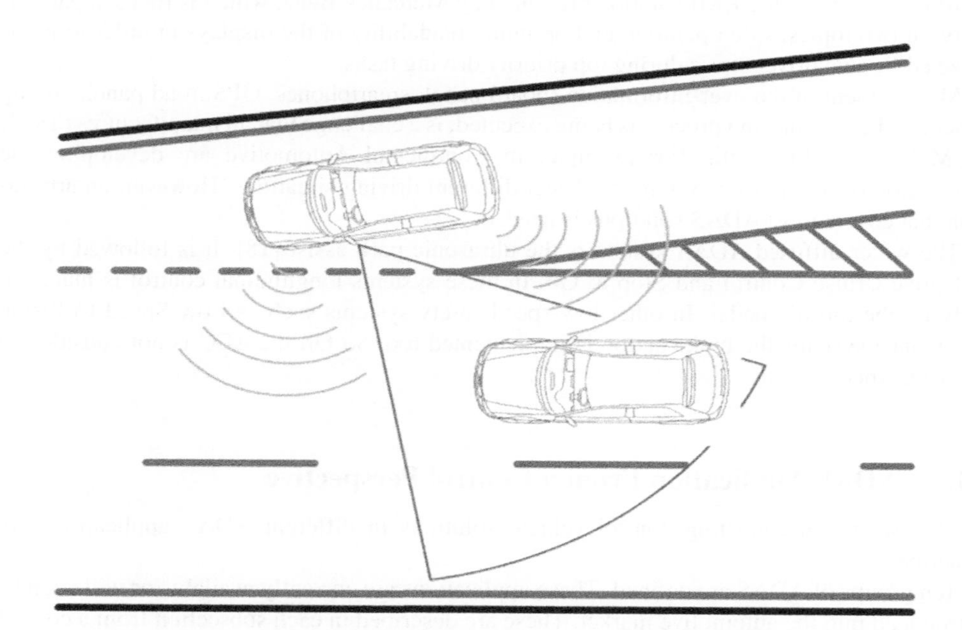

Figure 11.1 Blind-spot assistant ultrasonic sensors for safe lane changes.

The first pedestrian detection systems were based on stereo-vision cameras [25,26]. Different techniques were used: motion-based object detection and pedestrian recognition, which provide suitable measurements of the time to collision (TTC). The pedestrian detection can be classified like a Collision Warning System (CWS). However, since the reaction time of the driver is slow (around 2 seconds), these systems usually have access to the brake system (longitudinal control).

Recently, Volvo has implemented this system in many vehicle models: the new Volvo V40, S60, V60, XC60, V70, XC70 and S80. The vehicle can stop when a pedestrian or cyclists is detected in the vehicle path ahead.

These systems use Information Warning Intervention platforms to warn the driver before acting on the pedals.

11.6.3 Forward-looking Systems

These systems are related to forward detection, mainly using radars, lasers, cameras, infrared sensors and, in some cases, fusing different sensors. Although most of them have automatic actions, it is not mandatory. The forward-looking system involves, at least, three main modules: object detection, decision making and actuation [25]. The first one is related to perception tasks, which mean the analysis of environmental information obtained by one or more sensors. The decision-making system estimates when and how collisions can be avoided, or what kind of object is detected. Finally, the actuation stage adapts the target commands generated by the previous stage and transforms these commands to low-level control signals needed by the respective actuators. In some ADAS applications, the action of the steering has recently been considered. However, most of them use only access to the brake.

The most known and implemented forward-looking systems are the Collision Warning System (CWS) and Low Speed Collision Avoidance System (LSCAS). A brief description of CWS and LSCAS is presented in [18]). The LSCASs available in the market are usually limited to 30 kph. Other forward-looking systems are as follows:

- pre safe system;
- collision avoidance system;
- ahead emergency braking;
- electronic emergency brake light;
- intelligent intersection (emergency vehicle detection);
- rear approaching vehicle (although the rearview camera is most common in park assistant systems);
- queue warning.

Perception of the environment is one of the most important aspects of the forward-looking system for vehicle control.

11.6.4 Adaptive Light Control

These systems are recently used in commercial vehicles. The main advantage is that the lights can be adapted to the scenario (straight and curve roads). For example, Adaptive High Beam

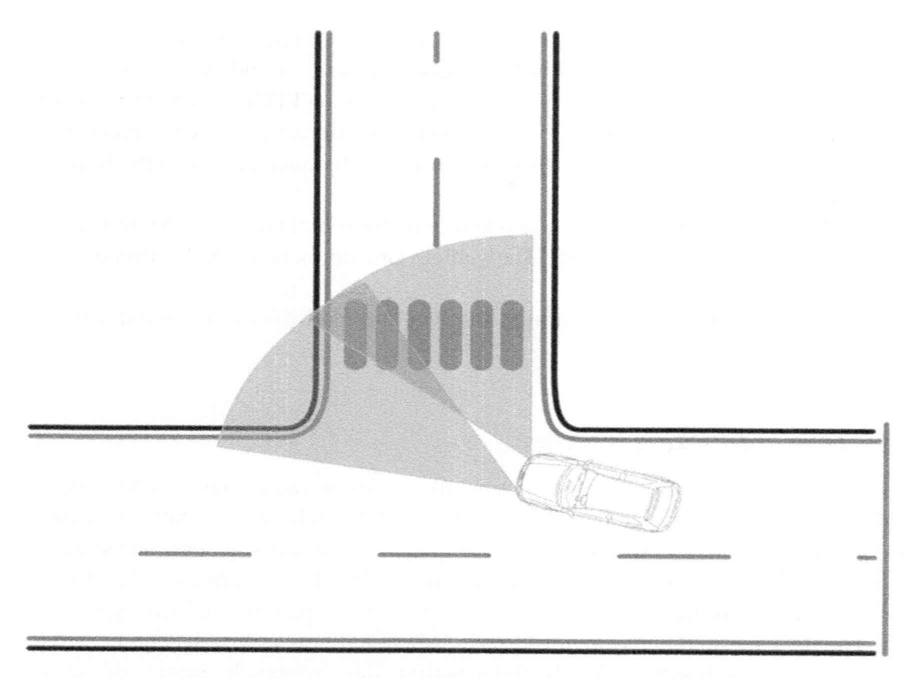

Figure 11.2 Example of the Adaptive Light Control.

can turn in the sense of the curves, anticipating possible undesirable obstacles (e.g. a bicycle or pedestrians) (Figure 11.2).

The map supported frontal lightings are based on the ability to adapt the headlamps dynamically, turning the reflectors, according to the environment condition and using GPS information.

Continental launched an updated integrated camera-LIDAR module, the SRLCAM400. It can be adjusted to three levels: 'Entry', 'Basic' and 'Premium', depending on the number of front windshields ADAS features (Adaptive Front-lighting System, Distance Warning or AEBS, LDWS and Traffic Sign Recognition).

These adaptive lights are a new control system on-board the vehicles to prevent dangerous situations.

11.6.5 Park Assistant

It is probably the most used (and demanded) ADAS today. Ultrasonic park assist systems have evolved from high-end to ordinary vehicles in few years. These systems can help in the parking manoeuvre in close-fitting spaces, by alerting the driver of rear obstacles and their distance to the vehicle.

The intelligent park assist provides easy parking by identifying sufficient parking spaces and steering the car into it. The system is always supervised by the driver, who can override the operation pushing the accelerator pedal or the brake pedal. Other parking assistant systems use rearview cameras instead of, or in addition to, the ultrasonic sensors. They provide a video image from the rear area of the vehicle. Lateral and longitudinal controllers are used simultaneously in these systems.

11.6.6 Night Vision Systems

Night Vision Systems (NVS) permit the drivers to see in low or difficult light conditions. When weather conditions are extreme, these systems can see beyond the range illuminated by the headlights of the vehicle. The technology is based on near and far infrared cameras, which permit illumination of the road ahead, along a spectrum invisible to the human eye.

Many manufactures are using these technologies (Mercedes-Benz, Toyota, Audi, BMW, among others). Recently, the night vision system came off second-best in preference for the car consumer option in Europe [27].

The night vision system has four basic functions: pedestrian detection, pedestrian collision warning, image display and sound warning. These systems use the information from an image which is composed with thermal radiation of objects.

Many premium vehicle brands offer different night vision systems. Most recent generation night vision systems have added pedestrian detection as a feature to assist drivers to avoid potential collisions. These are classified in near-infrared (NIR) and far-infrared (FIR) according to the regions of the electromagnetic spectrum [28].

Perception of the environment is one of the most important aspects of the night vision system for vehicle control in emergency situations.

11.6.7 Cruise Control System

Cruise Control (CC) systems are capable of automatically keeping the speed of the vehicle. The first CC implementations were based on controlling the accelerator pedal only (longitudinal control).

Adaptive Cruise Control (ACC), one of the most conventional forms of ADAS, was developed some years ago. It acts on the longitudinal control of the vehicle, permitting it to follow a leader – acting on the throttle and brake pedals autonomously – and to maintain a predefined headway with the vehicle in front. The next step in the evolution of this technology is based on cooperation among different vehicles in order to reduce this headway between vehicles and the accordion effect in traffic jams. This is known as Cooperative ACC (CACC) and is based on Vehicle to Vehicle (V2V) and in some cases Vehicle to Infrastructure (V2I) communications.

A feature of the ACC is to maintain constant speeds in motorway driving and continuously monitor the vehicle in front, depending on traffic conditions, and start following the traffic automatically. These kinds of systems are also known as Stop & Go for low speed or urban applications.

Recently, some manufactures (e.g. Volvo) have implemented the ACC in many of their models. The goal of this system is to maintain a set time interval (or speed) to the vehicle ahead. It is primarily intended for use on long straight roads in steady traffic, such as on highways and other main roads. This system has some limitation at lowest speed than 18 mph (30 kph). The distance to the vehicle ahead (in the same lane) is monitored by a radar sensor. The vehicle speed is regulated by accelerating and braking (Figure 11.3).

11.6.8 Traffic Sign and Traffic Light Recognition

Traffic sign and traffic light recognition deals with outdoor images, considering different techniques used in image processing and segmentation (i.e. colour analysis or shape analysis) for

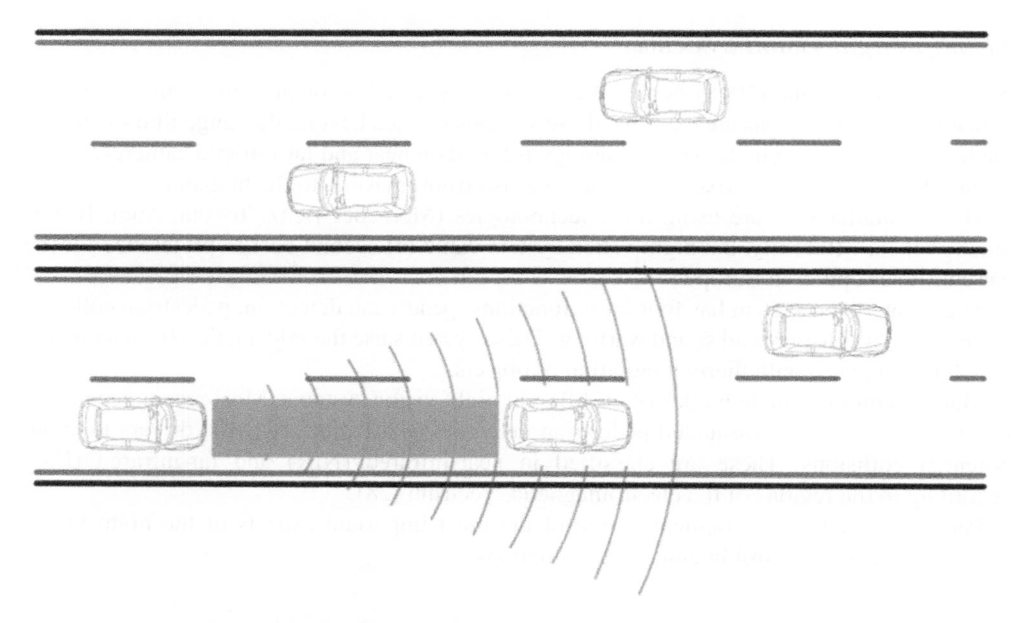

Figure 11.3 Volvo S60 Full Speed Range Adaptive Cruise Control and a Collision Warning and Mitigation system.

the recognition of traffic signs with daylight conditions in real scenarios [29] (Perception of the environment).

Artificial Intelligence techniques (such as neuronal networks and fuzzy logic) have been widely used in the recognition and classification processes of traffic signs. Some other techniques such as template matching or more classical learning based techniques using classifiers (Adaboost or Support Vector Machines) have also been used. Road and traffic sign recognition is one of the important fields in ITS, due to the visual language that drivers can understand on the road. Sometimes, these signals may be occluded by other objects, and may suffer from different problems like fading of colours, disorientation, and variations in shape and size, especially in images captured at night, in the rain and in sunny day conditions [30].

Different Traffic Sign Recognition (TSR) products have been available since 2008 on the BMW 7 Series based on a vision system or GNSS (Global Navigation Satellite Systems) or the fusion of both [31]. The system can help drivers to maintain a legal speed, obey local traffic instructions, or urban restrictions. Typical information obtained includes: speed limit, no-overtaking, prohibited access, among others. The information is shown in the control panel of the vehicle.

11.6.9 Map Supported Systems

Recently, another tendency has been the use of digital maps to support the driving process. The architectures of Map-Supported ADAS are described in [32]. They explain that Digital Map data can be classified in three levels: Nonmap ADAS, Map-Enhanced ADAS and Map-Enabled ADAS. An example of the first one is the ultrasonic parking distance control, whereas those for the second ones are the ACC and Speed Limit Info (SLI). These systems work without Digital Map information, but their functionalities can be improved with the addition of Digital Map data. Curve Speed Warning (CSW) and Dynamic Pass Predictor (DPP) are examples of systems that need digital map inputs.

One of the commercialized applications is the map supported for Advanced Front Lighting Systems (AFLS). These are based on the ability to adapt the headlamps dynamically, by means of turning reflectors, according to the current driving situation and the environment using data stored in the map database. The other is the CC Map adaptive, which attends to automatic speed and distance control, based on the preceding vehicle in the same lane and also based on predictive information from the navigation system. Finally, the map supported Lane Keeping System (Departure Warning System) allows the car to stay within the existing lane, based on on-board sensor inputs as well as the navigation system.

11.6.10 Vehicle Interior Observation

Driver drowsiness is one of the major causes of road accidents. For this reason, several driver drowsiness detection systems have been implemented to warn the driver in this dangerous situation. Some of them are based on iris (eye) detection and also gaze detection (direction) [33].

Vision-based eye tracking is one of the most commonly used techniques. Other researches use electrooculogram (EOG) as an alternative to video-based systems in detecting eye activities caused by drowsiness [34]. The problem of the EOG is the difficulty to install it on the driver face (or on head) every time that they drive.

For this reason, driver drowsiness warning systems available in the market are based on vision (camera) systems that monitor the driver's eyelids to detect signs of weariness or drowsiness and alert the driver.

Other manufactures are using biomedical signals (e.g. FICOSA), which permit the driving process to be characterized in order to set different fatigue and somnolence alarms depending on the driver [35].

11.7 Conclusions

This chapter provides an overview of existing vehicle control solutions applied in the field of Advanced Driver Assistance Systems (ADAS), from the functional requirements perspective.

In particular the main vehicle control software modules have been identified within the basic software architecture on three layers: Perception, Application and Information Warning Intervention (IWI) platforms. A classification of the different control levels, based on driver actions and fully autonomous vehicle functions.

Advances in vehicle control applications have been described, considering the 10 groups of DAS with 33 applications that are currently available or will soon be introduced in the automotive market.

The next step is to define a generic ADAS control architecture to be used in different platforms. The lateral and longitudinal controllers have to be considered for the arbitration and control of the vehicles. Most of the solutions consider the driver in the control loop; therefore future developments will focus on modelling the shared control between the vehicle and the driver (semi-automated and highly automated control level).

References

1. B. Ulmer (1992) VITA – an autonomous road vehicle (ARV) for collision avoidance in traffic, *Proceedings of the Intelligent Vehicles '92 Symposium, 29 June–1 July 1992*, pp. 36–41.
2. M. Lu, K. Wevers, R. van der Heijden and T. Heijer (2004) Adas applications for improving traffic safety, *Proceedings of the IEEE International Systems, Man and Cybernetics Conference, vol.* **4**, 2004, pp. 3995–4002.
3. S.E. Shladover (2006) *PATH at 20 – history and major milestones, Intelligent Transportation Systems Conference, 2006. ITSC '06. IEEE, 17–20 Sept. 2006*, pp. 1.22–1.29.
4. HAVEit (Highly automated vehicles for intelligent transport) project Deliverable D61.1, Final Report, 2011.
5. http://archive.darpa.mil/grandchallenge (last accessed 10 May 2015).
6. http://www.google.com/about/jobs/lifeatgoogle/self-driving-car-test-steve-mahan.html (last accessed 30 April 2015).
7. Annual Work Programme 2011 section 3.2: ASP1: Methods and processes for safety-relevant embedded systems.
8. Statistisches Bundesamt, Fachserie 8 – Reihe 7 – Verkehr – Verkehrsunfälle, 2010, p. 65.
9. F. Sugasawa, H. Ueno, M. Kaneda, *et al.* (1996) Development of Nissan's ASV, *Proceedings of IEEE Intelligent Vehicles Symposium, 19–20 Sept. 1996*, pp. 254–9.
10. S. Habenicht, H. Winner, S. Bone, *et al.* (2011) A maneuver-based lane change assistance system, *Intelligent Vehicles Symposium (IV), IEEE*, 5–9 June 2011, pp. 375–80.
11. D.F. Llorca, V. Milanes, I.P. Alonso, *et al.* (2011) Autonomous pedestrian collision avoidance using a fuzzy steering controller, *IEEE Transactions on Intelligent Transportation Systems* **12**(2): 390–401.
12. K. Gardels (1960) *Automatic car controls for electronic highways, General Motors Research Lab.*, Warren, MI, GMR276.
13. H. Fleyeh and M. Dougherty (2005) Road and traffic sign detection and recognition, *10th EWGT Meeting and 16th Mini-EURO Conference, 2005*, pp. 644–53.
14. Thum Chia Chieh, M.M. Mustafa, A. Hussain, *et al.* (2005) Development of vehicle driver drowsiness detection system using electrooculogram (EOG), *1st International Conference on Computers, Communications, & Signal Processing with Special Track on Biomedical Engineering, 14–16 Nov. 2005*, pp. 165–8.
15. A. Puthon, F. Nashashibi, B. Bradai (2011) A complete system to determine the speed limit by fusing a GIS and a camera, *IEEE International Conference on Intelligent Transportation Systems (ITSC), 5–7 October 2011*, Washington DC, USA, pp. 1686–91.
16. S. Leekam, S. Baron-Cohen, D. Perrett, *et al.* (1997) Eye-direction detection: A dissociation between geometric and joint attention skills in autism. *British Journal of Developmental Psychology*, pp. 77–95.
17. S. Durekovic and N. Smith (2011) *Architectures of map-supported ADAS, Intelligent Vehicles Symposium (IV), 5–9 June 2011, IEEE*, pp. 207–11.
18. http://gossip.libero.it/focus/25516094/anche-bosch-testa-l-auto-che-si-guida-da-sola/bosch-auto/?type= (last accessed 10 May 2015).
19. European Commission (2010) *Definition of Necessary Vehicle and Infrastructure Systems for Automated Driving: Study Report*, SMART 2010/0064.
20. Transportation Research Board (2014) *The Safety Promise and Challenge of Automotive Electronics, Insights from Unintended Acceleration*, National Research Council of the National Academies, Special Report 308.
21. L. Bi, X. Yang and C. Wang (2013) Inferring driver intentions using a driver model based on queuing network, *2013 IEEE Intelligent Vehicles Symposium (IV) 23–26 June 2013, Gold Coast,* Australia, pp. 1387–91.
22. Euro NCAP, *Moving Forward, 2010–1015 Strategic Roadmap*, December 2009.
23. *D12.1 – Development Platform Requirements, SP1, DESERVE Project*, 2013.
24. F. Beruscha, K. Augsburg and D. Manstetten, D. (2011) Haptic warning signals at the steering wheel: A literature survey regarding lane departure warning systems, *Haptics-e, The Electronic Journal of Haptic Research* **4**(5): 1–6.
25. M. Miyaji, K. Oguri and H.K. Kawanaka (2011) Study on psychosomatic features of driver's distraction based on pattern recognition, *11th International Conference on ITS Telecommunications (ITST), 23–25 Aug. 2011*, pp. 97–102.
26. http://vislab.it/proud-en (last accessed 30 April 2015).
27. S. Lee and Y. Yu (2012) Study of the night vision system in vehicle, Vehicle Power and Propulsion Conference (VPPC), *IEEE, 9–12 Oct. 2012*, pp. 1312–16.
28. M.A. Garcia-Garrido, M.A. Sotelo and E. Martin-Gorostiza (2006) Fast traffic sign detection and recognition under changing lighting conditions, Intelligent Transportation Systems Conference, ITSC '06. *IEEE, 17–20 Sept. 2006*, pp. 811–16.

29. Yun Luo, J. Remillard and D. Hoetzer (2010) Pedestrian detection in near-infrared night vision system, *Intelligent Vehicles Symposium (IV), IEEE, 21–24 June 2010*, pp. 51–8.
30. D. Barrick (1962) Automatic steering techniques, *IRE International Convention Record* **10**(2): 166–78.
31. M. Hanson (1966) *Projest Metran: An Integrated, Evolutionary Transportation System for Urban Areas*. Cambridge, MA: MIT Press.
32. Frost & Sullivan (2010) *2009 European Consumers' Desirability and Willingness to Pay for Advanced Safety and Driver Assistance Systems*, Frost & Sullivan.
33. C.G. Keller, Thao Dang, H. Fritz, *et al.* (2011) Active pedestrian safety by automatic braking and evasive steering, *IEEE Transactions on Intelligent Transportation Systems* 12(4): 1292–1304.
34. European Commission (2007) *Green Paper: Towards a New Culture for Urban Mobility*, Com (2007) 551.
35. P. Hrubes, J. Faber and M. Novak (2004) Analysis of EEG signals during micro-sleeps, *IEEE International Conference on Systems, Man and Cybernetics, 10–13 Oct. 2004*, vol. 4, pp. 3775–80.

12

Review of Legal Aspects Relating to Advanced Driver Assistance Systems

Alastair R. Ruddle and Lester Low
MIRA Ltd, Nuneaton, UK

12.1 Introduction

The adoption of intelligent transport systems (ITS) technologies is expected to bring many benefits to society in the areas of transport safety, mobility, and sustainability. However, the use of increasingly complex technological solutions to improve the safety of road users is accompanied by considerable uncertainty regarding legal liability for injuries and other damage that could result from defects of such systems (e.g. failures or malfunctions). Meanwhile, the increasing exploitation of personal data in intelligent mobility systems, such as e-Call and road tolling schemes, also raises concerns relating to data protection and the need to ensure the privacy of individuals. Furthermore, these uncertainties are exacerbated by the need to satisfy a range of different legal frameworks, which vary from country to country, and the mismatch between the pace of technological change and the rate at which suitable legislation can be developed. For example, a consultation carried out in Europe in 2008 [1] found that 61% of the respondents believed that uncertainty relating to liability and data protection aspects were major issues for the development of ITS, with only 15% considering that these issues were not significant obstacles to ITS development.

The term ITS covers a wide variety of applications ranging from traffic information systems, traffic signals, road tolling systems, and traffic management systems through to vehicle control systems. Furthermore, emerging two-way communications capabilities, both between vehicles and between vehicles and roadside infrastructure, will increase the quality and reliability of data relating to vehicle status, location, and local environment, including road and traffic

Intelligent Transport Systems: Technologies and Applications, First Edition. Asier Perallos, Unai Hernandez-Jayo, Enrique Onieva and Ignacio Julio García-Zuazola.

conditions. This will enable the development of more sophisticated cooperative ITS applications, which are expected to facilitate new and improved services for road users and traffic management authorities, further enhancing transport efficiency and safety. The work of the EC-supported collaborative research project EVITA [2] included a review [3] of the legal aspects pertaining to ITS in general. The scope of this chapter, however, is restricted to the subset of ITS functions relating to vehicle control, focusing on the rapidly developing trend towards increasing automation of vehicle driving functions and the associated regulatory implications, primarily from a European perspective. This review updates and extends the analysis of legal aspects relating to advanced driver assistance systems that was previously reported in [3].

12.2 Vehicle Type Approval

The importance and impact of vehicles on society are such that road vehicles have long been subject to specific certification and approval systems. In Europe, the EC Whole Vehicle Type Approval (ECWVTA) Directive (2007/46/EC [4]) provides for type approval of whole vehicles, vehicle systems and components. This directive references other EC directives, as well as regulations that have been developed by the United Nations Economic Commission for Europe (UNECE), which define technical requirements relating to specific aspects of vehicle safety (including lighting, braking, electromagnetic compatibility and many others).

The aim of ECWVTA is to avoid possible trade barriers whilst ensuring compliance with regulatory safety and environmental requirements for vehicles as well as their subsystems and components. If a production intent prototype passes the tests and the production arrangements also pass inspection, then other vehicles, subsystems or components of the same type are deemed to be approved for production and sale within Europe. This significantly reduces certification costs and time, providing benefits to manufacturers, importers and consumers. Although ECWVTA initially only applied to passenger cars (from 29 April 2009), the intention is that all new road vehicles and trailers will be covered from 29 October 2014. In addition, Directive 2007/46/EC also covers national schemes for small series vehicles (limited production) and individual approvals, although different directives apply to agricultural vehicles, quadricycles and tracked vehicles.

Those vehicles that are within the scope of 2007/46/EC are described in terms of a number of different categories and subcategories. For example, the category M encompasses passenger vehicles with at least four wheels, of which vehicles with no more than eight passenger seats (i.e. in addition to the driver's seat) fall into the subcategory M_1 (e.g. passenger cars). Further subcategories of passenger vehicles (category M) address those with more than eight passenger seats (i.e. buses and coaches), comprising M_2 (with masses not exceeding 5 tonnes) and M_3 (with masses exceeding 5 tonnes).

Other vehicle categories defined in 2007/46/EC include vehicles designed and constructed for the carriage of goods (category N, with three subcategories that are based on mass alone), trailers and semi-trailers (category O, which has four subcategories based on mass alone), and off-road vehicles (category G, with no subcategories). The latter comprises a subset of vehicles of classes M and N that also satisfy a number of other criteria relating to mass and particular design and performance features.

Detailed definitions for various subcategories are summarized in Table 12.1 below, for the vehicle categories M, N and O.

Table 12.1 Summary of vehicle subcategory definitions of EC Directive 2007/46/EC [4].

Vehicle category characteristics	Vehicle subcategory characteristics			
	1	2	3	4
M – motor vehicles with at least four wheels designed and constructed for the carriage of passengers	Passenger seats ≤ 8	Passenger seats > 8 Mass ≤ 5 tonnes	Passenger seats > 8 Mass > 5 tonnes	(Not used)
N – motor vehicles with at least four wheels designed and constructed for the carriage of goods	Mass ≤ 3.5 tonnes	Mass > 3.5 tonnes Mass ≤ 12 tonnes	Mass > 12 tonnes	(Not used)
O – trailers and semi-trailers	Mass ≤ 0.75 tonnes	Mass > 0.75 tonnes Mass ≤ 3.5 tonnes	Mass > 3.5 tonnes Mass ≤ 10 tonnes	Mass > 10 tonnes

The requirements for ECWVTA focus on the key safety and environmental issues which have been recognized over the years. The specific type approval requirements vary between the various categories of vehicle defined in 2007/46/EC, as specified in the tables of Annex IV Part 1 of this Directive, reflecting the differing operational roles of these vehicle categories.

Vehicle type approval is a formal process that is generally based on third-party testing by a recognized Technical Service and approval by an appointed Type Approval Body. In addition, type approval includes a Conformity of Production element, based around established quality systems principles (e.g. ISO 9001 [5]). An ECWVTA approval that is issued by one authority will be accepted in all of the other member states of the EC.

The main steps involved in the WVTA process are:

- application by the vehicle or component manufacturer;
- appropriate testing by a recognized technical service;
- granting of the approval by an authorized Approval Authority;
- Conformity of Production established by the manufacturer in agreement with the Approval Authority;
- Certificate of Conformity by the manufacturer for the end-user.

There are multiple methods available for type approval. For whole vehicles, manufacturers may select one of the following:

- *Step-by-step type approval*: A vehicle approval procedure consisting in the step-by-step collection of the whole set of EC type-approval certificates for the systems, components and separate technical units relating to the vehicle, and which leads, at the final stage, to the approval of the whole vehicle.
- *Single-step type approval*: A procedure consisting in the approval of a vehicle as a whole by means of a single operation.
- *Mixed type approval*: A step-by-step type approval procedure for which one or more system approvals are achieved during the final stage of the approval of the whole

vehicle, without it being necessary to issue the EC Type Approval certificates for those systems.

- *Multistage type approval*: A procedure whereby one or more Member States certify that, depending on the state of completion, an incomplete or completed type of vehicle satisfies the relevant administrative provisions and technical requirements of this Directive.

The multistage type-approval approach may be used for complete vehicles that are converted or modified by another manufacturer.

12.3 Trends in Vehicle Automation

12.3.1 EU Policy

Directions for future automotive policy were investigated by the CARS 21 High Level Group, which brought together the main stakeholders (including member states, industry, nongovernmental organizations and MEPs) in 2005 with the aim of examining the main policy areas impacting on the European automotive industry and making recommendations for future public policy and regulatory framework.

Specific aims that were identified by the CARS 21 High Level Group which are of relevance to vehicle automation include the following [6]:

- To investigate the costs, benefits and feasibility of introducing 'emergency braking systems' (EBS) in vehicles (particularly heavy-duty vehicles).
- Proposals to make the inclusion of 'electronic stability control' (ESC) mandatory, starting with heavy-duty vehicles and followed by passenger cars and light-duty vehicles.
- To continue efforts to promote the development, deployment and use of active in-vehicle safety systems and vehicle-infrastructure cooperative systems in the framework of the i2010 Intelligent Car Initiative [7].
- To encourage and support the conditioning of Community financing in the road sector to projects which follow best practice in road safety.
- Call on the Member States to further improve the enforcement of bans on drunk driving, enforcement of speed limits, enforcement of motor-cycle helmet use and to promote and enforce seat-belt use.

The pursuit of these aims has already resulted in amendments to Directive 2007/46/EC, and more can be expected in the future. In particular, a requirement for Brake Assist Systems (BAS) was introduced in EC Regulation 78/2009 [8] in order to enhance the protection of pedestrians and other vulnerable road users. Furthermore, Annex V of EC Regulation 78/2009 also amends Directive 2007/46/EC to include the requirements of EC Regulation 78/2009 as an additional type approval topic under the heading 'Pedestrian Protection'.

12.3.2 Brake Assist Systems

In order to support the implementation of EC Regulation 78/2009, specifications and test methods for BAS are described in EC Regulation 631/2009 [9]. The latter describes three slightly different categories of BAS, which are defined as follows:

- *Category A*: detects an emergency braking condition based on the brake pedal force applied by the driver;
- *Category B*: detects an emergency braking condition based on the brake pedal speed applied by the driver;
- *Category C*: detects an emergency braking condition based on multiple criteria, one of which shall be the rate at which the brake pedal is applied.

The required performance characteristic for BAS of Category A is that when an emergency condition has been sensed by a relatively high pedal force, the additional pedal force to cause full cycling of the ABS (i.e. Anti-lock Braking System) shall be reduced compared to the pedal force required without the BAS in operation. For categories B and C, when an emergency condition has been sensed, at least by a very fast application of the brake pedal, the BAS shall raise the pressure to deliver the maximum achievable braking rate or to cause full cycling of the ABS.

All three BAS categories require the driver to be involved in the braking action, and only unusual brake pedal demand can activate BAS operation for categories A and B. Although the description of Category C offers the opportunity to include inputs from other sources, it would appear that activation of Category C BAS functions must be instigated by driver activity (since the driver is still required to be making an unusual brake pedal demand). Thus, the driver remains the initiator of the braking action, although the nature of his brake pedal demand (and possibly other information sources, for BAS of Category C) may result in the braking system providing different performance characteristics than those that result under less extreme conditions of brake pedal demand.

The provisions of EC Regulation 78/2009 include requirements for all new vehicles of class M_1 (i.e. vehicles designed to carry no more than 8 passengers), as well as N_1 vehicles (i.e. goods vehicles up to 3.5 tonnes) that are derived from M_1 vehicles, to provide BAS functionality from 24 November 2009.

Among the motivations for EC Regulation 661/2009 [10] it is noted that:

- 'Technical progress in the area of advanced vehicle safety systems offers new possibilities for casualty reduction. In order to minimise the number of casualties, it is necessary to introduce some of the relevant new technologies.'
- 'The timetable for the introduction of specific new requirements for the type-approval of vehicles should take into account the technical feasibility of those requirements. In general, the requirements should initially apply only to new types of vehicle. Existing types of vehicle should be allowed an additional time period to comply with the requirements. Furthermore, mandatory installation of tyre pressure monitoring systems should initially apply only to passenger cars. Mandatory installation of other advanced safety features should initially apply only to heavy goods vehicles.'

The new requirements of EC Regulation 661/2009 include provisions for a number of such new technologies. In particular, Article 10 concerning 'advanced vehicle systems' (AVS) requires vehicles in the categories M_2, M_3, N_2 and N_3 (i.e. buses capable of carrying more than 8 passengers, and goods vehicles exceeding 3.5 tonnes) should be equipped with an 'advanced emergency braking system' (AEBS) and a 'lane departure warning system' (LDWS). Furthermore, Article 12 requires vehicles of a wide range of vehicle classes, including the more numerous M_1 and N_1 types, to be equipped with 'electronic stability control' (ESC). The requirement for 'tyre pressure monitoring systems' (TPMS), initially for vehicles of category M_1, is set out in Article 9 of EC Regulation 661/2009.

Such developments are not unique to the EU. Installation of TPMS has been mandated by the USA's National Highway Traffic Safety Administration (NHTSA) for all new light motor vehicles since September 2007 [11]. A requirement for ESC systems [12] has also been issued by the NHTSA, to be implemented from 2012 for a number of vehicle classes including passenger cars and buses. Similar measures have already been announced for a number of other counties, including Australia, Canada, Japan, Korea, and New Zealand.

Compliance with the requirements of Article 12 of EC Regulation 661/2009 (concerning ESC) is required for M_1 and N_1 vehicles from 1 November 2011. The schedule for introducing ESC in other vehicle classes (which is set out in Annex V of EC Regulation 661/2009) is variable, ranging from 1 November 2011 to 11 July 2016. Detailed rules concerning the specific procedures, tests and technical requirements for type-approval of systems relating to Article 10 (i.e. AEBS and LDWS) are required to be adopted by 31 December 2011, and installation of these systems is to be mandatory for new vehicles in classes M_2, N_2, M_3 and N_3 from 1 November 2015 (Article 13.13 of [10]). Compliance with the requirements of Article 10 of EC Regulation 661/2009 regarding TPMS for category M_1 vehicles (i.e. passenger cars) has been required in Europe since 1 November 2012.

12.3.3 Advanced Vehicle Systems

The advanced vehicle systems that are specifically mentioned in EC Regulation 661/2009 are defined as follows:

- *Advanced Emergency Braking System (AEBS)*: a system which can automatically detect an emergency situation and activate the vehicle braking system to decelerate the vehicle with the purpose of avoiding or mitigating a collision;
- *Electronic Stability Control (ESC)*: an electronic control function which improves the dynamic stability of the vehicle;
- *Lane Departure Warning System (LDWS)*: a system to warn the driver of unintentional drift of the vehicle out of its travel lane.

The implication of these descriptions is that, unlike the BAS described in EC Regulation 631/2009, the driver is not necessarily involved in initiating the actions of the AEBS or of the ESC. Technical requirements and test methods for ESC systems to be used in lighter vehicles (such as classes M_1 and N_1) are already described in Annex 9 of UNECE Regulation No. 13-H (introduced in Amendment 2 to Revision 1 of Regulation No. 13-H [13]). However, security against malicious interference is not included in these specifications. Regulations relating to AEBS and LDWS systems have already been developed by UNECE, and adopted by the EC [14].

A report commissioned by the EC on the subject of automated emergency braking systems [15] identifies three categories of such systems:

- *Collision Avoidance Systems (CAS)*: Sensors detect a potential collision and take action to avoid it entirely, taking control away from the driver.
- *Collision Mitigation Braking Systems (CMBS)*: Sensors detect a potential collision but take no immediate action to avoid it until it becomes unavoidable, at which point automatic braking is applied (independent of driver action) in order to reduce the speed, and hence the severity,

of the inevitable collision. Such systems may also trigger additional actions, such as pre-optimization of occupant restraints.
- *Forward Collision Warning (FCW)*: Sensors detect a potential collision and take action to warn the driver. Such systems could also be used to optimize occupant restraints.

Systems providing forward collision warning functions have been available on some EU vehicles since 1999. However, only the 'collision avoidance' and 'collision mitigation braking systems' outlined above correspond to the concept of AEBS as defined in EC Regulation 661/2009. Nonetheless, 'forward collision warning' may perhaps be a necessary adjunct to 'collision avoidance' in order to avoid contravening other legal requirements concerning the control of vehicles (see Section 12.4 below).

12.3.4 Advanced Driving Assistance Systems

Proposals concerning road safety policy directions for 2011–2020 [16] also call for the possibility of widening the deployment of 'advanced driving assistance systems' (ADAS) by retrofitting them to existing commercial and/or private vehicles to be further assessed. In addition, the role of vehicle technology in enforcing speed limits is also discussed, although only in the context of speed limiters for light commercial vehicles. The latter appears to be prompted by as much environmental concerns [17] as by road safety considerations. Thus, there would appear to be no current plans for actively promoting the deployment of 'intelligent speed adaptation' (ISA) systems, and presumably no intention to extend the existing ECWVTA legislation to include such systems in the near future.

The future deployment of 'collision avoidance systems' (CAS) is already anticipated in EC Regulation 78/2009, which notes that, subject to assessment by the Commission, vehicles that are equipped with CAS may be exempted from a certain subset of the type approval test requirements. The requirements that are alluded to are intended to establish the performance of vehicle structural features that should help to reduce the severity of injuries to pedestrians and other vulnerable road users that might arise from accidental impacts. The implication here is that it is expected that at least some CAS will be able to detect and avoid potential collisions with pedestrians and other vulnerable road users (i.e. cyclists, motorcyclists, horses and their riders, as well as infirm and disabled users of low-speed personal mobility vehicles), thus obviating the need to employ structural design measures to limit the severity of impacts between the vehicle and human bodies. However, these structural measures aim primarily to reduce injury to pedestrians and may not actually meet the needs of cyclists [18] and other vulnerable road users.

Further developments are also anticipated for the future in EC Regulation 661/2009, which notes that:

- 'The Commission should assess the feasibility of extending the mandatory installation of tyre pressure monitoring systems, lane departure warning systems and advanced emergency braking systems to other categories of vehicle and, if appropriate, propose an amendment to this Regulation.'
- 'The Commission should continue to assess the technical and economic feasibility and market maturity of other advanced safety features, and present a report, including, if appropriate, proposals for amendment to this Regulation, by 1 December 2012, and every three years thereafter.'

The intention behind such legislation is to promote the deployment of further advanced safety features (that are deemed to be sufficiently mature) in future vehicles with mandatory installation that will increase the pace of market penetration, which is likely to lead to a reduction in costs, thereby ensuring that society benefits (sooner rather than later) from the improvements in road safety, enhanced transport efficiency and reduced environmental impacts that new technologies are expected to deliver. A possible concern, however, is that reliance on such technology may also result in more reckless drivers being willing to take ever greater risks, which could potentially have the effect of negating some of the benefits for road safety.

12.3.5 Categorization of Vehicle Automation Levels

In response to the emergence of increasingly sophisticated driver assistance systems, a systematic categorization of vehicle automation levels was proposed [19] by the USA's NHTSA. This scheme, which is summarized in Table 12.2, takes account the distribution of responsibilities for key driving activities, including the fall-back position in case of system failures. Five levels of automation are described, ranging from 'no automation', in which the human driver is responsible for all driving tasks and environment monitoring activities, through to 'full automation', where the vehicle system assumes responsibility for all driving and related activities.

A similar but more structured approach has been proposed in SAE J3016 [20]. This scheme, which is summarized in Table 12.3, takes account of the distribution of responsibilities for key driving activities, including the fall-back position in case of system failures. Six levels of automation are described, ranging from 'no automation', in which the human driver is responsible for all driving tasks and environment monitoring activities, through to 'full automation', where the vehicle system assumes responsibility for all driving and related activities.

Another such classification approach [21] has been developed by an expert group established by the German road transport institution BASt (Bundesanstalt für Strassenwesen). This scheme, which is summarized in Table 12.4, not only categorizes the degree of automation, but also differentiates between different application classes. Specific applications can then be categorized by placing them in a matrix that reflects both their automation and application characteristics.

These categorizations of vehicle automation are useful in considering how different levels of automation may relate to other relevant legislation with which such vehicles may need to comply, such as international treaties concerning road traffic.

12.4 Vienna Convention on Road Traffic

The 1968 Vienna Convention on Road Traffic [22] is an international treaty that aims to facilitate international road traffic and to increase road safety through the adoption of uniform road traffic rules. In the signatory countries it replaces previous road traffic conventions, most notably the 1949 Geneva Convention on Road Traffic [23]. However, a number of countries (most notably Australia, China, India, New Zealand and the USA) are not signatories to the 1968 Vienna Convention, with the result that the 1949 Geneva Convention still applies in these regions.

With regard to vehicle automation, it should be noted that the Vienna Convention requires (see Article 8 of [22]) that:

Table 12.2 Levels of vehicle automation, summarizing NHTSA definitions [19].

Level	Nature	Description
0	No automation	Driver has complete and sole control of primary vehicle controls (i.e. brake, steering, throttle, and motive power) at all times. Driver is solely responsible for monitoring the road environment and for safe operation of all vehicle controls.
1	Function-specific automation	Involves automation of one or more specific control functions; if multiple functions are automated, they operate independently from each other. Driver has overall control, and is solely responsible for safe operation, but can choose to cede limited authority over a primary control (as in adaptive cruise control). Vehicle can automatically assume limited authority over a primary control (as in electronic stability control), or the automated system can provide added control to aid the driver in certain normal driving or crash-imminent situations (e.g. dynamic brake support in emergencies). Vehicle may have multiple capabilities combining individual driver support and crash avoidance technologies, but does not replace driver vigilance and does not take driving responsibility from the driver. Vehicle's automated system may assist or augment the driver in operating only one of the primary controls – either steering or braking/throttle controls (but not both). No combination of vehicle control systems working in unison enables driver to be disengaged from physically operating the vehicle by having hands off the steering wheel and feet off the pedals at the same time.
2	Combined function automation	Automation of at least two primary control functions designed to work in unison to relieve the driver of control of those functions. Vehicles at this level of automation can utilize shared authority when driver cedes active primary control in certain limited driving situations. Driver remains responsible for monitoring the road environment and ensuring safe operation, and is expected to be available to take control at all times and at short notice. System can relinquish control with no advance warning and the driver must be ready to take control and operate the vehicle safely.
3	Limited self-driving automation	Vehicle enables the driver to cede full control of all safety-critical functions under certain traffic or environmental conditions and in those conditions to rely heavily on the vehicle to monitor for changes in those conditions requiring transition back to driver control. Driver is expected to be available to take occasional control, but with sufficiently comfortable transition time. Vehicle is designed to ensure safe operation during automated driving.
4	Full self-driving automation	Vehicle is designed to perform all safety-critical driving functions and monitor road conditions for an entire trip. Driver provides destination or navigation input, but is not expected to be available for control at any time during the trip. This class includes both occupied and unoccupied vehicles. By design, responsibility for safe operation rests solely with the automated vehicle system.

Table 12.3 Levels of vehicle automation, summarizing SAE J3016 definitions [20].

Degree of automation		Distribution of responsibilities			System capability (driving modes)
Level	Nature	Steering, acceleration and braking	Monitoring driving environment	Fall-back position for driving tasks	
0	No automation	Driver	Driver	Driver	Not applicable
1	Driver assistance	Driver and system	Driver	Driver	Some modes
2	Partial automation	Vehicle system	Driver	Driver	Some modes
3	Conditional automation	Vehicle system	Vehicle system	Driver	Some modes
4	High automation	Vehicle system	Vehicle system	Vehicle system	Some modes
5	Full automation	Vehicle system	Vehicle system	Vehicle system	All modes

Table 12.4 Characteristics of vehicle automation and application categories proposed by BASt [21].

Automation level	Characteristics	Application type	Characteristics
Partly automated	Driver must supervise at all times. No nondriving activities may be undertaken.	Low speed manoeuvring	At low speeds (<10 km/hour), with minimal risk of injury.
Highly automated	System is aware of its performance limits and will request driver to take control with sufficient warning. Nondriving activities possible to same degree.	Manoeuvres of limited duration	Functions that can be completed within a few seconds.
Fully automated	No driver supervision is necessary. Nondriving activities are possible.	Permanent driving	Functions that are active for more extended periods, or permanently.

- 'Every moving vehicle or combination of vehicles shall have a driver.'
- 'Every driver shall at all times be able to control his vehicle or to guide his animals.'
- 'A driver of a vehicle shall at all times minimize any activity other than driving.'

Similar provisions are also to be found in the 1949 Geneva Convention (see Article 8 of [23]).

The objective of minimizing driver distraction has resulted in an additional requirement to prohibit the use of handheld mobile phones while driving, which (along with a number of other amendments [24]) was adopted in 2005.

Other relevant definitions (from Article 1 of [22]) are as follows:

- 'Combination of vehicles means coupled vehicles which travel on the road as a unit.'
- 'Driver means any person who drives a motor vehicle or other vehicle (including a cycle), or who guides cattle, singly or in herds, or flocks, or draught, pack or saddle animals on a road.'

Thus, in terms of the Vienna Convention, a driver must be a person, not a system, and must always be able to control the vehicle, or combination of coupled vehicles (or animals).

12.4.1 Implications for Driving Assistance Systems

The ABS and BAS systems provide enhanced braking support, but both of these functions require the driver to initiate them by applying the brakes. Thus, they can be considered as 'driving assistance systems' (DAS), with the driver remaining in control. The situation is similar for 'cruise control' (CC), which is used to maintain a fixed speed, and 'adaptive cruise control' (ACC), which tracks the speed of the vehicle in front by means of a radar system. These systems are manually engaged by the driver, who continues to drive the vehicle, and either manually disengaged or automatically disengaged (e.g. if the foot pedals are depressed, or if the speed of the vehicle in front falls below a threshold level). Thus, the driver remains in overall control of the vehicle and is able to override these systems when necessary.

However, the driver is already no longer in control, by definition, in situations where a collision has become unavoidable. Thus, the use of ADAS, which can provide some degree of mitigation in circumstances that are beyond the control of the driver, is probably justifiable as not contravening the requirements of the Vienna Convention.

The concept of 'controllability' for automotive applications (see Table 12.5) was originally developed by the EU project 'DRIVE Safely' [25], subsequently adopted by MISRA for on-board software development [26,27], and is now used as a qualitative probability measure in safety risk analysis methods that are applied to vehicle functional safety engineering (Part 3 of [28]). Approaches for evaluating controllability and undertaking risk assessments in the development of driver assistance systems have been proposed in the Code of Practice developed by the EU project 'RESPONSE 3' [29].

It has been suggested [30] that, in terms of this driver controllability criterion, while CMBS do not contravene the Vienna Convention this is not so for CAS, where the objective of the system is to take control from the driver before a collision becomes unavoidable (i.e. while the situation is still judged to be controllable). In order to avoid contravention of the Vienna Convention by DAS, it has also been proposed [31] that:

> 'The system must only "override" the driver if the latter is unable to intervene (e.g. loss of consciousness) and this is evident from the driver's failure to respond to certain information provided by the system. Automatic instant emergency braking initiated by a braking assistant in a speeding situation could impact vehicle handling and lead to the wrong reactions.'

Table 12.5 Controllability of automotive safety hazards [28].

Controllability class	Meaning
0	Controllable in general
1	Simply controllable
2	Normally controllable
3	Difficult to control or uncontrollable

In order to comply with this position, a CAS would need to warn the driver of the impending hazard and only take action if the driver fails to respond within a reasonable period of time (thereby demonstrating a lack of driver control of the situation, due to inattention or some form of physical incapacity). This would give a driver who is able to control the situation the opportunity to override the action of the CAS.

However, ESC systems, which may apply increased or decreased braking pressure amongst other actions (rather than just reduced braking pressure as in ABS), are intended to operate automatically without direct driver initiation at vehicle speeds in excess of 20 kph unless the driver has disabled the system or the vehicle is being driven in reverse. Such systems would comply with the Vienna Convention if, like the CMBS, they are only activated when the vehicle is no longer controllable by the driver, or after the driver has been warned of the threat but has failed to take action that would override the action of the ESC system.

12.4.2 Proposed Amendments

Proposals for possible amendments to the Vienna Convention have been made by UNECE, with the aim of ensuring that systems that are type approved under UNECE regulations are also accepted as complying with the Vienna Convention. These proposals [32] included a definition of a Driving Assistance System as follows:

> 'Driving Assistance System means a built-in system intended to help the driver in performing his driving task and which have an influence on the way the vehicle is driven, especially aimed at the prevention of road accidents.'

In addition, the following paragraph was proposed [32] as an addition to Article 13 (which is concerned with speed and distance between vehicles):

> 'Driving assistance systems shall not be considered contrary to the principles mentioned in paragraph 1 of this Article and mentioned in paragraphs 1 and 5 of Article 8 as well, provided that:
>
> - either these systems are overridable at any time or can be switched off,
> - or they only optimise at technical level some functions which operating depends only on the driver,
> - or they operate in case of emergency when the driver lost or is about to lose the control of the vehicle,
> - or the intervention of these systems is identical with a usual property of a motor vehicle (e.g. speed limiting device).'

The UNECE Inland Transport Committee Working Party of Road Traffic Safety recommended that these criteria should be observed when establishing rules for the design of a given DAS. The paragraph 1 and 5 of Article 8 that is referred to above are the requirement that every moving vehicle or combination of vehicles shall have a driver (Article 8.1) and that every driver shall at all times be able to control his vehicle (Article 8.5). The paragraph 1 of Article 13 that is referred to states that:

> 'Every driver of a vehicle shall in all circumstances have his vehicle under control so as to be able to exercise due and proper care and to be at all times in a position to perform all manoeuvres required of him. He shall, when adjusting the speed of his vehicle, pay constant regard to the

circumstances, in particular the lie of the land, the state of the road, the condition and load of his vehicle, the weather conditions and the density of traffic, so as to be able to stop his vehicle within his range of forward vision and short of any foreseeable obstruction. He shall slow down and if necessary stop whenever circumstances so require, and particularly when visibility is not good.'

A number of vehicle manufacturers now offer automatic parking systems, which will autonomously manoeuvre the car into a selected parking space with the aid of on-board sensors to identify the positions of nearby obstacles. In most such systems the driver still controls the speed of the vehicle with the accelerator and brake pedals, but any intervention with the steering process causes the vehicle to return to full control to the driver. Systems of this type would therefore comply with the first of the criteria proposed by UNECE for DAS to be acceptable under the Vienna Convention. Vehicles with automatic parking capability (which depends on the availability of electric power steering) may also offer an active lane keeping support function, in which the steering will be adjusted if the vehicle is determined to be on course to leave the current traffic lane without the relevant indicator being activated (rather than just issuing a warning to the driver, as in LDWS). Automatic suspension under driver intervention and the option to turn off this feature would also enable such systems to satisfy the first of the proposed acceptable DAS criteria.

Systems providing 'intelligent speed adaptation' (ISA) have been widely studied (including in practical field trials), and are regarded by safety organizations such as the UK Royal Society for the Prevention of Accidents [33] and the European Transport Safety Council [34] as offering significant potential to reduce both the occurrence and the severity of road accidents. The last of the DAS criteria proposed by UNECE for inclusion in the Vienna Convention would also permit the adoption of ISA systems that actively restrict the maximum speed of vehicles according to prevailing local limits (other ISA systems simply provide warnings to the driver). However, there are other types of system under investigation that are probably still outside the scope of these proposed amendments.

More recent proposals [35] include the addition of points (a) and (b) below to paragraph 5 of Article 13:

'5. Every driver shall at all times be able to control his vehicle or to guide his animals.

(a) Vehicle systems which influence the way vehicles are driven shall be deemed to be in conformity with the first sentence of this paragraph and with paragraph 1 of Article 13, when they are in conformity with the conditions of construction, fitting and utilization according to international legal instruments concerning wheeled vehicles, equipment and parts which can be fitted and/or be used on wheeled vehicles.

(b) Vehicle systems which influence the way vehicles are driven and are not in conformity with the aforementioned conditions of construction, fitting and utilization, shall be deemed to be in conformity with the first sentence of this paragraph and with paragraph 1 of Article 13, when such systems can be overridden or switched off by the driver.'

Corresponding changes have also been proposed [36] in relation to the 1949 Geneva Convention on Road Traffic, as some countries that have not ratified the Vienna Convention may still be parties to the 1949 Convention.

The aim of these proposed amendments is to remove legal obstacles to the deployment of more automated driving systems, allowing a car to drive itself provided that the system can be overridden or switched off by the driver. A driver must therefore be present and able to take control at any time.

12.4.3 Implications for Autonomous Driving

A modified Toyota Prius has been reported [37] to have been operating autonomously on public roads in California, although with a human co-pilot constantly monitoring performance and ready to take manual control if needed (this is reported to have been necessary when an earlier vehicle unexpectedly veered off a road in 2005 [38]). In this case, therefore, the driver has voluntarily given up control to an on-board system in circumstances that are clearly not uncontrollable, although retaining a supervisory role.

However, such a scheme is of limited practical benefit to the driver, as the supervision activity will require the same level of attention as when actually driving, but is probably more difficult to maintain while not actively involved in the driving task. It is more likely that the use of a supervising driver is simply a stepping stone towards the ultimate objective of fully autonomous driving without reliance on human supervision (i.e. the highest automation levels outlined in Section 12.3.5). In the USA, laws permitting the use of automated cars have been in force in California, Florida and Nevada since 2012 [39], although a driver is still required for supervision and back-up purposes. Trials involving 100 such vehicles have been initiated in Sweden [40], and the national governments are seeking to modify traffic legislation to allow similar demonstrations in other countries. An experimental fully autonomous vehicle [41], without a steering wheel and having no brake or accelerator controls, has been developed but is currently limited to a maximum speed of 25 mph and is not yet permitted to operate on public roads.

The concept of 'platooning', in which a number of vehicles travel as an ensemble for some period of time (also described as 'road-trains'), has been investigated in a number of collaborative research projects supported with EU and national funding. The perceived benefits of such schemes include improved traffic flow, higher vehicle density on the road, reduced engine emissions, and improvements in road safety. In the EU project SARTRE [41,42] the platoon is envisaged as comprising a 'lead vehicle' that is driven by a trained, professional driver, together with one or more 'following vehicles' that are being driven autonomously (but linked to the 'lead vehicle' via wireless communication), thus allowing the drivers of the 'following vehicles' to perform tasks other than driving their vehicles. Thus, the 'following vehicles' would not be under the control of their drivers whilst part of the platoon, despite the fact that the driving situation is not expected to be uncontrollable. Nonetheless, they do have the ability to choose to join or leave the platoon, and therefore have the opportunity to override the external control of the driver of the 'lead vehicle'. However, the override capability of the drivers of individual 'following vehicles' may need to be limited under some circumstances in order to avoid compromising the safety of other platoon members.

The platoon concept described above does not comply with the view that driver assistance systems should override the driver only if the latter is unable to maintain control. In the platoon scenario the 'following vehicle' drivers have opted to give up direct control of their vehicle to an autonomous system, although the autonomous systems are linked by wireless communication to a 'lead vehicle' that is under the control of a driver. It may also not fully comply with the amendments proposed by UNECE, which require that it must be possible to override DAS at any time. Furthermore, it is possible that the 'lead vehicle' may also encounter situations in which on-board systems take control from the driver. In such circumstances the drivers of the 'following vehicles' cannot be expected to immediately resume control of their own vehicles, since they have given up control in order to undertake other activities whilst part of the platoon, so the automatic systems of the 'lead vehicle' would effectively be in control of the 'following vehicles' in the platoon as well.

Although the Vienna Convention does allow for coupled vehicles which travel on the road as a single unit, provided that they are controlled by a driver, references to coupling elsewhere in the text clearly indicate that this was expected to be a mechanical coupling (e.g. for trailers and articulated vehicles). In addition, even the proposed amendments to the Vienna Convention would not support fully autonomous driving. Consequently, on-going amendment and clarification of the Vienna Convention will be necessary in order to ensure that it takes account of recent and anticipated technological developments.

12.5 Liability Issues

Historically, the responsibility for road accidents and failure to comply with traffic regulations has most commonly been attributed to human errors. Typical examples include the following:

- failure to pay full attention while driving;
- failure to follow the accepted rules;
- failure to maintain the vehicle correctly;
- failure to take adequate account of local traffic and/or environmental conditions.

Less frequently, the cause may be attributed to failures or defects of specific vehicle parts or systems, or perhaps due to some shortcoming of the road management (e.g. inadequate road signs or poor junction design). With the introduction of ADAS, however, this situation is likely to become increasingly complex.

12.5.1 Identifying Responsibilities

The basic manual braking system has already evolved over recent years, with the addition of on-board sensors, actuators and more sophisticated control algorithms, through enhancing the performance of braking actions instigated by the driver (i.e. ABS and BAS) and on to automatically supporting the driving process (i.e. ESC). These systems enhance safety and are now mandatory for new vehicles. At present, however, the driver still remains responsible for the driving activity and for sensing and processing information received from outside the vehicle (such as local traffic and weather conditions, prevailing speed limits, etc.). Nonetheless, failure to ensure that the user is fully informed of the features and limitations of the available driving support functions may lead to increased accidents if the driver mistakenly believes that the system can be relied upon to mitigate the potential effects of poor driving practices. Behavioural adaptation of this kind is reported to have been observed in connection with a number of different driver support systems [43].

However, the 'presentation' of a product is an important factor in relation to EU Directive 2001/95/EC [44] on general product safety and Directive 85/374/EEC [45] on liability for defective products. For example, inadequate instructions or misleading advertisements regarding the use of ADAS equipment could be regarded as making the system 'defective' through inappropriate influence on customer expectations. It is essential for the user to have a correct understanding of the operational characteristics and limitations of such systems in order to ensure that they can be used in a safe manner. Nonetheless, warnings do not mitigate the impact of safety limitations that could have been avoided using an alternative design that was economically viable [46].

Further considerations of the Product Safety and Product Liability Directives include 'the use to which it could reasonably be expected that the product would be put' (Article 6.1(b) of [45]) and 'normal or reasonably foreseeable conditions of use' (Article 2(b) of [44]). Thus, there is an obligation on the product producer to take account of the possible impact of foreseeable misuse and nonideal operating conditions when developing their products. There is already widespread experience of malicious interference with computing systems via the Internet, including attacks against individual home computers as well as institutional networks. Thus, the requirements of the Product Safety and Product Liability Directives would lead to the expectation that ADAS producers should anticipate the possibility of attacks on the security of in-vehicle assets, which could exploit the new wireless communications channels that are now beginning to be provided in modern vehicles. Consequently, failure to address the security of on-board vehicle networks, which may have potential safety implications as well as other possible impacts, could also be considered as a product defect.

The ADAS producer could therefore be held liable for any deaths or injuries that could be attributed to such defects, as well as compensation for associated physical damage sustained by other products (although not the defective product itself) provided that they are intended for private use (see below: the scope of Product Liability is generally restricted to private use). Other possible types of damage that might result (i.e. nonmaterial damage types, such as financial losses or loss of reputation) are not covered by the Product Safety and Product Liability directives and would need to be pursued under the applicable national laws.

The EU-supported project RESPONSE 3 classified ADAS products in terms of three generic types [47], drawing conclusions about the associated liability issues as follows:

- *Information and warning systems* – where liability generally remains with the driver, who remains in full control although the ADAS producer or distributor may be liable if incorrect or inaccurate information is provided by the system.
- *Intervention systems for which driver override is possible at any time* – where the driver retains overall responsibility and may therefore be liable, depending on the circumstances, although system malfunctions may also lead to liability for the ADAS producer or distributor.
- *Intervention systems which the driver either cannot override, or where override is impracticable (because of human reaction time)* – where ADAS producers and distributors are likely to be liable as the driver is not in control.

The functionality that is envisaged for future vehicle systems will be increasingly dependent on inputs from a variety of external systems (e.g. positioning and navigation signals, and messages from other vehicles or roadside infrastructure), as well as a widening array of on-board sensors, actuators and electronic control capabilities. Such systems may diminish the driver's current role, and perhaps ultimately replace the driver with fully autonomous driving systems. In these scenarios the quality of information received from outside the car, the reliability of wireless communication channels, and the dependability of the on-board systems will be increasingly significant factors for successful and safe operation. Consequently, responsibility for accidents might be expected to shift away from the driver towards vehicle manufacturers and their on-board systems suppliers, as well as to external information providers.

Under fully autonomous operation there is no driver involvement, but determining whether responsibility lies with the on-board systems or the external information sources may not be easy to establish. However, it may prove difficult to establish the responsibility of actors other than the driver in circumstances where the driver still has a role. This was demonstrated by the recent investigation into unintended acceleration in Toyota vehicles, which was carried out by NHTSA. The origin of these behavioural anomalies was widely debated, with driver error, electromagnetic compatibility, mechanical issues and software defects all mooted as possible causes [48]. The associated NASA report [49] concluded that although the unintended acceleration events were unlikely to have been caused by the electronic systems, this was not considered to be impossible. The NASA investigations were unable to demonstrate that the unintended acceleration events were due to unexpected behaviour of the electronic systems. However, exhaustive evaluations were not feasible due to the very large number of possible combinations of system inputs. Thus, the absence of evidence of such effects cannot be assumed to be evidence of their absence. Consequently, the possibility that electronic systems defects could have caused the unintended acceleration events cannot be ruled out based on the available data.

12.5.2 Event Data Recorders

Showing that a mechanical part has broken, perhaps then resulting in the failure of a safety-related function such as the braking system, should be relatively straightforward. However, establishing that a vehicle control system responded in an unexpected way to a particular combination of transient inputs is likely to be extremely difficult. For this reason, it has been suggested that there should perhaps be an obligation to install an event data recorder (similar to the so-called 'black box', which has been used in aircraft for many years) when ADAS are more widely deployed [46]. The first standard for such a device, known as a 'motor vehicle event data recorder' (MVEDR), was developed by the IEEE in 2004 (IEEE 1616 [50]), and has been amended in 2010 [51] to address potential security issues associated with MVEDRs, including:

- *data tampering* – modification, removal, erasure of, or otherwise rendering in-operative, any device or element, including MVEDRs;
- *VIN theft* – duplication and transfer of unique vehicle identification numbers, enabling stolen cars to be passed off as nonstolen;
- *odometer fraud* – rolling back of vehicle odometers, reducing the reported total distance travelled by the vehicle;
- *privacy* – prevention of the misuse of collected data relating to vehicle owners.

The availability of MVEDR data could raise a number of possible privacy and liability issues. For example, insurance companies may have an interest in using such data to influence vehicle insurance premiums. In the event of an accident occurring, they might perhaps wish to try to use MVEDR data in order to attempt shifting liability towards:

- the driver, if the data suggest that the driver has been behaving recklessly;
- the ADAS producer, if the data suggest that the accident could be attributed to a defect in the performance of the electronic systems;
- organizations providing information to the vehicle, if the data suggest that erroneous information caused or contributed to the accident.

The United States NHTSA and Federal Motor Carrier Safety Administration (FMCSA) both take the position that the MVEDR and its data belong to the vehicle owner [52], with the implication that no private party could force the vehicle owner to relinquish that data without consent. However, it is conceivable that insurance companies could perhaps require the vehicle owner to provide consent as a condition of the insurance policy, or alternatively offer an incentive such as reduced insurance premiums in return for such consent. The latter approach has already been adopted in the USA [53]. The monitoring device installed in the car does not track where people drive, but only their driving patterns. A similar scheme (recording speed and acceleration) has recently been launched in the UK targeted at young and inexperienced drivers, for whom car insurance costs are becoming prohibitive [54].

Vehicle manufacturers in the USA have been voluntarily installing MVEDRs as part of car and light truck airbag modules since 1996. These devices are triggered by conditions such as rapid changes in vehicle speed in order to collect a variety of data during crash and near-crash events. The data collected typically includes speed at time of impact, steering angle, whether brakes were applied, and seatbelt usage during the crash. The United States NHTSA requires MVEDRs, where voluntarily fitted, to meet specific data collection standards from September 2010 [55] for light vehicles. Furthermore, the findings of the NHTSA-NASA investigation of unintended acceleration in Toyota vehicles [56] include (amongst others) recommendations to:

- consider initiating rulemaking to;
- require brake override systems (to ensure that the brake has priority over the throttle);
- standardize operation of keyless ignition systems (so that drivers know how to stop the engine quickly);
- require the installation of MVEDRs in all passenger vehicles;
- begin broad research on the reliability and security of electronic control systems for vehicles by examining existing industry and international standards for best practices and relevance to automotive applications.

In this proposed reliability and security research, NHTSA plan to give full consideration to NASA's recommendation that NHTSA should consider controls for managing safety critical functions in vehicles, based on those currently applied to the rail, aerospace, military, and medical industries.

12.6 Best Practice for Complex Systems Development

The difficulties involved in developing and demonstrating the reliability of ADAS derive from their inherent complexity. The following definitions, which derive from the defence systems domain [57], serve to demonstrate this issue.

- 'Simple. A hardware item may be classified as "simple" if its design is suitable for exhaustive simulation and test.'
- 'Complex. The degree to which a system or component has a design or implementation that is difficult to understand and verify. For the purposes of this document "complex" is defined as "unsuited to the application of exhaustive test."'
- 'Exhaustive test. Thorough exercising of a component through test or analysis using values applied at its terminals. The aim is to exercise all possible combinations. The phrase 100% test is not used because the number of possible tests is infinite, taking account of all physical properties. Judgement is involved which needs to be justified in a safety case.'

A consequence of complexity is that the traditional test-based methods used in establishing compliance with WVTA requirements, which have been developed for validating the safety performance of relatively simple and largely independent mechanical and electrical systems, are unlikely to be suitable for increasingly sophisticated mechatronic vehicle systems with significant software content and widespread inter-dependencies. Exhaustive testing is not practicable for such complex systems because the number of possible system states (i.e. combinations of inputs) is extremely large. Furthermore, in complex, software based systems it is systematic, rather than random, faults that predominate, with the result that testing to establish probabilistic failure rates is also likely to be of impracticable duration.

12.6.1 Safety Case

The recommended approach for establishing the safety of complex electronic control systems, based on experience in safety-critical applications found in the aerospace, defence, nuclear, rail and off-shore oil industries, is to create a safety argument to show that the system is acceptably safe for the intended application and for the intended operating environment. The important points here are that complete safety is recognized as unachievable, although mitigation measures must be implemented as necessary to ensure that any residual risks are deemed to be acceptable, and that the safety argument only applies to the intended application and operating environment. For networked vehicles, however, the operating environment is known to include hackers and criminals, who are already actively engaged in security attacks against existing computer networks. Thus, a safety case for such applications should also take account of safety-related security threats.

The safety argument and supporting evidence should be documented in a 'safety case', which should [58]:

- make an explicit set of claims about the properties of the system;
- identify the supporting evidence (i.e. facts, assumptions, or subclaims derived from lower-level arguments);
- provide a set of safety arguments that link the claims to the evidence;
- make clear the assumptions and judgements underlying the arguments;
- allow for different viewpoints and levels of detail.

The safety case should be subject to independent assessment and audit by a suitably qualified third party. Constructing the safety case in the form of a relatively simple top-level safety claim supported by a hierarchy of subclaims makes it easier to understand the main arguments and to partition the safety case development activities. Claims can be made more robust by using independent evidence and more than one argument to support the claim, ideally with different styles of safety argument. A catalogue of generic patterns for a number of canonical safety argument types that could be used in this have recently been described and applied to an automotive case study [59].

The EC WVTA legislation is also moving in the direction of a Safety Case approach for vehicle systems based on complex electronic controls. Extensions have already been added to the UNECE regulations concerning braking [60] and steering [61], which detail special requirements to be applied to the safety aspects of complex electronic vehicle control systems. It is expected that similar extensions will eventually be added to all regulations concerning vehicle systems that may involve complex electronic control systems.

12.6.2 *Safety Development Processes*

The international standard IEC 61508 [62], concerning the functional safety of safety-related electrical, electronic or programmable electronic systems, provides a basic functional safety standard applicable to all kinds of industry. It has its origins in the process control industry sector, but is also intended to provide a basis for the development of sector-specific safety standards. In particular, IEC 61508 reflects the following views on safety risks:

- zero risk is unachievable;
- safety must be considered from the outset;
- unacceptable risks must be reduced 'as low as reasonably practicable' (ALARP).

Consequently, hazard identification, analysis of safety risks and assessment of the need for measures to mitigate such risks are key elements of the IEC 61508 approach.

Another important aspect of the IEC 61508 approach is the requirement for increasingly rigorous development processes to be applied for more critical safety functions, which is intended to provide greater confidence in the reliability of complex systems that are not amenable to exhaustive testing. The safety requirements are described in terms of 'safety functional requirements' (i.e. what the function should do) and 'safety integrity requirements" (i.e. the likelihood that the safety function will be carried out satisfactorily). The safety integrity requirements of the safety functions are specified in terms of a number of discrete levels, known as 'safety integrity levels' (SILs), which are related to the risk level and range from SIL1 to SIL4. The SILs reflect requirements for increasingly rigorous processes to be applied in a range of development activities, ranging from specification and design, through configuration management, testing, validation and verification, to independent assessment. The safety argument for the achievement of a particular SIL should be as follows [63]:

> The requirement was for a SIL X system, and good practice decreed that I adhered to the standard's processes for a SIL X system. In doing so, I have generated the evidence appropriate to a SIL X system, and assessment of the evidence has found that I have adhered to the defined processes.

An interpretation of the IEC 61508 approach that has been developed specifically for the automotive industry is provided by ISO 26262 [28], which:

- provides an automotive safety lifecycle (management, development, production, operation, service, decommissioning) and supports tailoring of the necessary activities during these lifecycle phases;
- covers functional safety aspects of the entire development process (including such activities as requirements specification as well as system design, implementation, integration, verification, validation, and configuration);
- provides an automotive-specific and risk-based approach for determining risk classes ('automotive safety integrity levels', ASILs) that are analogous the IEC 61508 SILs;
- uses ASILs for specifying the necessary safety integrity requirements for the safety goals that are identified as needed in order to achieve an acceptable level of residual risk, where class D represents the highest integrity category and class A is the lowest;
- provides suitable requirements for validation and confirmation measures to ensure that a sufficient and acceptable level of safety is achieved.

The main difference between the ASILs of ISO 26262 and the SILs of IEC 61508 is that the latter employ quantitative target probability values, while the ASILs are based on qualitative measures.

Related guidance regarding safety analysis for vehicle based programmable systems has also been developed by the Motor Industry Software Reliability Association (MISRA) [29]. This is based on an iterative process, starting with a Preliminary Safety Analysis (PSA) carried out at the system concept stage. This is subsequently refined through more comprehensive Detailed Safety Analysis (DSA) activities as the system design and development activities progress. Thus, the MISRA safety engineering process is expected to be an iterative activity that is developed and refined as the system evolves and matures.

12.6.3 ECWVTA Requirements

The ECWVTA Directive [4] also contains requirements related to 'Complex Electronic Systems'. Unlike the definition of complexity reported at the beginning of this section, complexity in the context of these regulations refers to a hierarchy of control, where driver demands on the basic control functions of a vehicle, such as braking and steering, can be overridden or modified by an electronic system. Systems that fall within the scope of the regulations include functionality such as ABS and ESC, as well as driver assistance functions such as LDWS, AEBS and adaptive cruise control. These requirements are part of UNECE Regulations 13 [64] and 13-H [13] on braking, Regulation 79 on steering [65], and of EC Regulation 661/2009 [10] mandating the fitment of AEBS to commercial vehicles.

The 'Complex Electronics Systems' requirements are effectively a very high level regulatory check of an implicit safety case for the functionality. The applicant is required to submit a set of information that includes a description of the development process used for software and the safety strategies (safety concept) that indicates how the system is intended to perform in the presence of faults. The approval authority then conducts testing where faults are injected into the system and the performance in accordance with the safety concept verified.

The requirements do not currently refer to ISO 26262, but it should be trivial to demonstrate compliance with the regulation if an ISO 26262 development process has been followed. At present there are a wide range of interpretations of the requirements, with some approval authorities accepting the minimum documentation requirements as per the letter of the regulation and others insisting on application of IEC 61508 or ISO 26262.

12.6.4 Cyber Security Issues

The need to consider cyber security issues in relation to ADAS has motivated a number of European collaborative research projects (e.g. EVITA [2], SeVeCom [66]) and is also identified in the NHTSA's Automated Vehicles Policy [19]. As with safety, zero security risk is in practice unachievable and a similar risk-based approach is needed in order to evaluate potential security threats and to identify requirements to mitigate those threats for which the level of risk is judged to be unacceptable.

The standard IEC 15408 [67] is concerned with security evaluation for IT products, but does not explicitly address the possible safety implications of security breaches for safety-critical control systems. A further limitation is that it does not provide a framework for risk analysis. Methods for evaluating the probability of a successful attack (described as 'attack potential') are described in IEC 18045 [68], but the severity of the impact is not evaluated to allow risk to be assessed. Risk analysis in an IT security context is outlined in ISO/IEC TR 15446 [69] and described in more detail elsewhere (e.g. ISO/IEC 13335 [70], NIST IT Security Handbook [71]).

In IEC 15408 the concept of 'evaluation assurance levels' (EAL) has a similar role for security considerations to the SIL and ASIL categories used in the safety context. The EALs are similarly associated with graded levels of increasing development rigour, ranging from functional testing where the security threat is not deemed to be serious (EAL1), through to formally (i.e. mathematically) verified design and testing for cases where the security risks are judged to be extremely high (EAL7). The similarities between the EAL and SIL/ASIL concepts suggest the potential for developing a unified approach for automotive safety and security [72]. Similar observations have also been made with regard to the security and safety of mobile ad-hoc network applications [73]. Unifying safety and security engineering processes offers potential benefits in terms of reduced costs through sharing of evidence and risk analysis for those applications where security may also have possible safety implications.

The MISRA 'Development Guidelines for Vehicle Based Software' [26,74] identified the need to protect vehicle software from unauthorized access that could compromise software, or to provide detection of tampering. It was also noted that unauthorized reprogramming of vehicle control systems (so-called 'chipping') may cause the vehicle manufacturer to become legally liable in some countries.

The 2010 edition of IEC 61508 (2nd Edition [75]) now includes consideration of security issues with regard to their potential impact on safety. Possible malevolent and unauthorized actions are required to be addressed during the hazard and risk analysis. If a security threat is seen as being reasonably foreseeable, then a security threat analysis should be carried out and if security threats have been identified then a vulnerability analysis (i.e. security risk analysis) should be undertaken in order to specify corresponding security requirements. However, security threats that are not safety-related, such as those affecting privacy or financial security, are beyond the scope of IEC 61508.

For the purposes of the EVITA project a risk analysis approach [76] was developed from the IEC 61508-based concepts of ISO 26262 and MISRA, which were extended to encompass nonsafety aspects of security threats in a unified manner, with security-related risks assessed using the attack potential concept of IEC 15408 and IEC18045 (see also Chapter 5).

Given that the safety case concept has been widely adopted in many safety-related industrial sectors, it seems logical to consider developing an analogous 'security assurance case' [77] to present the security argument for security-related applications, particularly for those where security may also have potential safety implications. Furthermore, it would be also desirable that such a security case should be subject to independent assessment and audit by a suitably qualified third party, as with the safety case.

12.7 Conclusions

The deployment of ADAS, and particularly the development of fully automated driving systems, are currently hindered by uncertainties relating to the legal framework. The most notable issues concerning liability aspects and the significant variations that are found between national legislations. In addition, further complexities result from the complex web of different stakeholders and interdependencies that may be involved in the provision of ITS functions and services.

The functionality that is envisaged for future vehicle systems will be increasingly dependent on inputs from a variety of external systems (e.g. positioning and navigation signals, and messages from other vehicles or roadside infrastructure), as well as a widening array of on-board sensors, actuators and electronic control capabilities. Such systems may diminish the driver's current role, and perhaps ultimately replace the driver with fully autonomous driving systems. In these scenarios the quality of information received from outside the car, the reliability of wireless communication channels, and the dependability of the on-board systems will be increasingly significant factors for successful and safe operation. Consequently, responsibility for accidents might be expected to shift away from the driver towards vehicle manufacturers and their on-board systems suppliers, as well as to external information providers.

Under fully autonomous operation there would be no driver involvement, but determining whether responsibility lies with the on-board systems or the external information sources may not be easy to establish. Furthermore, it may prove difficult to establish the responsibility of actors other than the driver in circumstances where the driver still has a role. Establishing that a vehicle control system responded in an unexpected way to a particular combination of transient inputs is likely to be extremely difficult. For this reason, it has been suggested that there should perhaps be an obligation to install an event data recorder (similar to the well-known 'black box' used in aircraft) when ADAS implementing highly automated driving are more widely deployed.

Given the technical complexity of ADAS, and the difficulties of validating the performance characteristics of such systems reliably, it will be increasingly important to adopt system design processes that implement best practice in the design and analysis of complex electronic control systems. This increasingly needs to take account of possible cyber security threats, as well as more established vehicle safety engineering considerations.

Although the Vienna Convention allows for coupled vehicles which travel on the road as a single unit, provided that they are controlled by a driver, references to coupling elsewhere in the text clearly indicate that this was expected to be a mechanical coupling (e.g. for trailers and articulated vehicles). In addition, even the currently proposed amendments to the Vienna Convention would not support fully autonomous driving. Consequently, on-going amendment and clarification of the Vienna Convention will be necessary in order to ensure that it takes account of recent and anticipated technological developments.

Ongoing changes in automotive safety requirements and technology will also have an impact on the vehicle type approval legislation, as increasingly sophisticated ADAS become mandatory equipment in future vehicles. Consequently, further development of the necessary UNECE regulations to specify the technical requirements and appropriate validation procedures will also be required.

Acknowledgements

Part of the research leading to these results received funding from the European Community's Framework Programme (FP7/2007–2013) under grant agreement FP7-ICT-224275 (EVITA). The authors are grateful for the contributions of the other EVITA project participants, from BMW Group (Germany), Continental Teves (Germany), escrypt (Germany), EURECOM (France), Fraunhofer Institute for Secure Information Technology (Germany), Fraunhofer Institute for Systems and Innovation Research (Germany), Fujitsu (Sweden, Austria, and Germany), Infineon Technologies (Germany), Institut Telecom (France), Katholieke Universiteit Leuven (Belgium), MIRA (UK), Robert Bosch (Germany), and Trialog (France).

References

1. European Commission (2008) *Results of Consultation ITS, 26/03/2008*. Available online at: http://ec.europa.eu/transport/road/consultations/doc/2008_03_26_its_results.pdf (last accessed 1 May 2015).
2. EVITA project overview. Available online at: http://www.evita-project.org (last accessed 1 May 2015).
3. J. Dumortier, C. Geuens, A.R. Ruddle and L. Low (2011) *Legal Framework and Requirements of Automotive On-board Networks*, EVITA Deliverable D2.4, 19 September 2011. Available online at: http://www.evita-project.org (last accessed 1 May 2015).
4. Directive 2007/46/EC of The European Parliament and of the Council of 5 September 2007 establishing a framework for the approval of motor vehicles and their trailers, and of systems, components and separate technical units intended for such vehicles, *Official Journal of the European Union*, L 263, 9 October 2007, pp. 1–160.
5. ISO 9001:2008, *Quality Management Systems – Requirements*.
6. COM (2007) 22, *Communication from the Commission to the European Parliament and Council: A Competitive Automotive Regulatory Framework for the 21st Century – Commission's position on the CARS 21 High Level Group Final Report*, 7 February 2007.
7. COM (2006) 59, *Communication from the Commission to the Council, the European Parliament, the European Economic and Social Committee and the Committee of the Regions: On the Intelligent Car Initiative – Raising Awareness of ICT for Smarter, Safer and Cleaner Vehicles*, 15 February 2006.
8. Commission Regulation (EC) No 78/2009 of the European Parliament and of the Council of 14 January 2009 on the type-approval of motor vehicles with regard to the protection of pedestrians and other vulnerable road users, amending Directive 2007/46/EC and repealing Directives 2003/102/EC and 2005/66/EC, *Official Journal of the European Union*, L35, 4 February 2009, pp. 1–31.
9. Commission Regulation (EC) No. 631/2009 of 22 July 2009 laying down detailed rules for the implementation of Annex I to Regulation (EC) No 78/2009 of the European Parliament and of the Council on the type-approval of motor vehicles with regard to the protection of pedestrians and other vulnerable road users, amending Directive 2007/46/EC and repealing Directives 2003/102/EC and 2005/66/EC, *Official Journal of the European Union*, L 195, 25 July 2009, pp. 1–60.
10. Commission Regulation (EC) No. 661/2009 of 13 July 2009 concerning type-approval requirements for the general safety of motor vehicles, their trailers and systems, components and separate technical units intended therefor, *Official Journal of the European Union*, L 200, 31/7/2009, pp. 1–24.
11. US Department of Transportation, National Highway Traffic Safety Administration, Federal Motor Vehicle Safety Standard FMVSS No. 138, 49 CFR, Parts 571 & 585: Tire Pressure Monitoring Systems.

12. US. Department of Transportation, National Highway Traffic Safety Administration, Federal Motor Vehicle Safety Standard FMVSS No. 126, 49 CFR, Parts 571 & 585: Electronic Stability Control Systems.
13. UNECE Regulation No. 13-H, Uniform provisions concerning the approval of passenger cars with regard to braking, Revision 1 – Amendment 2, 11 November 2009.
14. Ares(2014)818137, Status of EU accession to UN ECE regulations in the area of vehicle approval as of 31 December 2013, 20 March 2014.
15. C. Grover, I. Knight, F. Okoro, *et al.* (2008) *Automated Emergency Braking Systems: Technical Requirements, Costs and Benefits*, TRL (UK), Report PPR 227, April.
16. COM (2010) 389, *Communication from the Commission to the Council, the European Parliament, the European Economic and Social Committee and the Committee of the Regions: Towards a European Road Safety Area – Policy Orientations on Road Safety 2011–2020*, 20 July 2010.
17. COM (2009) 593, *Proposal for a Regulation of the European Parliament and of the Council Setting Emission Performance Standards for New Light Commercial Vehicles as Part of the Community's Integrated Approach to Reduce CO2 Emissions from Light-duty Vehicles*, 28 October 2009.
18. SWOV Fact sheet R-2003-33, *Fietser-autofront*, 4 July 2004.
19. US Department of Transportation, National Highway Traffic Safety Administration, *Preliminary Statement of Policy Concerning Automated Vehicles*, 30 May 2013.
20. SAE J3016, *Taxonomy and Definitions for Terms Related to On-Road Motor Vehicle Automated Driving Systems*, 16 January 2014.
21. J. Schwarz (2012) Designing safe automated driving functions – challenges from the legal framework, *EC ITS Workshop on Liability Aspects Related to ITS Applications*, 13 June 2012. Available online at: http://ec.europa.eu/transport/themes/its/events/doc/2012-06-13-workshop/8-schwarz_daimler_highly_automated_driving_legal_aspects_2011_06_13.pdf (last accessed 1 May 2015).
22. *Convention on Road Traffic*, Vienna, 8 November 1968, as amended on 3 September 1993 and 28 March 2006.
23. *Convention on Road Traffic*, Geneva, 19 September 1949. Available online at: http://en.wikisource.org/wiki/Geneva_Convention_on_Road_Traffic (last accessed 1 May 2015).
24. UNECE Inland Transport Committee Working Party of Road Traffic Safety (2004) *Proposals for amendments to the Vienna Convention on Road Traffic*, TRANS/WP.1/2003/1/Rev.4, 23 April 2004.
25. *Towards a European Standard: The Development of Safe Road Transport Informatics Systems*, Draft 2, DRIVE Safely (DRIVE I Project V1051), March 1992.
26. MISRA (1994) *Development Guidelines for Vehicle Based Software*, MIRA Ltd, November 1994.
27. *MISRA Guidelines for safety analysis of vehicle based programmable systems*, MIRA, 2007.
28. ISO 26262-1:2011, *Road Vehicles – Functional Safety* (10 parts).
29. *Code of Practice for the Design and Evaluation of ADAS*, Deliverable 11.2, RESPONSE 3 (a sub-project of the 'PREVENT Integrated' Project. Available online at: http://citeseerx.ist.psu.edu/viewdoc/download?doi=10.1.1.174.4717&rep=rep1&type=pdf (10 May 2015).
30. J. Schwarz (2007) Legal problems and suggested solutions in connection with the development of Driver Assistance Systems, *German Presidency eSafety Conference*, Berlin, June 2007.
31. W. Botman (2007) Potential benefits of active driver assistance systems and the legal context, *German Presidency eSafety Conference*, Berlin, June 2007.
32. UNECE Inland Transport Committee Working Party of Road Traffic Safety (2011) Consistency between the Convention on Road Traffic, 1968, and the vehicle technical regulations, Informal document No. 1, March 2011.
33. Royal Society for the Prevention of Accidents (UK) (2007) *Cars in the Future*, Policy Paper, January 2007.
34. European Transport Safety Council (2011) *ETSC MEP Briefing: European Parliament Own Initiative Report on Road Safety*, 4 March 2011.
35. UNECE Inland Transport Committee Working Party of Road Traffic Safety (2014) *Consistency between the Convention on Road Traffic, 1968, and the Vehicle Technical Regulations*, ECE/TRANS/WP.1/2014/1, March 2014.
36. UNECE Inland Transport Committee Working Party of Road Traffic Safety (2014) *Consistency between the Convention on Road Traffic, 1949, and the Vehicle Technical Regulations*, ECE/TRANS/WP.1/2014/4, March 2014.
37. J. Markoff (2010) Google cars drive themselves, in traffic, *New York Times*, 9/10/2010. Available online at: http://www.nytimes.com/2010/10/10/science/10google.html?pagewanted=1&_r=2 (last accessed 1 May 2015).
38. J. Markoff (2010) Guided by computers and sensors, a smooth ride at 60 miles per hour, *New York Times*, 10/10/2010. Available online at: http://www.nytimes.com/2010/10/10/science/10googleside.html?ref=science (last accessed 1 May 2015).

39. B. Walker Smith (2014) Automated vehicles are probably legal in the United States, *1 Tex.* A&M L. Rev. 411.
40. Volvo Press Release, *Volvo Car Group's first self-driving Autopilot cars test on public roads around Gothenburg*, 29 April 2014. Available online at: https://www.media.volvocars.com/global/en-gb/media/pressreleases/145619/volvo-car-groups-first-self-driving-autopilot-cars-test-on-public-roads-around-gothenburg (last accessed 1 May 2015).
41. L. Gannes (2014) *Google's new self-driving car ditches the steering wheel*, 27 May 2014. Available online at: http://recode.net/2014/05/27/googles-new-self-driving-car-ditches-the-steering-wheel (last accessed 1 May 2015).
42. T. Robinson, E. Chan and E. Coelingh (2010) Operating platoons on public motorways: an introduction to the SARTRE platooning programme, *17th World Congress on Intelligent Transport Systems, October 2010*, Busan, Korea, pp. 1–11.
43. K. van Wees and K. Brookhuis (2005) Product liability for ADAS; legal and human factors perspectives, *European Journal of Transport and Infrastructure Research* **5**(4): 357–72.
44. Directive 2001/95/EC of the European Parliament and the Council of 2 December 2001 on general product safety, *Official Journal of the European Communities*, L11, 15 January 2002, pp. 4–17.
45. Council Directive 85/374/EEC of 25 July 1985 on the approximation of the laws, regulations and administrative provisions of the Member States concerning liability for defective products, *Official Journal of the European Communities*, L210, 7 August 1985, pp. 29–33.
46. R. van der Heijden and K. van Wees (2001) Introducing advanced driver assistance systems: some legal issues, European Journal of Transport and Infrastructure Research **1**(3): 309–26.
47. J. Schwarz (2005) Code of practice for development, validation and market introduction of ADAS, *5th European Congress on ITS, Hannover, Germany, 3 June 2005*. Available online at: http://www.ftm.mw.tum.de/uploads/media/09g_schwarz.pdf (10 May 2015).
48. F. Ahrens (2010) Why it's so hard for Toyota to find out what's wrong with its vehicles, *The Washington Post*, 4/3/2010. Available online at: http://voices.washingtonpost.com/economy-watch/2010/03/i_wont_lie_to_you.html (last accessed 1 May 2015).
49. NASA Engineering and Safety Centre (2011) *National Highway Traffic Safety Administration: Toyota Unintended Acceleration Investigation*, NESC Assessment Report TI-10-00618, January 2011. Available online at: http://www.nhtsa.gov/staticfiles/nvs/pdf/NASA-UA_report.pdf (last accessed 1 May 2015).
50. IEEE 1616:2004, *Standard for Motor Vehicle Event Data Recorder (MVEDR)*.
51. IEEE 1616a:2010, *Standard for Motor Vehicle Event Data Recorders (MVEDRs) – Amendment 1: Motor Vehicle Event Data Recorder Connector Lockout Apparatus (MVEDRCLA)*.
52. T.M. Kowalick (2005) *Fatal Exit: The Automotive Black Box Debate*, Hoboken, NJ: IEEE Press, p. 277.
53. J. Lendino (2008) *Progressive uses 'black box' to monitor drivers*, 31/07/2008. Available online at: http://www.pcmag.com/article2/0,2817,2326909,00.asp (last accessed 1 May 2015).
54. N. Lyndon (2011) Black box technology to monitor young drivers, *The Telegraph*, 18/07/2011. Available online at: http://www.telegraph.co.uk/motoring/columnists/neil-lyndon/8458515/Black-box-technology-to-monitor-young-drivers.html (last accessed 1 May 2015).
55. US Department of Transportation, National Highway Traffic Safety Administration, 49 CFR, Part 563: *Event Data Recorders – EDRs in Vehicles*. Available online at: http://www.nhtsa.gov/DOT/NHTSA/Rulemaking/Rules/Associated%20Files/EDRFinalRule_Aug2006.pdf (last accessed 1 May 2015).
56. US Department of Transportation, National Highway Traffic Safety Administration, *Technical Assessment of Toyota Electronic Throttle Control (ETC) Systems*, February 2011. Available online at: http://www.nhtsa.gov/staticfiles/nvs/pdf/NHTSA-UA_report.pdf (last accessed 1 May 2015).
57. Def Stan 00-54 (1999) *Requirements for Safety Related Electronic Hardware in Defence Equipment*, UK Ministry of Defence.
58. P. Bishop and R. Bloomfield (1998) A methodology for safety case development. In F. Redmill and T. Anderson (eds), *Industrial Perspectives of Safety-Critical Systems: Proceedings of the 6th Safety-critical Systems Symposium, Birmingham, UK, February 1998*. New York: Springer.
59. R. Palin and I. Habli (2010) Assurance of automotive safety – a safety case approach, *Proceedings of the 29th International Conference on Computer Safety, Reliability and Security, Vienna, Austria, September 2010*, pp. 82–96.
60. UNECE Regulation No. 13, *Uniform Provisions Concerning the Approval of Vehicles of Categories M, N and O with Regard to Braking: Annex 18 – Special Requirement to be Applied to the Safety Aspects of Complex Electronic Vehicle Control Systems*, Revision 5, 08/10/2004.

61. UNECE Regulation No. 79, *Uniform Provisions Concerning the Approval of Vehicles with Regard to Steering Equipment: Annex 6 – Special Requirement to be Applied to the Safety Aspects of Complex Electronic Vehicle Control Systems*, Revision 2, 21/10/2005.
62. IEC 61508: 1998–2005, *Functional Safety of Electrical/Electronic/Programmable Electronic Safety-related Systems* (8 parts).
63. F. Redmill (2000) Understanding the use, misuse and abuse of Safety Integrity Levels, *Proceedings of the 8th Safety-critical Systems Symposium, Southampton, UK, February 2000*. Available online at: http://www.csr.ncl.ac.uk/FELIX_Web/3A.SILs.pdf (last accessed 1 May 2015).
64. UNECE Regulation No. 13, *Uniform Provisions Concerning the Approval of Vehicles of Categories M, N and O with Regard to Braking*, Revision 8 – Amendment 1, 9 October 2014.
65. UNECE Regulation No. 79 (2014) *Uniform Provisions Concerning the Approval of with Regard to Steering Equipment*, Revision 2 – Amendment 1, 13 February 2014.
66. SeVeCom project overview. Available online at: http://www.transport-research.info/web/projects/project_details.cfm?id=46017 (10 May 2015).
67. ISO/IEC 15408:2005, *Information Technology – Security Techniques – Evaluation Criteria for IT Security* (3 parts), 2nd Edition, 01/10/2005.
68. ISO/IEC 18045:2008, *Information Technology – Security Techniques – Methodology for IT Security Evaluation*, 2nd Edition, 15/08/2008.
69. ISO/IEC TR 15446:2004, *Information Technology – Security Techniques Guide for the Production of Protection Profiles and Security Targets*, Technical report, 01/07/2004.
70. ISO/IEC 13335, *Information Technology — Security Techniques — Management of Information and Communications Technology Security*.
71. NIST Special Publication 800-12, *An Introduction to Computer Security: The NIST Handbook*. October 1995. Available online at: http://csrc.nist.gov/publications/nistpubs/800-12/handbook.pdf (last accessed 1 May 2015).
72. P.H. Jesty and D.D. Ward (2007) Towards a unified approach to safety and security in automotive systems, *Proceedings of 15th Safety-critical Systems Symposium*, Bristol, UK, February 2007, pp. 21–35.
73. J.A. Clark, H.R. Chivers, J. Murdoch and J.A. McDermid, *Unifying MANET Safety and Security*, International Technology Alliance in Network-Centric Systems, Report ITA/TR/2007/02 V. 1.0, 06/11/2007. Available online at: http://www.usukita.org/papers/3155/ITA-TR-2007-02-v1.0_0.pdf (last accessed 1 May 2015).
74. MISRA (1994) *Development Guidelines for Vehicle Based Software*, MIRA Ltd, p. 43.
75. IEC 61508:2010, *Functional Safety of Electrical/Electronic/Programmable Electronic Safety-related Systems*, 2nd Edition, 30/04/2010.
76. A.R. Ruddle, D. Ward, B. Weyl, *et al.* (2009) *Security Requirements for Automotive On-board Networks Based on Dark-side Scenarios*, EVITA Deliverable D2.3, 30 November 2009. Available online at: http://www.evita-project.org (last accessed 1 May 2015).
77. I. Ibarra and D.D. Ward (2013) Cyber-security as an attribute of active safety systems and their migration towards vehicle automation, *Proceedings of the 8th IET International System Safety Conference incorporating the Cyber Security Conference 2013* Cardiff, UK, October 2013, p. 2.2.

Part 5

Applications and Services for Users and Traffic Managers

13

Traffic Management Systems

António Amador, Rui Dias, Tiago Dias and Tomé Canas
Brisa Innovation, Porto Salvo, Portugal

13.1 Introduction

This chapter covers traffic management systems (TMS), one of the most important components of modern ITS. The main motivation for traffic management is the efficient management of a road network, optimizing the traffic flow while maximizing road users' safety and comfort. These systems involve several stakeholders but are mainly directed towards road users, helping them act efficiently and safely whilst on the road. This goal is achieved through different applications and services, by communicating, analysing and sharing information.

We start this chapter with the definition of a 'traffic management system', followed by a depiction of different traffic environments on which traffic management systems operate. Afterwards, a proposal for a conceptual framework of traffic management systems is presented, on which data collection (inputs), data processing and analysis (platform) and information dissemination and actuation (outputs) are established as the pillars for a TMS.

Although ITS involve multiple transport modes, the focus for this chapter will be brought on roads. Throughout the text, Traffic Management Centres (TMC) and Traffic Control Centres (TCC) will be referred to – although being very similar in meaning, the first term refers to a more generic management facility, whereas the second will usually refer to actual control centres used by road concessionaires and municipalities.

13.1.1 Objectives

Our goal in this chapter is to show the relevance for traffic management systems in different contexts and road environments, to propose a conceptual framework to address the different challenges and to present the several aspects of a typical Traffic Control Centre – which usually is the core embodiment of the traffic management system.

Intelligent Transport Systems: Technologies and Applications, First Edition. Asier Perallos, Unai Hernandez-Jayo, Enrique Onieva and Ignacio Julio García-Zuazola.

13.1.2 Traffic Management

Several definitions of what Traffic Management is can be found in modern literature and standards.

The ISO International Standard 14813-1 on ITS [1] describes Traffic Management as 'The management of the movement of vehicles, travellers and pedestrians throughout the road transport network'.

While broad, the ISO definition adequately defines the main targets of traffic management: vehicles, travellers and pedestrians and their scope in the road transport network.

Regarding the actual act of management it is important to further understand the objectives that drive traffic management efforts. In this case we have already stated that traffic management aims at being the efficient management and operation of a road network, optimizing the traffic flow while maximizing road users' safety and comfort. To this end, the management part is broken down into traffic management services that span several stages and create a cycle. There are three main stages where traffic management services fit:

- *Planning*: involving services like defining the road network layout and structure (or its evolution), planning multimodal interfaces, determining fixed signalling elements, the roadside traffic management systems and the actual traffic management procedures or public transport routes.
- *Operating*: involving activities like monitoring and patrolling, resolving traffic incidents or emergencies, managing demand, anticipating or predicting future situations, providing information and assistance to drivers, repairing damages to the infrastructure, managing parks and other facilities, enforcing and policing or managing multimodal public transport.
- *Analysing*: evaluating the results of the operating activities and, in case of need, determining if these can be positively influenced by applying changes to the results of the planning stage.

These stages define the traffic management loop as a positive feedback continuous enhancement loop. In terms of feedback from analysis to the planning stage, while changing the built infrastructure is hard, changes to the other assets of the planning stage become easier to adopt as we move further ahead in the planning services list, right until we reach the procedures and routes, these being the most agile and easy to change result of the planning stage.

While the main stakeholders that benefit from traffic management are drivers, public transport users and pedestrians, the main clients that deploy traffic management systems are the local and central traffic control agencies, highway concessionaires and municipalities.

Several advantages are normally linked with traffic management services – these allow us to:

- ensure the safety of road users through the effective monitoring of highway operating conditions;
- lower average travel times and carbon emissions by optimizing traffic flow and harmonizing speeds through variable speed limits;
- optimize traffic maintenance manpower through automated information system control;
- facilitate real-time traffic information flow and improve transportation system performance by coordinating and integrating cross-unit traffic information exchange;
- make overall network traffic distribution return to a state of equilibrium in the event of congestion or accident, by influencing driver behaviour and resolving situations in a timely fashion;
- provide appropriate and effective traffic control strategies in response to the actual demand from within different vehicle types.

13.1.3 Traffic Environments

Several traffic categories are managed with traffic management systems and can be clustered in several ways. In the present approach, we opted to do so according with the environment surrounding the road network, as this classification highlights the most significant differences between the different broad scenarios. As so, these traffic environments can be enumerated as follows:

- *Urban traffic environment*: This environment is found in cities, metropolitan areas and other urban sets. It is characterized by a network of streets and roads, with different capacities and different typical densities of traffic. Normally it includes road intersections with one or multiple vehicle lanes, one-way streets, different grade vertical bypasses, tunnels and bridges, cross-walks for pedestrians, traffic calming mechanisms, limited access zones, parking spots and lots, multimodal transportation interfaces, increasing use of bike lanes, among others. In this environment, data is gathered from multiple sources such as traffic counters and classifiers, increasingly Smartphone apps and in-car services [2], closed circuit television (CCTV) / video cameras, automatic number plate recognition, car parking systems, and meteorological and air quality monitoring stations. Traffic lights, access control systems, strategic variable message signs (VMS) and variable speed limit signs are among the most common systems used to control the flow of traffic in urban environments. Municipalities are the main traffic systems operator in urban environments with local, county or regional control centres. Frequently these traffic control centres are also involved in multimodal transport management.
- *Interurban traffic environment*: This environment is typically characterized by motorway and road networks (Figure 13.1) connecting cities. In motorways, it features dedicate lanes for vehicle traffic, exclusive access for certain classes of motor vehicles, no pedestrians and no road intersections at the same level, providing one of the most efficient infrastructures for road users. On the other hand, interurban roads have less intersections or pedestrians than in urban roads but may cross villages and is usually made up of single lanes that are also used by heavy goods vehicles or agricultural vehicles that might make a constant speed trip impossible.

  Whereas in interurban roads there are not many traffic management systems, aside from occasional speed limit systems in villages or high-risk road stretches, motorways usually rely heavily on traffic management systems. Many countries make the presence of electronic call boxes (ECB) along the motorways mandatory, rendering ECB one of the most widespread and numerous ITS in motorways throughout the world. Electronic call-boxes allow road users to communicate with the traffic control centres. Many other types of sensors, some of these similar to the ones used in the urban environments, collect traffic data in motorways. The most commonly used are traffic counters and classifiers, CCTV and weather stations.

  The multiple actuators, providing the way to influence the traffic in motorways include variable message signs, variable speed and lane control signals or traveller advisory services (such as radio or Smartphone apps). As in urban environments, these actuators can be used to provide a diverse set of traffic management services like incident warning and management, dynamic lane management, variable speed limits, ramp metering, hard shoulder running or specific overtaking bans.

  Typically, motorway traffic control centres are operated by the road concessionaires (either public or private) and span from tens to hundreds or thousands of kilometres operated in a single control centre.

Figure 13.1 Intersection between two highways in an Interurban traffic environment.

Sometimes motorways are directly or indirectly tolled. For such cases, specific tolling ITS are used, ranging from manual toll booths to fully automatic and transparent solutions like geotolling, shadow tolling or free-flow electronic tolling.

- *Long-distance (interurban/international) traffic environment*: Whereas interurban traffic environments deal with the perspective of the specific motorways operated by certain concessionaires, long distance needs to see beyond that perspective, focusing on networks of the different motorways and corridors of motorways, which sometimes span several countries. This allows a strategic management of these networks and corridors, implementing traffic management plans such as alternative routing in case of road closures or intense traffic. While data gathering and actuator systems remain pretty much the same as the ones used in interurban environments, it is the traffic management services that vary greatly in these environments, becoming complex interactions between several traffic control centres. These interactions are based on data exchange between the traffic control centres and on requests and responses regarding the implementation of strategic traffic management plans [3].

These scenarios are particularly challenging when involving different traffic control centres and organizations spanning different jurisdictions, frequently with different laws, traffic regulations, languages and even signalling.

In Europe, projects directed at integrating traffic control centres to manage international corridors date back to 1993 [4]. Currently the Trans-European Network is finally becoming a reality based on a strategic, coordinated approach to manage existing infrastructures is gradually pushing a move from road to rail and water transport, mostly for environmental reasons [5].

In the US, the notion of corridors is more associated with physical multimodal corridors, which is quite controversial [6].

13.2 Traffic Management Framework

To achieve adequate levels regarding traffic management indicators an appropriate framework must be made available. A 'traffic management framework' is understood as being the set of tools needed to achieve the desired efficiency goals. In general terms, a traffic management framework can be systematized as involving inputs, data processing, on one or more platforms, and outputs, as suggested in Figure 13.2.

A conceptual framework, by nature, represents a theoretical basis from which particular implementations may arise. This principle is particularly relevant for the matter at hand, since traffic management scenarios often present differences regarding their scope and focus, although usually having in mind similar goals, such as increased mobility or improved road safety. As so, when considering the implementation of a Traffic Management Centre, one must position itself in order to assess the best approach, e.g. if being a road operator or a municipality, roadside means are most likely available not only to provide data collection but also for influencing user behaviour; when being a fleet manager or service provider of some sort, interfacing with external entities gains a wider relevance. It is also important to state that these elements and interfaces have a strong dynamic nature – they have evolved tremendously in the past decades with several technology breakthroughs (especially in communication technologies, but not alone), roadside technologies and the advent of the internet, with its consequent immediate information availability; currently, with the wide dissemination of Global Navigation Satellite System (GNSS) and communication devices, especially Smartphones. Massive floating data (originated from on-board units or nomadic devices) is proving to be a major asset for assessing traffic conditions and multiple business models have grown around this concept – these models are usually positioned alongside with the 'connected cars' concept and range from traffic information services, insurance and behaviour services or even infotainment. In the next years, with the expected deployment of cooperative systems in vehicles and infrastructure, a potential game change leap may be on the verge of becoming, allowing for improved road safety, more accurate traffic flow assessment at lower costs and new services will become available for users.

In a simplistic approach, the framework diagram (Figure 13.2) illustrates the need for inputs, which are interfaced through an integration layer or data exchange gateway, processed according with their nature, purpose and the platform needs, and stored in a data repository. This data repository will then feed two main elements: business processes (operation) and traffic modelling, which is usually used for forecast and future planning, although sometimes traffic modelling can also feed business processes. From business processes and rules, action

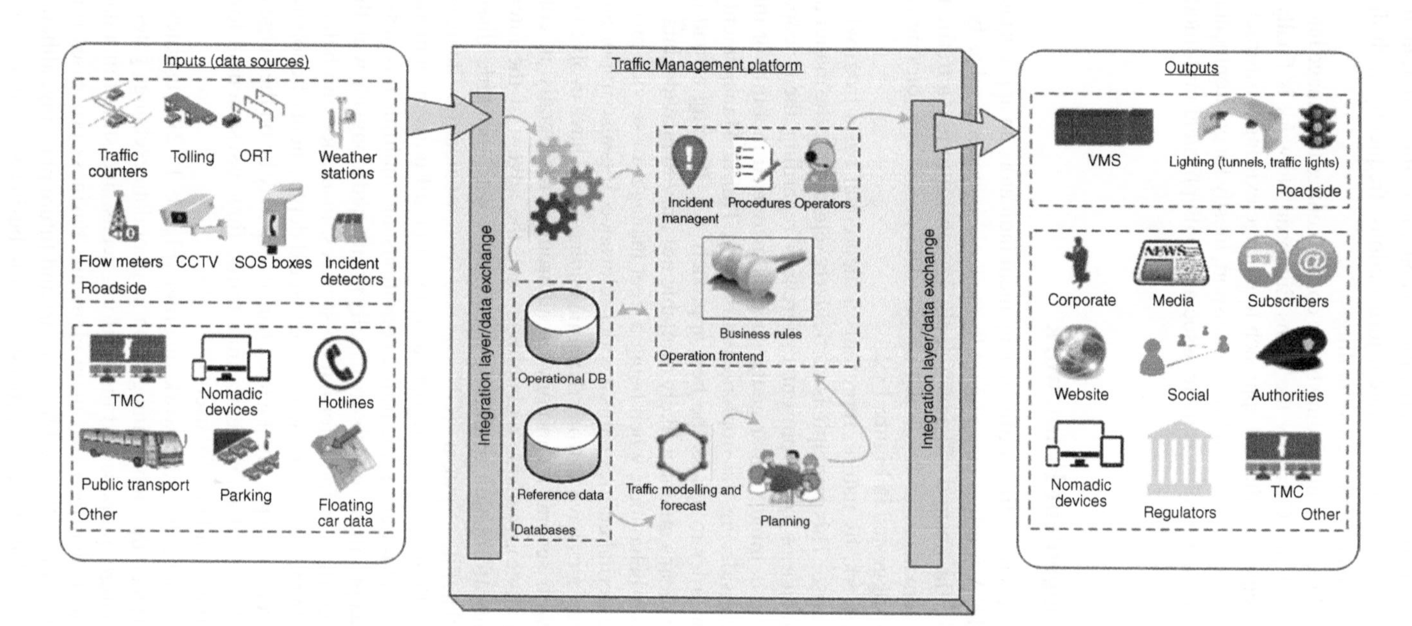

Figure 13.2 Example diagram for a conceptual traffic management framework.

over outputs is then performed, whether being acting on roadside means, providing data to third parties or issuing alerts to appropriate actors.

Some key elements of any system were intentionally left out of this scope, such as monitoring systems, maintenance management/ticketing systems or specific business rules and features for motorway or urban environments.

The present section will focus on the main potential elements for the conceptual framework presented in Figure 13.2, being sure that this exercise is not intended to be exhaustive on their enumeration.

13.2.1 Inputs

Regarding possible inputs within a given Traffic Management Framework, one can structure these in two groups: Roadside (deployed) and (other) Interfaces.

13.2.1.1 Roadside Inputs

Roadside inputs are usually managed by a single entity, which may eventually provide the acquired data to external parties through specific interfaces, after appropriate transformation, if applicable – thus becoming inputs on the second group for other traffic management implementations.

- *Traffic counters*: Traffic counters (Figure 13.3) are one of the oldest and most widespread approaches on traffic flow measurement. These systems are deployed for multiple purposes, which range from the actual measurement of traffic volume for appropriate planning and

Figure 13.3 Traffic counters.

forecasting, to shadow tolling charge models, on which revenue is based upon a volume and class schema. Being subjected to several decades of evolution and multiple purposes with diverse accuracy requirements, traffic counters are available in the market under a multitude of technologies, being the most common the inductive loop based one – in fact, almost every traffic counter design involves at least one inductive loop, and some designs rely on loops alone. Commonly, these systems use the magnetic signature of vehicles to assess their class within a given schema, using additional sensors to measure other features – axle counting, for example. It can consider weigh in motion (WIM) stations as a subset of this class of roadside systems (additionally to volume and class assessment, WIM stations are able to measure each axle weight, thus obtaining the vehicle's total weight). Traffic counters are still one of the best performing and accurate approaches when it comes to assessing actual traffic in a discrete point, although being intrusive in its deployment (sensor installation) and carrying maintenance costs that vary according with the sensor layout. Common data elements obtained with these systems for discrete transactions are vehicle class, speed, number of axles, vehicle length, direction and time gap (from the previous vehicle); when considering data aggregation within a time period, other metrics may be obtained, such as volume (as a whole and/or per class), average speed, average vehicle length, average spacing (between vehicles) or occupancy.

- *Tolling systems*: Tolling systems are deployed when a given business model for a particular infrastructure operation demands and allows so – their purpose is to provide revenue for subsidizing the infrastructure's maintenance and assure an adequate (and sometimes contracted) service level. Tolling systems have very strict mechanisms to provide revenue assurance and these always involve a rigorous control of the passing vehicles and, usually, their class assessment (depending on the charging model applied). As such, these systems provide a very good data input for assessing traffic volume, although they are not a good measure point when it comes to road occupancy, for example, since the presence of a toll plaza commonly affects several relevant metrics to take into account when evaluating a road section – an exception to this principle are the Open Road Tolling systems (ORT), which discard the need for a toll plaza as their charge point. Recently, new geotolling charging models are being deployed (especially in Europe) – these charging models often involve the use of a GNSS on-board unit and route data collection. In some of these models it is possible to feed this (floating vehicle) data to Traffic Management Systems, enriching its assessment on traffic conditions of a wide part of the road network.
- *Flow meters*: Flow meters refer to traffic counting systems on which only statistical relevance is important, e.g. traffic volume does not need to be accurately assessed, nor does the vehicle classes; instead, a statistically relevant sample must be able to be measured and from it several metrics be obtained – for example, average speed, which enables congestion detection. These systems present themselves over a wide range of technologies and approaches to measurements, of which one can pinpoint some examples: radar based stations, Bluetooth beacons network, magnetometer based sensors (measurements of slight local variations to the earth's magnetic field originated by passing vehicles), Automatic Number Plate Recognition (ANPR) systems or even image based counters.
  Some of these systems may eventually achieve accuracy performance levels that come close to traffic counters.
- *Weather stations*: Weather has a profound influence in road conditions and safety [7]. Some weather phenomena are relatively self-evident (e.g. rain, snow, fog) but others are invisible to drivers (e.g. ice sheets/black ice, water accumulation). In traffic management, even the

self-evident situations need to be advised and alerted to drivers to promote safety through adequate driver behaviour.

Weather is usually monitored using standard Weather Stations (WS) and weather sensor technology with specific sensors used to gather information on road pavement conditions (pavement temperature, state and water or snow accumulation levels).

Weather stations provide relevant meteorological information for traffic management centres, since they are able to assess weather and pavement conditions on a discrete point of a road section. With these inputs, a traffic management centre is able to identify hazardous weather and, after acknowledgment, disseminate such information over the appropriate media – for example, informing road users directly through nearby VMS. Common data elements obtained from weather stations are visibility, air temperature, air humidity, pavement temperature, wind speed and direction, rainfall intensity, snow presence, ice formation, air pressure, dew point or underground temperature. Many of these stations are also able to issue alerts for particular hazardous conditions, such as black ice, very low visibility or other extreme weather conditions. The collection of such data also provides a useful database for post analysis and correlation with other metrics and events.

- *CCTV and Automatic Incident Detectors*: Video cameras are nowadays an invaluable tool for traffic management and are probably the most versatile piece of equipment in any ITS installation.

  Video cameras have gradually gained additional relevance through increased image quality and additional functions – these improvements have been made possible by the recent evolution of computer processing (particularly, image processing) and higher available bandwidths. Video cameras are key for the evaluation of traffic accidents and overall traffic flow conditions, as well as immediate assessment upon event triggering (for example, when a roadside system such as traffic counter detects wrong way drivers or presents metrics matching traffic congestion). Furthermore, video cameras can be associated with image processing algorithms targeting the identification of specific patterns such as stationary cars in a motorway, double parking in a city, or traffic offences – most of these automatic incident detections should always be validated by an operator, since false positives can be common, depending on the particular detection scenario.
- *Emergency Call Boxes*: Emergency/SOS call boxes (ECB) are placed along motorways and are an important input for emergency distress calls, although their usage has been declining with mobile phones' penetration. Nevertheless, the widespread and almost ubiquitous presence of these roadside systems throughout motorway networks may prove to be useful in the near future, since it represents a multitude of spots with the elementary needs for any kind of roadside system: power and communications.

13.2.1.2 Other Inputs

As 'other inputs', we refer to all external sources of information apart from the owned roadside ones. These inputs are often generated or managed by external entities that can either be private corporations, public organizations or even the users themselves. Since there is some specificity on these, a reduced set of the more common ones is commented on.

- *Hotlines*: Hotlines represent the nonroadside counterpart of emergency call boxes. Hotlines' importance as a source for incident detection has been growing in the past decades due to the wide usage of mobile phones, providing information about some events that other automated means are incapable or have difficulties on reporting, like animal in a motorway, for example. More and more, the adequate processing of information provided by users through hotlines is of the utmost importance for an efficient traffic management operation.

- *Other traffic management platforms and services*: As people, no traffic management system 'is as island' – there are multiple interfaces between traffic management platforms, being that some are institutional (e.g. when a motorway concessionaire is obliged to provide traffic information to a regulator or supervisor), other are mandatory within a legal framework (for example, providing immediate information to authorities) and others are driven by business requirements or decisions (such as traffic information interchange between road operators in order to optimize processes and service levels). At this level, information exchange standards play a key role. Any kind of information can be exchanged, although the most common are traffic flow metrics and road incidents (accidents, congestions, roadworks or extreme weather).
- *Nomadic devices*: As 'nomadic devices' we basically refer to two categories of devices: Smartphones and on-board units (OBUs). The most common one is obviously Smartphones, and in this context, besides their part on the hotline usage, Smartphones are also used as 'probes' – these devices include location information and communication abilities that allow them to supply travel information, as well as to receive any relevant information for the user. As long as the user is willing to use such kind of services, he will be able to provide the traffic management platform with his travel info whether in real time or in a batch fashion.

  One traffic management platform can easily integrate such information in a useful way, and to allow this channel to exist a mobile app will have to be made available. There are several known examples of such approaches, but their success rate depends a lot on the service they offer against the demanded device usage (data plan and battery usage) – for a successful implementation, two aspects should be taken into account: the appetency for users to adopt the service and the service cost (effective cost/fees, communication costs and battery). There are several examples of efficient floating data collection on Smartphones alone (as mentioned ahead).
- *Public transport multimodal management platforms and parking service providers*: Interfacing with public transport multimodal management and parking systems is essential in an urban environment. The holistic view on public transport schedules and parking availability in a city is a key concept to reduce congestion and improve mobility for citizens; in an urban traffic management platform this interfacing is a must and further interfacing should be considered with online navigation service providers in order to efficiently disseminate such information.
- *Floating car data providers*: Floating Car Data (FCD), also referred to as Floating Vehicle Data, or simply as floating data, consists on collecting travel time data from a statistically relevant number of vehicles in a particular region and using it to map the traffic conditions on that region, in real time – we are all pretty familiar with the green-yellow-red street markings in the morning news, or on more global web platforms, such as Google® Maps or Bing® Maps. Most of these data come from FCD. Currently, there are several approaches on FCD collection, and three are depicted below:
  - *Fleet managers* are companies that provide a fleet management service to others relying on installed OBUs. The customer usually pays a monthly fee per vehicle and another monthly fee according with the level of reporting it requires for its business and the location scope (transportation companies have different needs from, for example, maintenance

services companies). From their activity, fleet managers come into possession of a very large quantity of data that usually surpasses the needs of its customers – however, this data can be used as FCD if detailed enough and with enough samples; in this case, the fleet manager can find additional income either by selling this data to a third party (after customer privacy is assured), who will infer traffic conditions from it, or by becoming itself an FCD provider.

- *Insurance firms* are now using methods of defining insurance costs according with the driving behaviour of their customers (this model is usually referred to as PAYD insurance – Pay As You Drive – or PPUI – Pay Per Use Insurance); as in with fleet managers, they too can become FCD providers, as long as their customer privacy is safeguarded.
- *Social/community applications* are also a source for FCD; one rather well-known example is the Waze® social network which provides its users with traffic information and events, in exchange for the user's contribution for the network (FCD and event reporting). Other examples are online traffic services, such as those provided by TomTom® – relying on community supplied FCD to infer traffic conditions.

13.2.1.3 A Note on Cooperative Systems

Cooperative Systems (CoSy), V2V/V2I (Vehicle-to-Vehicle/Vehicle-to-Infrastructure) or V2X are some of the designations used for the vehicular communications over Dedicated Short-Range Communications (DSRC) 5.9 GHz. This technology is still emerging and many usage aspects of it are still in standardization process as of 2014, with a deep involvement of the automotive industry and road operators and regulators, as well as other Original Equipment Manufacturer (OEM) suppliers. In the associated model, cars are able to communicate with other cars or with the infrastructure (or vice versa), and a data element catalogue will allow for the implementation on many use case scenarios – maybe the most well-known one is the break warning scenario, on which a vehicle that breaks suddenly informs all nearby vehicles of that event, thus increasing road safety by early awareness on a potentially hazardous situation. However, many other use case scenarios are considered, and some of them may also constitute valuable data sources for traffic management centres, particularly, the vehicle-to-infrastructure (V2I) Probe Vehicle Data (PVD) scenario, on which a 5.9 GHz Roadside Unit is able to collect travel time data from all vehicles within a 500–1000 m radius, regardless of travel direction or even travelling road – thus becoming an efficient 'FCD collector'. Additional use cases can also be mentioned: Road Works Warning, on which vehicles are informed ahead of any roadworks on the infrastructure, and respective nature, In-Vehicle Signage, which intends to present users with information regarding the infrastructure on which they travel, Hazardous Locations Warning, to warn drivers of upcoming obstacles or dangerous weather conditions, or Signal Phase and Time, which addresses optimizations regarding travel speeds when approaching a traffic light, or even traffic light management according with surrounding traffic conditions.

Slow/Stationary Vehicle Warning, Traffic Jam Ahead Warning, Speed Management, Emergency Vehicle Warning or Motorcycle Approaching Warning are other examples of what can be considered the 'Day One' use case scenarios for Cooperative Systems – however, and although deployment plans are already being finalized in Europe, the key element for the success of Cooperative Systems will be the deployment of on-board units in vehicles, which is supposed to begin within this decade. The current scenario raises expectations on what Cooperative Systems may bring to road safety and road operations in the coming decade.

Some Cooperative Systems use case scenarios may eventually overlap 'Connected Cars' (internet) ones – however, Cooperative Systems aim mostly at road safety scenarios, on which latency must be minimal.

13.2.2 Analysis

Some of the data from inputs to traffic management processing/analysis will already provide direct event information (e.g. fire in the tunnel, wrong way driver) and details related to the event (e.g. the event location, direction). Other inputs will only provide raw data (e.g. traffic volumes, air temperature) that will require further processing in order to generate valid events (e.g. queue detected) or predictions (e.g. risk of black-ice in two hours) either for operation and/or dissemination to end-users.

In either case the event is a key concept. An event is an occurrence of some sort that can be categorized (what) and located in time (when) and space (where). It may have extra details like the origin system or an image captured of the event but the event is always identifiable by its category. As you will see in the next section, Traffic Management Systems forward and filter such events to traffic control operators and also to external systems targeted at end-users, such as websites and apps.

Events can be filtered using rule engines that define which events get forwarded to which operators or end-users. Rule engines can be used to define a rich set of business rules that use the event information like category and location to determine what is done with the event.

Sometimes using information from a single event is not enough. In some cases a single detection is not able to catch the operator's attention. In these cases correlation between events can be added to express more complex business rules (e.g. after two or more detections of this event in an interval below one hour, forward it to the active traffic control operator, marked as urgent).

Data from events is also used to determine the adequate response plan that can be used to handle the event. A response plan is a set of operations that determines results on the outputs in our traffic management framework, for example, by setting a warning message in VMS in response to a wrong way driver event.

We will now review some examples where processing is required to generate valid events or prediction. Most of these examples are mature fields in traffic management but where new approaches have been driving changes in the last few years. In the last group, the events and predictions are used to determine automatic outputs such as traffic light cycles or variable speed limits, creating a closed inputs → processing → outputs → impact that changes inputs loop.

13.2.2.1 Automatic Incident Detection

Automatic Incident Detection (AID) can be implemented either by image processing algorithms applied to the output of video cameras or by processing traffic flow data.

AID systems based on image processing (Figure 13.4) are capable of detecting incidents such as stopped vehicles, wrong-way drivers, queuing, slow-moving vehicles, fire and smoke.

Figure 13.4 AID system detecting a wrong way incident.

Nevertheless AID offers these detections with accuracy limitations, that is, there can be 'false positive' detections (detection of an incident that is not actually occurring) and detection omissions (an incident occurs but it is not detected). These limitations are inherent to the difficulty of implementing algorithms that can mimic the human eye (and brain) to detect and identify complex patterns in moving images [8].

There are different approaches among different AID algorithms. The principles behind AID algorithms are not complex and we can analyse some examples of two different detection types:

- Wrong-way drivers [9]
  - In order to detect wrong-way drivers, AID algorithms will need to know or learn the normal direction of the flow of traffic.
  - By determining the current flow of the vehicles and checking it against the normal one it will be possible to detect flows in relatively opposite directions which are potential wrong-way drivers.
- Stopped vehicles [10]
  - To detect stopped vehicles, AID algorithms will require an image without any vehicles (background model), either by learning or by a manual configuration process.
  - Each current image can be compared against the background model in order to detect differences; differences that stay in the image for long are candidates for stopped vehicles.

Although the principles can be very simple, AID algorithms face a set of complex challenges that directly influence the accuracy of results. First of all, the moving elements in the image may be other than road vehicles (e.g. airplanes in the sky, birds, insects or even vehicles on other roads not being monitored). To cope with this, most AID systems limit the areas being processed, either through calibration, using a mask of the image areas that are processed, or by determining the road lane lines and producing that mask automatically [11]. Another step to solve this problem is to distinguish between vehicles and other objects [10].

Other challenges [11,12] may be:

- the change of the image background over time, e.g. as the night sets, this requires an online background model update;
- shadows that are considered as moving objects require being detected and removed;
- abrupt contrast or brightness changes;
- occlusions by other vehicles.

The way in which AID algorithms cope with these challenges will determine the effectiveness of the algorithms. Whether the algorithms require initial or periodic calibration or can learn more elements on their own will determine the ease of use and versatility of the algorithms.

Video-based AID accuracy can be enhanced and has been getting better as image processing algorithms are further developed. AID accuracy is also influenced by the image quality – having low noise in the acquired images is essential. Depending on the algorithms used and hardware computation capabilities of AID devices, using a high-definition video signal might also enhance accuracy.

In AID systems based on processing traffic flow data, the same technology used to gather traffic flow information is also capable of detecting some incidents like queuing, slow driving vehicles and wrong-way drivers, in most cases with almost 100% accuracy. There are some limitations in these technologies: incidents like fire or smoke are not detected at all, stopped cars can only be detected if they stop at the point of measurement and wrong-way drivers using the hard shoulder, when there are no sensors there, are not detected.

Several algorithms for traffic flow based on AID systems exist. Some algorithms, like the TVS-system, are based on microscopic analysis while most are based on pattern recognition [13]. There are different types of pattern recognition algorithms [14]. Comparative algorithms like the California, PATREG and APID compare the measured values with pre-defined threshold values. Statistical algorithms like SND and Bayesian algorithms use standard statistical techniques to determine if the measured values differ statistically from estimated or predicted traffic characteristics. The time series algorithms family, which includes ARIMA and HIOCC, goes one step further than comparative algorithms and determines threshold values for each time period automatically. Smoothing and Filtering Algorithms like DES, LPF and DWT-LDA remove short-term noises from traffic to avoid false alarms and enhance the true traffic pattern to better detect incidents. Traffic Modelling algorithms include dynamic models (MM, GLRS), catastrophe theory models (McMaster) and low-volume (LV). These algorithms create traffic predictions/models and compare them against the actual flow in order to detect incidents.

Most of these algorithms have the need for very fine-grained data, ideally from sensors deployed every 100–1000 m. Successful deployments of pattern recognition algorithms include Expressway Monitoring and Advisory System (EMAS) in Singapore [15], the Freeway

Traffic Management System (COMPASS) in Canada [16], which relies on the McMaster algorithm, and the Highway Incident Detection and Automatic Signalling (MIDAS) project in the United Kingdom, which uses the HIOCC algorithm and is now covering 1368 km (48%) of the English highway network [17]. The MIDAS project goes one step further by going from analysis to outputs, in this case as automatic speed limit control. This establishes a closed loop automated control system that is reviewed further ahead.

New trends in flow-based incident detection algorithms [14] are coming from artificial intelligence, which applies 'black box' approaches. Nomadic devices are also having a strong influence in the field of incident detection with MIT, ADVANCE, TTI, UCB and TRANSMIT algorithms being applied to probe-data from nomadic sources. Sensor fusion-based algorithms are also emerging, relying on a mix of several different sources (e.g. nomadic and traffic flow data), sometimes also combining more classic approaches with artificial intelligence ones.

AID solutions are typically evaluated according to three parameters [18,19]:

- detection rate – percentage of real incidents detected by the system;
- false alarm rate – percentage of detections by the systems that do not correspond to real incidents;
- time to detection – average time required to detect an incident.

Typically video-based AID solutions are more versatile, detecting more kinds of incidents, but at the cost of higher rates of false positive detections (in the order of 4–25% in open roads, less than 1% in tunnels [13]) than traffic-flow-based AID solutions (usually less than 2% [18]). In terms of time video-based AID commonly has average detection times under 5 seconds if the incident is covered by a camera, while flow-based AID solutions range from 20 seconds to 4 minutes [14] when there is flow information every 100 meters [18]. If the distance between flow data collection is higher the time to detect will also become higher.

There is not one winner solution, algorithm or approach for automatic incident detection. The different approaches have been applied successfully in different scenarios. Usual difficulties come from false positives that cause negative reactions from operators and from difficulties in maintaining solutions that require frequent calibration.

13.2.2.2 Travel Time Estimation

Travel time estimation (TTE) is a traffic management service by which estimated values for travel times to different destinations are made available to drivers. The effects of such information are not fully predictable and depend on the individual reaction, but, overall, may help relieving congestion by having drivers choose alternative routes.

TTE is the estimated duration of travel for a specific origin-destination pair (e.g. 'Lyon to Paris 5 h'). The estimation can be made from the current situation, in such case it may be very imprecise, or use prediction algorithms to forecast future conditions, far enough in time to be able to offer a reliable estimation.

TTE algorithms usually rely on historical data and range from statistical techniques to artificial intelligence and simulation [20].

TTE algorithms are typically based on route travel time data (obtained from trips between two points). This involves being able to identify the exact timestamp of the same vehicle in these two points (either by licence plate recognition or by some kind of on-board device – a mobile phone, OBU, RFID, etc.). TTE approaches based on single point data also exist and can take advantage of existing infrastructure, such as traffic counters, but have a higher error margin than route based data. TTE can use the same set of data sources as AID (some provide point data, e.g. traffic counters, others route data, e.g. nomadic devices).

Some examples of travel time information deployments in Europe are in the UK (Leeds, all motorways and major roads), Finland (3300 km around the largest cities) and Germany (Bavaria) [17].

13.2.2.3 Closed Loop Automated Control Systems

If the processing of inputs is used to automatically set outputs and the result from these outputs then influences the same inputs we have a closed loop automatic control system. This is the case of traffic lights and variable speed limit.

(a) Adaptive Traffic Signal Control

If traffic lights are controlled by processing data from traffic flow sources in order to optimize throughput, the resulting optimized traffic light cycles will in turn cause changes to the traffic flow that will again be fed as input to the process and end-up further influencing the traffic light cycles.

The basic principle behind adaptive traffic signal control is to avoid having green traffic signals without cars passing. This can be done locally at a single traffic sign or, usually, coordinated across areas or an entire city.

Adaptive traffic signal control systems vary in several respects, namely in the type of sensor used to detect vehicles waiting for a traffic signal, distributed or centralized architecture, timing between processing cycles, what is the mathematical goal of the system, among several other respects that influence the process and the end results obtained.

Several adaptive signal control systems exist worldwide [21,22]. Split Cycle Offset Optimization Technique (SCOOT) developed in the UK, Sydney Coordinated Adaptive Traffic System (SCATS) in Australia, MOTION in Germany, Real Time Hierarchical Optimized Distributed Effective System (RHODES) [24] in the US, UTOPIA in Italy and Optimized Policies for Adaptive Control (OPAC) in the US. SCOOT is currently the most common system in the US [22]. Most of the systems developed in countries other than the US are also currently deployed in the US [21,22].

(b) Variable Speed Limit Control

Another example of closed loop automated control systems is Variable Speed Limit Control (VSLC) [24,25]. VSLC uses traffic flow data to feed real-time algorithms that determine the speed limits that are then displayed through variable speed signs on the road.

The basic analysed parameters of traffic flow are speed, density and flow. Based on collected parameters, real-time algorithms calculate the optimum maximal speed of the vehicles. VSLC algorithms can also incorporate other environmental elements like weather and road surface conditions to provide maximized safety through the speed limit values.

The main purpose of VSLC is to homogenize traffic flow, resulting in a unique optimum mean speed of all vehicles on all lanes, thus reducing the risk of the so-called ‘shock waves’ and reducing the potential for traffic accidents and congestion. The principle is simple: during congestions, when density and flow are high, vehicles are packed closer together – under these conditions, if vehicle speeds differ a lot from vehicle to vehicle it will be more difficult to keep the traffic flow equilibrium, frequent lane changes will possibly occur and there will be a greater risk of accident.

VSLC can also be used strategically to reduce the influx of vehicles on a critical section (a congested section or one where an accident just occurred) by reducing the speed on prior sections.

VSLC algorithms range from theoretical methods based on microscopic traffic flow models, practical methods based on heuristic control logic, methods based on artificial intelligence and even methods that use predictions of the current potential of accidents to determine the optimized flow that lowers the risk of accident.

Whereas Adaptive Traffic Light Control is an urban ITS system used mostly in the core network of the cities, Variable Speed Limit Control is mostly found in inter-urban or urban ring roads.

VSLC is extensively deployed. Some examples deployments in Europe [17] are in the UK managed motorways (M1, M25, M42, M60), Sweden (19 intersections), Netherlands (A1, A12, A58, A13 motorways), Germany on over 1200 km of autobahn, Austria: several motorways and expressways (about 450 km), Italy motorways and France (A7, A9, A13 motorways).

13.2.3 Outputs

The data collected (inputs) can be stored, analysed and transformed into useful information that can be consumed either by operation staff, by drivers or by external entities such as security or emergency forces.

13.2.3.1 Variable Message Signs

VMSs are a common way to transmit information about road conditions to drivers. They can inform about circulation conditions at the approaching cross sections, weather conditions, collisions and other event notifications, recommended speed for better traffic flow and, in case of road blocks or major congestion, they may suggest alternative routes.

13.2.3.2 Traffic Lights

These components are used since 1868 but even though they are in principle a simple solution they provide a very effective solution for controlling competing traffic flows in road intersections. The effectiveness of the solution is enabled by enforcement and by applying significant penalties to transgressors.

Traffic lights can be controlled locally, coordinating only a set of visible lights in a single intersection. Timing for each direction can be fixed accordingly to the usual flow of that direction. As seen in the previous section, traffic lights can also be coordinated using Adaptive Traffic Signal Control systems.

13.3 Key Stakeholders

In a traffic management operation, several key stakeholders must be taken into account, according to their roles and responsibilities.

- *Operator*: The operator is in charge of the infrastructure or service and commonly has to guarantee a minimum set of service levels that can range from keeping fatalities and injuries under given values, maintaining defined levels of availability of the roads, the timely provision of information to regulators or the public or repairing damages in the infrastructure within a given time-frame. Usually some of these obligations are bound in contracts, but others are intrinsic to either the role that the operator performs or come from the mission the operator defines for himself. The operator can be a private company or a public one, and its scope can either be over an urban (usually municipalities or public companies) or interurban scenarios. As the operator will be the manager of the operation centre, it becomes the main stakeholder regarding the business rules definition, and thus, the traffic management platform design and implementation.
- *Regulator*: The regulator, or road authority, is usually a government dependent body which defines the terms of concession and some service levels to road users and monitors their execution. The regulator acts within the legal framework of the country and often accumulates competences over road signage or roadside device approvals/characteristics, or participates in standardization committees, for example. The regulator usually has a responsibility before the general public regarding road safety aspects. Depending on the particular national juridical framework, these competences can be concentrated on a single body or in multiple ones.
- *Road users*: Road users are key players when it comes to traffic conditions. Road users are in the 'frontline' and are the primary interested parties when it comes to road safety. The interests of road users are the main reason for traffic management, with particular focus on road users' safety and mobility.
- *Operator service suppliers*: To assure the contracted (or desired) service levels, the operator needs to manage a diversity of service providers that include, but may not be limited to, road assistance and maintenance vehicles, roadworks subcontractors or insurance providers. The coordination of these players is essential to achieve optimal efficiency, whether they are within the organization or not, and the operation centre must somehow address their management and allocation.
- *Authorities and assistance means*: Authorities, like police or the military can ask for access to road related information – more commonly, police forces are a close partner on road operation due to their enforcement and road signalling competences. Assistance means such as Emergency Assistance means play a critical role in road user assistance in the event of accidents – sometimes, the assistance means allocation time can be life or death defining, and any road operator must have lean and efficient business processes to interface with such entities, supported on their Traffic Management platform.

13.4 Traffic Management Centres

Traffic Management Centres (TMC) – or Traffic Control Centres – are currently the central point where decisions are made and traffic management is actually performed (Figure 13.5). TMC are supported by the information from the roadside (inputs) and rely on specific software

Figure 13.5 Brisa's Traffic Management Centre (example of a centralized traffic management centre).

to deal with such information (analysis) and support the traffic management operation process. These systems are known as Traffic Management Systems (TMS) and play a critical role in the business of Traffic Management Centres and the entire structure of the entities managing the road network. The TMS information life-cycle is not limited to the road operation alone, as it interfaces with back office and administrative processes that are vital to the organizations.

13.4.1 Scope

TMC are responsible for analysing, monitoring and receiving information about the road conditions such as traffic status, weather, incident and accident occurrence and details. TMC then use this information to influence behaviour and change conditions on the road, namely by sending messages to Variable Message Signs (VMS), deploying assistance vehicles or disseminating traffic information. TMC usually follow a set of operational procedures that define how to address and follow-up different situations and scenarios.

In some cases the TMC are also responsible for managing road patrolling and assistance and/or maintenance vehicle fleets.

Many of TMC business processes will not end when the operational situations are resolved. For instance, there will be insurance claims to manage, payment processes for entities involved in road repairs and the need for situation information review and completion of reports.

Traffic management is a heavily information-oriented activity, producing a lot of automatic and manual information on traffic flow and situations. This information is extremely valuable for operational and strategic management of the highway operator (when considering the typical interurban scenario). This is where a Business Intelligence component can deliver very important business value; guiding the highway operator in regards to service levels and towards continuous evaluation and enhancement. The operator service levels can be monitored and controlled using a Business Intelligence component.

13.4.2 *Operation Platforms*

A Traffic Management System (TMS) acts as a bridge between the roadside and the Traffic Management Centre (TMC). A TMS should handle two main function groups regarding the roadside:

- gather information (e.g. traffic data, weather data, reports regarding incidents and accidents);
- influence behaviour (e.g. send messages to VMS, disseminate traffic information).

A TMS (Figure 13.6) should not just be confined to providing ways of viewing the gathered information and controlling equipment. Traffic management is all about maximizing the traffic flow and increasing safety. This is achieved by identifying or anticipating any impediments to the traffic flow or safety concerns and adequately reacting to them.

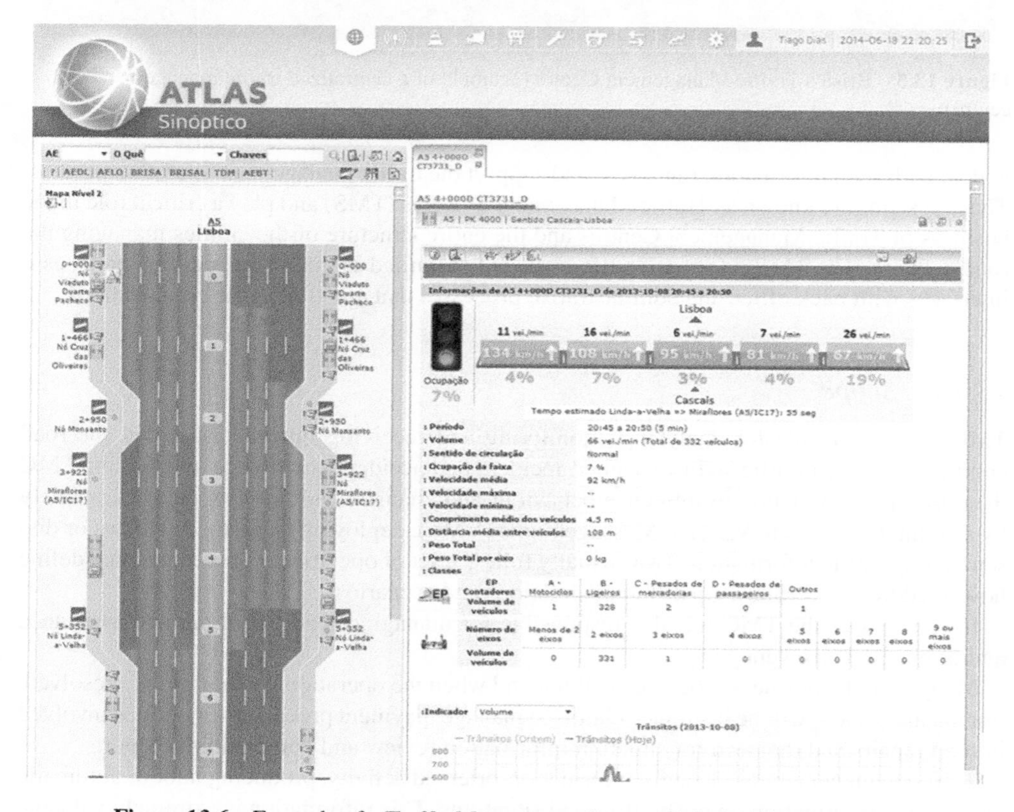

Figure 13.6 Example of a Traffic Management System (Brisa's Atlas Platform).

The TMS is expected to do more than just provide the information gathered from the roadside equipment to the operator. The TMS is expected to add value to that information by processing, correlating and filtering it for the most relevant elements for operation. Finally, the TMS should present that processed information in an intuitive and simple form for human operator analysis.

The TMS supports the TMC operator analysis of road situations by providing adequate access to more details (data or live video) about the identified situations/incidents.

To help with the response, the TMS should also offer guidance on how to address the different situations, usually by having embedded in its business rules originated on the operator. The TMS will need to address two main fields of the response:

- active traffic management, by offering a high-level control of the roadside equipment, for example, determining the appropriate VMS messages for the appropriate VMS location according to a situation that is occurring;
- identifying, dispatching and managing assistance entities that will intervene in the situation.

The TMS will then need to offer help in monitoring the evolution of situations/incidents. The TMS should also provide support to manage other activities on the road that contribute or are in some way related to the traffic flow. It should help manage road patrolling activities as well as roadworks.

As stated before, TMC business processes are not limited to the operational scope. The TMS should be able to support the continuity of all business processes or otherwise be integrated with other systems that will support that business continuity. Namely by supporting insurance claims, payment processes and situation information review and reporting.

The different layers described for the TMS can be summarized by six different TMS levels:

Level 1 – Physical equipment;
Level 2 – Local equipment control and automation;
Level 3 – Remote equipment control (telematics);
Level 4 – Operations support (response business processes);
Level 5 – Nonoperational support (follow-up work flows of incidents and roadworks);
Level 6 – Business intelligence.

The TMS will also need to act as a traffic information hub and be able to share or distribute that information. Sharing can be done by directly supporting human access to information, for instance with an Internet traffic information component, or by acting as an integration point for other applications that need information.

13.4.2.1 Traffic Management System Architectures

Two main TMS architectures exist; both are client-server based. An event-based TMS gathers data and events from the roadside, transforms them on the fly if necessary, and delivers them directly to a client component that is used by human operators in the TMC. The client holds in memory all the data received, filters it and displays it to the operator. The event-based architecture has the advantage of being able to provide real-time information. It has the disadvantage of not storing the information received centrally for historical analysis.

The second architecture is the data-store. Data-store based TMS are still client-server based but the server stores all data in a persistent fashion and the client only requests the data it needs. The major disadvantage of such architecture is that clients will not receive real-time information. The advantages are to be more lightweight clients and the persistence of all the necessary information for historical analysis.

Due to the disadvantages of both architectures, more commonly, TMS implement a hybrid solution either by adopting a special client to maintain all the data in event-based TMS or by adopting an event channel from the server to the client in data-store based TMS.

When operating large road networks, operators need to decide whether to implement several dispersed smaller TMC or one centralized TMC. In some cases there may be no communications between different points of the network, in which case the centralized option is not possible.

In a centralized, larger TMC, it is easier to adopt uniform procedures and applications. Workloads must be divided between TMC operators. This can be done either by function or by geographic location. In the centralized approach it is easier to optimize the use of human, hardware and software resources. Nevertheless, dispersed TMC can also be uniform, using the same hardware, software and procedures (or not), providing some of the benefits of the centralized TMC approach.

Many TMC create their TMS by adopting several different applications for different subsystems like video-cameras, emergency calls, variable message signs, incident management. In this case some kind of integration between the different applications is sometimes necessary to achieve added benefits (e.g. traffic flow alerts with video-camera images, VMS control based on incident information).

Unified TMS platforms, on the other hand, tend to integrate all the different subsystems in one single application. On top of this, unified TMS platforms can offer integrated services, like automatic or semi-automatic response plans involving outputs in different subsystems (e.g. VMS, traffic lights or tunnel systems), or crossing information from different subsystems (e.g. video-cameras, traffic counters, assistance vehicle positioning). Unified TMS platforms can offer cross subsystem functionalities like auditing or alarms with the added value of using the same user interface.

The main disadvantage with unified TMS platforms is when the TMC needs to integrate a subsystem that the TMS platform does not support. In this case the TMC should adopt the new subsystems software but integration in the TMS may provide added benefits.

13.5 Conclusions

By defining and contextualizing traffic management, in this chapter we were able to present a traffic management framework. This framework can be structured as involving inputs, data analysis and outputs. Several inputs sources could be used: traffic counters, tolling systems, flow meters, weather stations, CCTV and nomadic devices, among others. The core where all is coordinated is the TMC, a real time management component of a much wider and longer running process. TMS is the platform that provides access to inputs, results of analysis and allows sending outputs/actions. Finally we overviewed two opposite architectures commonly used for TMS platforms, both client-server based: event based and data-store.

References

1. ISO 14813-1:2007. Intelligent transport systems – Reference model architecture(s) for the ITS sector – Part 1: ITS service domains, service groups and services. ISO, 2007.
2. K. Friso, K. Zantema and E. Mein (2013) On-line traffic modelling in Assen: the sensor city, *Proceedings of 3th International Conference on Models and Technologies for Intelligent Transportation Systems 2013, Dresden*, 1–10.
3. EasyWay Project. TMS-DG07. Traffic management plan service for corridors and networks. Deployment guideline. Version 02-00-00, December 2012.
4. I. Fraser, N. Hoose and J.-M. Hotteau (1993) The PLEIADES project: the design of inter-urban traffic management field trials on the Paris – London – Brussels corridor, *Proceedings of the IEEE-IEE: Vehicle Navigation and Information Systems Conference, 1993*, Dept. of Transport, Bristol, pp. 213–16.
5. C. Guasco (2013) Trans-European transport network and cross- border governance, *Selected Proceedings from the Annual Transport Conference at Aalborg University*, ISSN 1903-1092, 10/2013, pp. 1–13.
6. Trans-Texas Corridor, from Wikipedia <http://en.wikipedia.org/wiki/Trans-Texas_Corridor> (last accessed 2 May 2015).
7. T.H. Maze, M. Agarwai and G. Burchett (2006) Whether weather matters to traffic demand, traffic safety, and traffic operations and flow, *Transportation Research Record: Journal of the Transportation Research Board* **1948**: 170–6.
8. M.S. Shehata, Jun Cai, W.M. Badawy, *et al.* (2008) Video-based automatic incident detection for smart roads: the outdoor environmental challenges regarding false alarms, *IEEE Transactions on Intelligent Transportation Systems:* **9**(2): 349–60.
9. CTC & Associates LLC, *Automated Video Incident Detection Systems*, Caltrans Division of Research and Innovation, 28 October 2012.
10. G. Monteiro, M. Ribeiro, J. Marcos and J. Batista (2007) A framework for wrong way driver detection using optical flow, *Lecture Notes in Computer Science*, Vol. **4633**. Berlin: Springer.
11. G. Monteiro, M. Ribeiro, J. Marcos and J. Batista (2008) Robust segmentation for outdoor traffic surveillance, *IEEE ICIP 2008, San Diego, California, USA*, pp. 2652–5.
12. G. Monteiro (2009) Traffic Video Surveillance for Automatic Incident Detection on Highways, Master Thesis.
13. J. Michek (2013) Automatic incident detection. Lecture notes K620ARR, Czech Technical University in Prague.
14. E. Parkany and C. Xie (2005) *A Complete Review of Incident Detection Algorithms & Their Deployment: What Works and What Doesn't*, The New England Transportation Consortium, 7 February.
15. Expressway Monitoring and Advisory System, from Wikipedia <http://en.wikipedia.org/wiki/Expressway_Monitoring_and_Advisory_System> (last accessed 2 May 2015).
16. Freeway Traffic Management System, from Wikipedia <http://en.wikipedia.org/wiki/Freeway_Traffic_Management_System> (last accessed 2 May 2015).
17. P. Kroon (2012) Traffic management to reduce congestion, *Conference of European Directores of Road, 29 June 2012*.
18. J. Ozbay and P. Kachroo (1999) *Incident Management in Intelligent Transporation Systems*. Artech House Intelligent Transportation Systems Library, Artech House.
19. R. Brydia, J. Johnson and K. Balke (2005) *An Investigation into the Evaluation and Optimization of the Automatic Incident Detection Algorithm Used in TxDOT Traffic Management Systems*, Texas Transportation Institute, October.
20. H. Lin, R. Zito and M. Taylor (2005) A review of travel-time prediction in transport and logistics, *Proceedings of the Eastern Asia Society for Transportation Studies* **5**: 1433–48.
21. Adaptive Signal Control, Federal Highway Administration, 2013. <http://www.fhwa.dot.gov/everydaycounts/technology/adsc> (last accessed 2 May 2015).
22. A. Stevanovic (2010) *Adaptive Traffic Control Systems: Domestic and Foreign State of Practice*, National Cooperative Highway Research Program, Synthesis 403, Washington DC.
23. P. Mirchandani and L. Head (2001) A real-time traffic signal control system: architecture, algorithms, and analysis, *Transportation Research Part C: Emerging Technologies* **9**(6): 415–32.
24. J. Youngtae, K. Yoon and J. Inbum (2012) Variable speed limit to improve safety near traffic congestion on urban freeways, *International Journal of Fuzzy Systems* **14**(2): 43–50.
25. P.S. Virginia (2001) Variable speed control: technologies and practice, *Proceedings of the 11th Annual Meeting of ITS America*, pp. 1–11.

14

The Use of Cooperative ITS in Urban Traffic Management

Sadko Mandžuka, Edouard Ivanjko, Miroslav Vujić, Pero Škorput and Martin Gregurić
Intelligent Transport System Department, Faculty of Traffic and Transport Science, University of Zagreb, Zagreb, Croatia

14.1 Introduction

Traffic and transport engineering of the 21st century requires a new approach in order to make transport safer, more efficient and reliable, with minimal impact on the environment and society. Basic characteristics of this new approach are increasing demands followed by the imperative of lower costs. Traffic safety is especially very important. The challenge is to increase traffic safety by 50% until 2020 in order to reach the main goal of zero fatalities in 2050. The goal for more efficient traffic system is set for reducing congestion – 2% Gross Domestic Product (GDP) – and 60% energy efficiency and emissions until 2050 [1]. The expectations for urban traffic are especially high with demands in integration of different transport modes and significant reduction of noise and air pollution. The European Union has recognized intelligent transport systems (ITS) as the technological basis for accomplishing these goals. ITS is a holistic, control and information communication technologies (ICT) upgrade of classic traffic and transport systems which significantly improves system performance, traffic safety, efficiency in transportation of goods and passengers, increases passenger protection and comfort, reduces pollution, etc. [2]. The possibilities of use of modern ITS in traffic safety improvements can be divided into the following:

1. systems related to infrastructure (roads, bridges, tunnels…);
2. systems related to vehicles;
3. systems based on cooperation.

Intelligent Transport Systems: Technologies and Applications, First Edition. Asier Perallos, Unai Hernandez-Jayo, Enrique Onieva and Ignacio Julio García-Zuazola.

A particularly potent approach was recognized in the possibility of application of cooperative systems in traffic. Cooperation can be viewed as a basic form of organization, in a broad sense, and as a problem of communication of a moving entity (vehicle) with road infrastructure and/or other moving entities, in a narrow sense. In the broad sense Ramage defined the cooperative system as a combination of technologies, people and organizations which enables communication and coordination required for achieving a common goal of a certain group which performs various activities, for the benefit of all participants [3]. In the narrow sense of the cooperation definition, the following communications were recognized: V2V – vehicle to vehicle (link 1), V2I – vehicle to infrastructure (link 2), V2P – vehicle to pedestrian (link 3), I2P – infrastructure to pedestrian (link 4), see Figure. 14.1. Dedicated short-range communications (DSRC) is one of basic vehicular communications technology used for these applications.

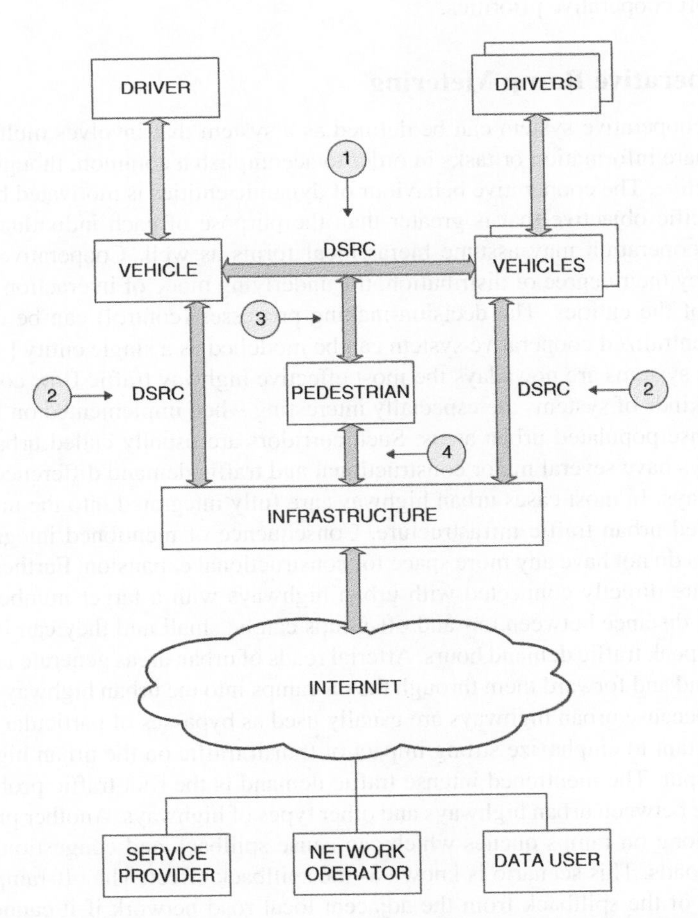

Figure 14.1 Basic topology of cooperative system in traffic and transportation.

The main characteristics of a cooperative approach are:

1. consider the driver, vehicle, infrastructure and other road users as a unique system;
2. consider operational and management needs of the entire system;
3. integrated approach to safety of traffic and all participants;
4. apply technology in a coherent manner in order to support overall integration of system parts.

Currently we recognize the following systems onto which the cooperative approach can be successfully applied: navigation systems and travel information systems, warning systems, emergency services' vehicle management, priority management in urban public transport, intelligent systems for speed management, support systems for endangered transport users and others.

The following examples of the cooperative approach application in urban traffic are described in this chapter: cooperative ramp metering, incident management in urban areas and public transport cooperative priorities.

14.2 Cooperative Ramp Metering

In general, a cooperative system can be defined as a system that involves multiple dynamic entities that share information or tasks in order to accomplish a common, though perhaps not singular, objective. The cooperative behaviour of dynamic entities is motivated by the need to achieve a specific objective that is greater than the purpose of each individual entity. This implies that cooperation may assume hierarchical forms as well. Cooperative systems are characterized by their degree of distribution, the underlying mode of interaction and the level of autonomy of the entities. The decision-making processes (control) can be distributed or centralized. Centralized cooperative system can be modelled as a single entity [4].

Cooperative systems are nowadays the most effective highway traffic flow control method in ITS. These kinds of systems are especially interesting when implemented on highway corridors near dense populated urban areas. Such corridors are usually called urban highways. Urban highways have several major constructional and traffic demand differences in contrast to other highways. In most cases urban highways are fully integrated into the urban environment and related urban traffic infrastructure. Consequence of mentioned integration is that urban highways do not have any more space for constructional expansion. Furthermore, urban arterial roads are directly connected with urban highways with a larger number of on- and off-ramps. The distance between on- and off-ramps can be small and they can influence one another during peak traffic demand hours. Arterial roads of urban areas generate large amounts of traffic demand and forward them through the on-ramps into the urban highway mainstream. This happens because urban highways are usually used as bypasses of particular urban areas. It is also important to emphasize strong impact of transit traffic on the urban highway mainstream throughput. The mentioned intense traffic demand is the root traffic problem and the main difference between urban highways and other types of highways. Another problem is the occurrence of long on-ramps queues which can cause spillback and congestion on adjacent local arterials roads. This scenario is known as the spillback effect. The off-ramp congestion area is a result of the spillback from the adjacent local road network if it cannot accept all vehicles leaving the highway.

On- and off-ramps are also places where it is possible to make a significant influence on highway mainstream. The control method that regulates the number of on-ramp vehicles allowed to enter mainstream flow (interaction between on-ramp and mainstream traffic flows) by means of a traffic light is called ramp metering. Implemented as a local responsive control method it is effective only for mitigation of moderate traffic congestion [5]. Cooperative approach in ramp metering is related to cooperation between several on-ramps in order to mitigate upstream congestion back-propagation [6]. Besides ramp metering, several other traffic control systems are developed in order to mitigate congestions at highways: variable speed limit control (VSLC), selectively prohibiting lane changes and various personal driver information systems [7]. Mostly, these control systems function as a standalone application. However, active research is being done with the goal of achieving cooperation between ramp metering and other traffic control systems [8,9].

14.2.1 Ramp Metering

The consequences of urban highway congestions are manifesting themselves with the following indicators: traffic demand exceeding road capacity, increased number of accidents and incidents, queues on arterial roads that spill over onto the highway, and peaks in traffic demand resulting from platooned vehicle entry from on-ramps [4]. Common source of congestions on urban highways is in the merging of flows from on-ramp and mainstream traffic. The area where these two types of traffic flows are actually coming in interaction is known as downstream bottleneck. In Figure 14.2 control system architecture for cooperative ramp metering with location of the downstream bottleneck close to the on-ramp and spillback effect is given.

Respectively, modern ramp metering systems control the number of vehicles entering the highway according to the current traffic conditions using traffic lights. Ramp metering algorithms compute the restriction rate of the total traffic flow entering the highway thereby using on-ramps as a temporary vehicle storing place. This process is known as 'access rate reduction'. The whole system is based on traffic data collected in real time by road sensors (inductive loops, cameras, etc.) and controllable traffic lights [5].

Generally it is possible to divide ramp metering algorithms into two major control categories or strategies: local (or isolated) and coordinated [10]. Local strategies include ramp metering algorithms which take into account only the traffic conditions on particular on-ramp and nearby segment of highway where they are applied. Linear overriding control of motorway on-ramp (ALINEA) algorithm is the most often used standard local ramp metering algorithm [10]. The reason for this lies in the algorithms optimal ratio between simplicity and efficiency. The core concept of ALINEA is to keep the downstream occupancy of the on-ramp at a specified level by adjusting the metering rate. The main disadvantages of ALINEA are: inability to resolve upstream congestions of the particular ramp and to locate optimal detector mounting zone.

Coordinated algorithms are taking into account the traffic situation on the overall highway transportation system. In literature these algorithms are further divided into: cooperative, competitive and integrated algorithms [5]. Ramp metering algorithms based on cooperation will be described in the next section with more details. In competitive algorithms, two or more sets of metering rates are computed based on different algorithms types. The more restrictive metering rate is selected as the final one. Typical representative of competitive ramp metering

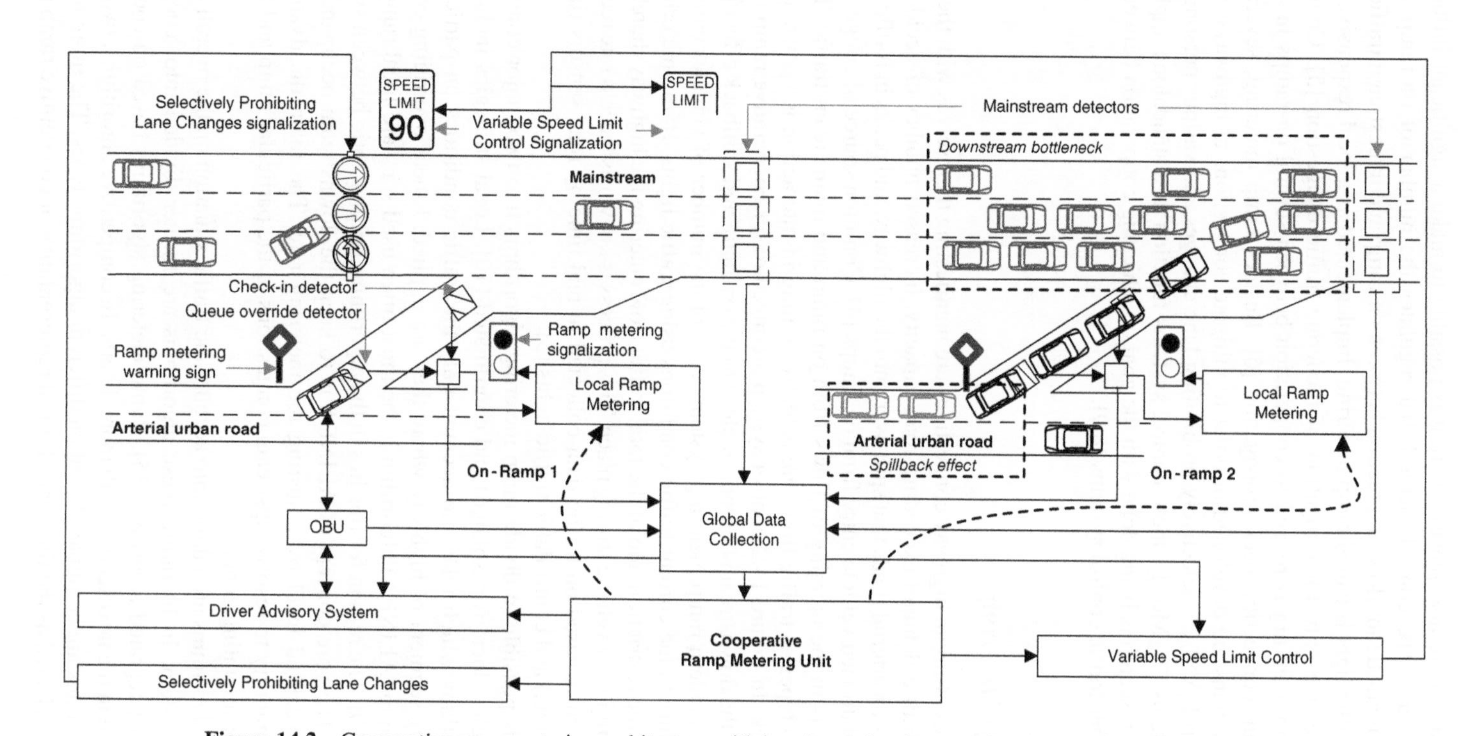

Figure 14.2 Cooperative ramp metering architecture with locations of mainstream bottleneck and spill-back effect.

algorithms is SWARM [10]. The SWARM algorithm has two major modules. First module performs local, while second module conducts global predictive ramp metering based on the same traffic control scenario. As final metering rate minimal/restrictive solution for particular traffic scenario is selected. Integrated algorithms compute the ramp metering rates through optimizing an objective such as specific level of service (LoS) value while considering system constraints like maximum allowable ramp queue, bottleneck capacity, etc. Fuzzy logic based algorithms are the most representative members of integrated algorithms.

Modern approaches in ramp metering include use of artificial intelligence in order to perform adaptive mitigation of congestion which is varying in strength and in time. Recent work described in [11] includes use of an adaptive neural-fuzzy inference system (ANFIS) algorithm. Artificial neural network (ANN) as the part of ANFIS algorithm is trained by learning data set which is obtained from the simulation results of the several different ramp metering algorithms. ANN modifies parameters of ANFIS fuzzy inference system according to acquired knowledge from learning data set in order to cover a wide range of traffic scenarios on a particular highway.

14.2.2 Cooperation between Local Ramp Meters

Ramp metering algorithms based on cooperation are also working in two phases. In the first phase the metering rate for each on-ramp is computed. Usually this is done by a local ramp metering algorithm. In the second phase further adjustment of each local metering rate is done based on system-wide information about the traffic situation on the whole ramp metered highway segment. Adjustment criterion is to prevent both traffic problems, congestion at the downstream traffic bottleneck and spill-back at critical on-ramps in upstream traffic.

HELPER, as the most prominent cooperative ramp metering algorithm, uses local traffic-responsive algorithms with the added feature of central override control [12]. Override control can adjust local metering rates if queue threshold is reached. A particular on-ramp is categorized as ‘master’ if it is operating at the minimum metering rate, and occupancy on the queue detector exceeds a pre-determined threshold value. The centralized module increases the metering rate at the ‘master’ on-ramp by one level and reduces the rate for the several upstream on-ramps by one level. These upstream on-ramps are categorized as ‘slave’ on-ramps. The main idea is to exploit their queue capacity in order to mitigate downstream congestions at mainstream.

The working principle of cooperative ramp metering algorithms can be very complex and sensitive considering fluctuations in traffic parameters. Because of the above-mentioned reasons it is imperative to conduct simulations in order to analyse the impact of cooperative ramp metering on the traffic flows. One appropriate simulator is CTMSIM [13]. It is an open source interactive simulator based on macroscopic traffic models specifically designed for highway traffic flows simulations. The highway model used in CTMSIM is based on the Asymmetric Cell Transmission Model (ACTM) and allows implementation and development of user-pluggable on-ramp flow and queue controllers.

In order to analyse cooperative ramp metering, an appropriate high-level traffic management framework was developed as augmentation to the original CTMSIM simulator [9]. Cooperative ramp metering algorithm HELPER was created with the use of the above-mentioned cooperative framework. The second step in the augmentation process was to develop an additional highway subsystem which will potentially provide cooperative support to the HELPER algorithm. VSLC is selected as an adequate highway management subsystem and implemented into CTMSIM as the second augmentation.

Ramp metering control systems are evaluated using different measures of services. The basic measure of service quality for ramp metering is travel time (TT) and delay. TT provides an answer to how much time one vehicle needs to travel through observed highway segment. If this measure is unusually high, it is a clear sign of LoS quality drop for the observed highway segment. Delay is the difference between the actual time spent by all vehicles on a congested highway and the time spent in case they have travelled at free flow speed. Delay also considers vehicles which are waiting in on-ramp queues or at mainstream queues caused by the bottlenecks.

An important step is to select appropriate highway for modelling in a simulation environment. The urban highway section between the nodes Jankomir and Lučko of the Croatian city of Zagreb bypass is selected as the use case model. This section contains 70% of traffic generated by the nearby town Zagreb [14]. Additionally, this section has an increased traffic load during the whole day and a prominent effect of daily migrations on it.

14.2.3 Cooperation between Ramp Metering and Other Traffic Management Systems

It is possible to add several other highway management strategies into the cooperation with ramp metering algorithms. A higher number of management strategies included in cooperation with ramp metering contributes to the more comprehensive control over highway section. Management strategies which can be included in cooperation with ramp metering are: VSLC, selectively prohibiting lane changing, information systems for drivers and most recently semi-automatic control over vehicles. Basic functionality of cooperation between ramp metering and other highway management strategies can be seen in Figure 14.2.

One of the first research approaches in cooperation between several different highway traffic management systems is published in [15]. Research presents results in which cooperative system consisting of ALINEA and VSLC reduces TT around 1.62% compared to the ALINEA standalone implementation. Comparative analysis based on simulation results derived from the implemented use case shows that cooperation between HELPER ramp metering algorithm and VSLC produces the smallest TT [9]. Results of comparative analysis are shown in Figure 14.3 according to TT and delay.

According to Table 14.1 it is possible to conclude that cooperation between these two traffic control methods produces the best average TT results in comparison with other methods involved.

VSLC in cooperation with HELPER decreases speed of the incoming vehicles to the area between last ‘slave’ on-ramp and congested cell. It can be concluded that virtual queues provided by HELPER and speed reduction in area between last ‘slave’ cell and congested cell induced by VSLC significantly reduce traffic density upstream of the congested on-ramp. Lower upstream density of the congested on-ramp leaves additional mainstream capacity to accept vehicles coming from congestion back-propagation.

VSLC as part of HELPER and VSLC cooperation does not have strong influence on the delay as in the case of TT. According to Table 14.1 it can be concluded that VSLC has a lower

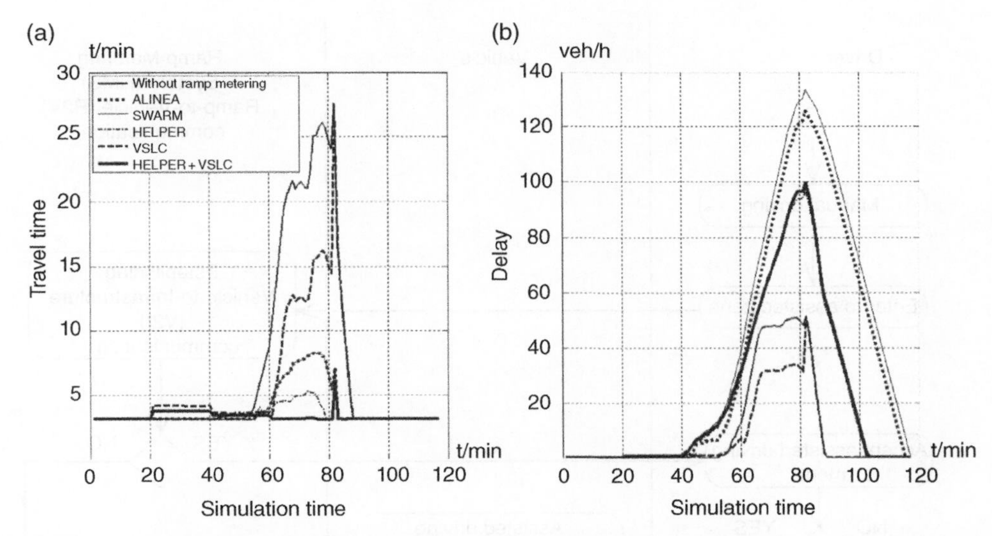

Figure 14.3 Results of comparative analysis between different ramp metering algorithms according to (a) TT and (b) Delay.

Table 14.1 Results of comparative analysis between different traffic control algorithms.

Traffic control algorithm	TT (min)	Delay (vehicle-hour)
None	7.06	15.87
ALINEA	3.90	36.88
SWARM	3.71	41.49
HELPER	3.40	22.63
VSLC	5.59	12.24
HELPER + VSLC	3.30	21.50

influence on delay if the mainstream density is decreased by HELPER's exploitation of on-ramp queues. Used simulation settings were set to enable immediately merging of all vehicles from the on-ramps with mainstream if in a current cell maximal mainstream capacity was not exceeded. This feature enables higher values of mainstream density and consequently lower delay values compared to the case of no ramp metering. Standalone VSLC produces the best delay results. VSLC firstly gradually decreases mainstream speed, but enables higher mainstream speed during the congestion period unlike the scenario without VSLC and ramp metering.

A special option is the use of a cooperative metering approach in the case of applying vehicle to infrastructure (V2I) communication (see Figure 14.4). Cooperation between vehicle and on-ramp control system can be established at the moment when the vehicle has stopped at the on-ramp end and is waiting for the green light. At the moment when the green light is turned on, the on-ramp control unit sends a message to the first vehicle in the queue which obtains throttle control. Similar cooperation between vehicle and infrastructure cooperation unit can be established with vehicles in the highway mainstream. In this case vehicle speed is

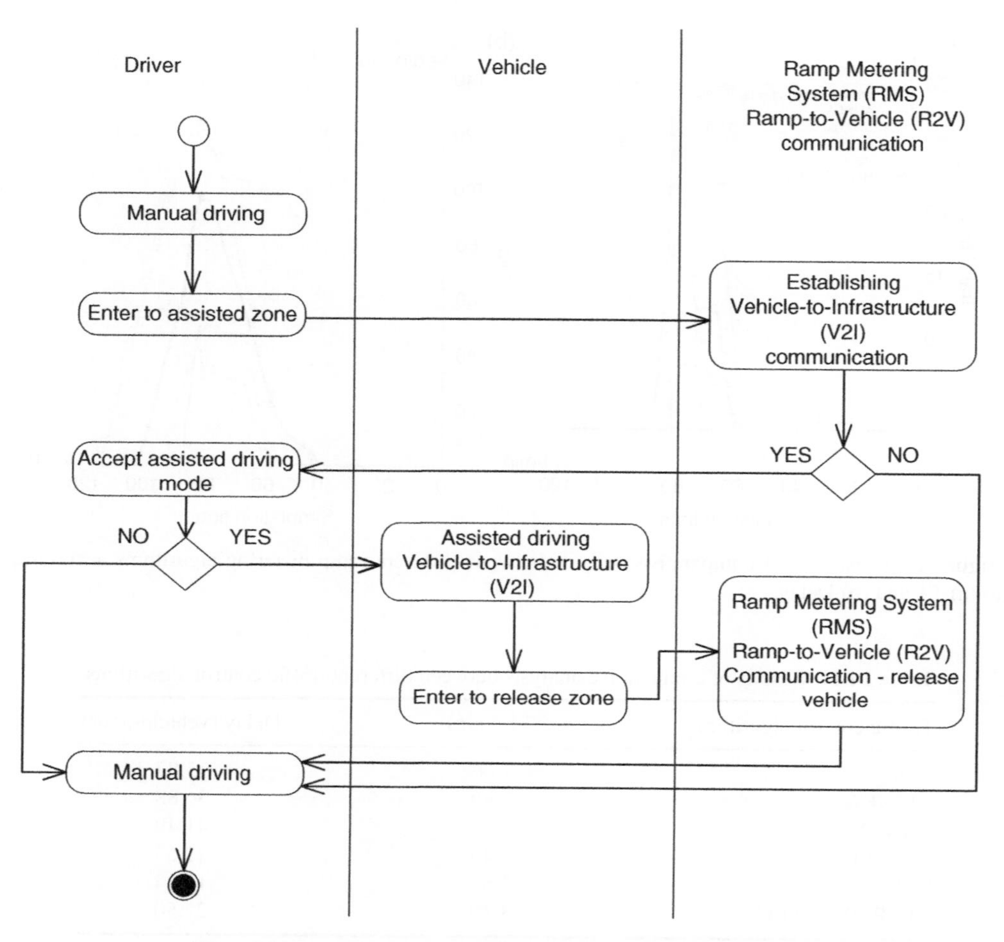

Figure 14.4 Activity diagram of V2I Ramp Metering System.

adapted (Intelligent Speed Adaptation) according to the current traffic state near an on- and off-ramp area. The result is safer highway traffic near inclusion of on-ramp vehicles into mainstream flow.

14.3 Incident Management in Urban Areas

Traffic incident is an unforeseen or unpredictable event that impacts on the safety and the capacity of the urban and other road networks and causes extra delay to road users. Traffic incidents which occur in the urban road network have larger consequences on regular traffic flow compared to ones that occur in other parts of the traffic network.

The incident management process is divided into different phases where each group has its own strategy and definitions for handling traffic incidents. In general, the incident management process is divided into incident detection and verification, incident response, clearance of the incident and recovery of normal traffic flow [16]. One of the major transport challenges becomes how to enhance mobility while at the same time reducing congestion, incidents and pollution. Across Europe, urban road incidents account for 10–25% of congestion.

There are several different events that influence the normal or desired traffic flow in urban road network. The following events which may lead to temporary reduction in road network capacity are identified (compared to requirement) [17]:

- vehicle-conditioned incidents, ranging from minor vehicle damage to multiple accidents with injuries and fatalities;
- debris/barriers on the road;
- maintenance activities;
- unpredicted congestions;
- any combination thereof.

Incident management could involve the use of cooperative systems based on V2V and V2I communication to improve road safety in urban areas. Cooperative vehicles and infrastructure are equipped with a significant amount of sensory and communication equipment. Such systems allow vehicles to 'feel' (e.g. radar, ultrasound sensors), see (e.g. camera, infrared camera) and communicate with each other and with the infrastructure.

Cooperative ITS is based on the principle that all cooperative participants exchange information between each other. Every actor evaluates the data received and considers the information for its data analysis and information provided to the driver. The incident management system (IMS), fleet management centres and other stakeholders as well as centralized or decentralized traffic control can be part of the cooperative systems network. Development of such systems involves a vehicle sensing platform and applications, infrastructure sensing platform and applications and innovative technologies. Based on the cooperative data received, the cooperative incident management system will be able to generate the up-to-date status of the traffic situations and its forecast for traffic control and traffic management in urban areas. Prioritization of emergency vehicles at traffic lights will be possible as well as adaptive 'green waves', smoothing traffic flows and reducing efforts for finding incident location, etc. Also, it is possible to prevent road incidents by developing cooperative incident management applications which detect and predict potentially dangerous situations and extend, in space and time, drivers' awareness of the surroundings. Cooperation is possible with vulnerable road users like pedestrians and bicyclists and can support the mobility of blind and visually impaired persons. Cooperative systems offer the potential to reduce these impacts by creating additional effective road network capacity and a more efficient utilization by vehicles. Furthermore, it involves development and deployment of core architecture, the business model and legal aspects.

Cooperative scenarios in incident management in urban areas can be shown by one example of traffic incident (see Figure 14.5). In that case, emergency vehicle that is approaching the location of incident sends V2I data set to road side unit (RSU). Afterwards, infrastructure part of system adapts various traffic lights, variable message signs (VMS), etc. Also, I2V data set is sent to other vehicles in neighbourhood of incident. Consequently driver's display shows recommended vehicle speed and alternative route, whereby driver adjusts the speed and passes without stopping.

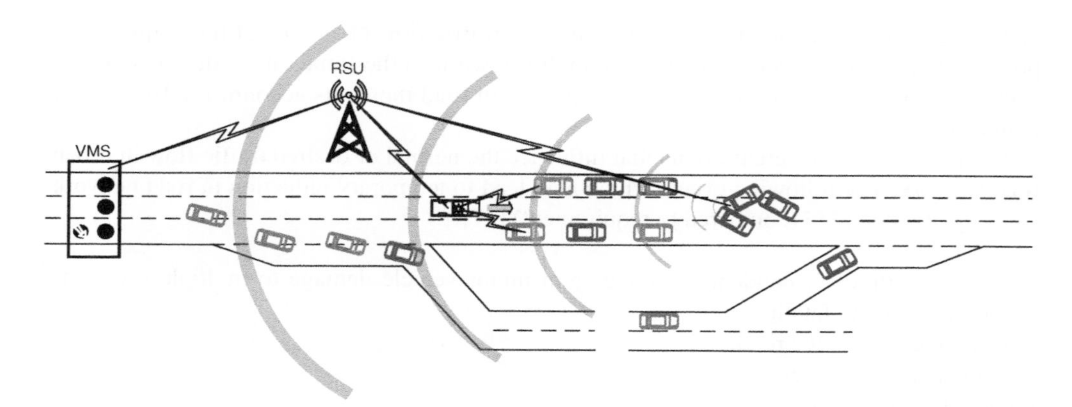

Figure 14.5 Cooperative systems and emergency vehicle.

Incident detection is a process of identifying the location and time of the incident event. Also, possible nature of the incident is very important. In cooperative ITS environment, cooperation among vehicles can detect traffic incidents or congestion on roads. Also, cooperative vehicle travelling in a group estimates traffic conditions separately. These estimations are joined together to collectively decide whether the roads are congested. Similarly, neighbouring vehicles in a V2V network can avoid traffic casualties by exchanging collision warnings in a cooperative manner. Based on V2V and V2I cooperative communication it is possible to develop a variety of applications such as:

- hazard warning for frontal collision prevention, ghost driver and obstacle detection;
- dangerous curve warning;
- vulnerable road user detection;
- reduced visibility warning;
- safety margin for assistance and emergency vehicle;
- intersection collision prevention;
- road condition status;
- roadworks status.

The European ITS Architecture define user-needs and functionality as Cooperative ITS services and applications. It is a set of high-level viewpoints that enable plans to be made for integrating ITS applications and services. It normally covers technical aspects, plus the related organizational, legal and business issues. ITS architecture supports harmonized deployment of ITS in Europe and enabling a framework for systematic planning ITS implementations, facilitates integration for multiple systems and ensures wide interoperability. ITS architecture is the conceptual design that defines the structure and behaviour of an integrated ITS. Architecture description is a formal description of a system, organized in a way that supports reasoning about the structural properties of the system. It defines the system components or building blocks and provides a plan from which products can be procured, and systems developed, that will work together to implement the overall system. This may enable managing investment in a way that meets business needs. In the ITS environment, incident management is the part of traffic management that provides functionality enabling the management of traffic in urban and inter-urban environments. In a wider sense, incident management also includes traffic control, demand management, road maintenance management and providing various information for other organizations [18].

The incident management system in urban areas shall provide facilities for the management of incidents that occur within an urban road network. Also, it shall integrate and involve other ITS applications in the incident management processes. The incident management system needs to perform the following internal functions or abilities in the cooperative ITS environment:

- provide operator interface for incident management;
- detect incidents from data;
- classify and identify incidents;
- send incident details to vehicles;
- provide incident mitigations to traffic management;
- send incident details to information providers or others;
- manage and store incident data.

Detecting incidents from data is the ability of the incident management system to analyse the data that it receives about traffic conditions in the road network in order to detect the possibility that incidents have occurred. It can also represent the ability to analyse all types of data for patterns that suggest the occurrence of an incident and the ability for such patterns to be linked to the same incident if they occur in adjacent sections of the road network. Classifying and identifying incidents is the ability to process data in order to identify and classify the particular type of incident that has been detected, according to the source, and using its own internal 'rules' that may relate to some form of approved standard. Sending an incident detail to vehicles is the system's ability to manage the output of instructions contained in an incident strategy to other functionality in the vehicle in response to incidents that have been detected by other functionalities. Providing incident mitigations to traffic management is the ability for the instructions included in the incident management strategies to be output to require the replacement of, or changes to, any traffic management strategies that are currently in operation. Assessing incidents and devising responses enable functions of managing the assessment of incident data and to create strategies in response to incidents that have been detected by other functionality. They can also periodically review the data that has been collected about incidents and decide if any mitigation action is needed by either using an existing incident management strategy or devising a new one. Sending incident details to information providers or others will provide facilities for managing the output of information to external service providers as part of an incident strategy. This is done in response to incidents that have been detected by other functionality and sent out to require the output of information to other functionality such as emergency support, public transport management and traveller assistance. Managing a store of incident data refers to taking responsibility for the management of data about incidents and the production of statistical reports. The efficient management of available information, data exchange as well as intelligent real-time decision-making can reduce the consequences of traffic incidents, especially preventing secondary incidents. Advanced inventive technologies and the approach based on the ITS paradigm significantly improve the system performances.

14.4 Public Transport Cooperative Priorities

Increased usage of traffic infrastructure leads to traffic flow congestions and increment of travel times for road vehicles. This problem of capacity congestion is particularly expressed in urban areas where possibilities of physical enlargement of urban roads capacity are limited or impossible. One approach to resolving the problem is significant usage of public transport (PT), which is achieved by applying certain measures that can improve the quality of public transport. Measures for public transport quality improvement can be divided into four basic categories [19–21]:

1. *Roadway (infrastructure) improvements* – the simplest type of measures that include minor changes to roadways, relocation of PT stops, improved traffic regulations, etc.).
2. *Improvement of PT system operations* – improved PT management centre, design of PT vehicles, modifications in payment methods.
3. *Administrative measures* – congestion charging, limitations for delivery vehicles in urban areas.
4. *Adaptive traffic control* – usage of control strategies on signalized intersections with PT priority techniques.

Adaptive traffic control is the most efficient measure for PT quality improvement. The basic principle of existing adaptive traffic control with PT priorities (without cooperation between vehicle, infrastructure and driver) is shown in Figure 14.6.

If the PT vehicle is behind a predefined timetable, it sends a PT priority request for upcoming signalized intersection (via detection sensors). In accordance with the active signal phase, a certain priority algorithm is activated. If PT vehicle cannot pass on its green phase, then the time for phase prolongation is selected based on the position of detection of the PT vehicle (in the case presented in Figure 14.6, the detector is located 50 m ahead of the intersection, driving speed is 10 m/s, so the PT vehicle needs a minimum of 5 seconds to arrive at the intersection). After the PT vehicle successfully passes through the intersection, a fixed signal timing plan is reactivated. In this case, the signal controller does not know if multiple

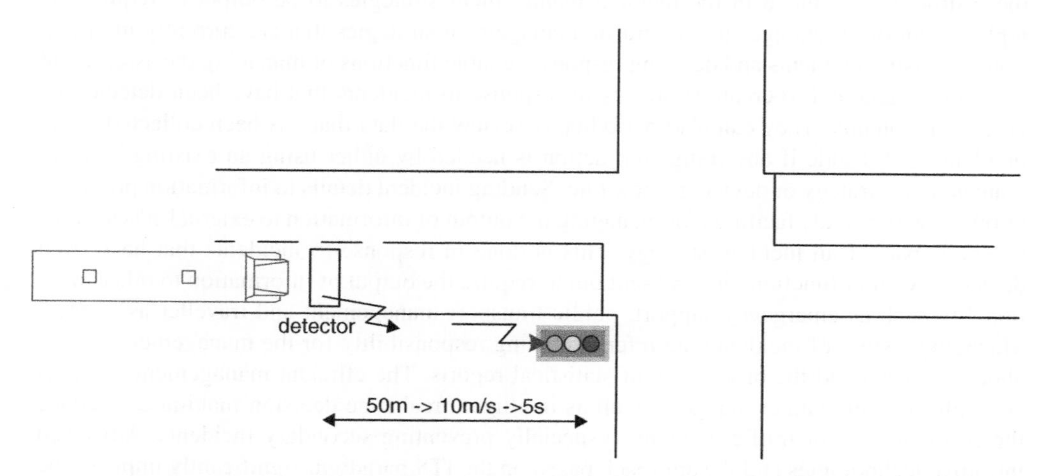

Figure 14.6 Basic principle of existing adaptive traffic control with PT priorities.

PT priorities were requested, or what to do if two PT vehicles arrive at the same intersection at the same time, etc.

Today's adaptive traffic control strategies rely on implementation of fundamental priority techniques based on a small amount of traffic data (PT vehicle's number, delay according to predefined timetables, etc.). New cooperative concept enables the exchange of extended traffic data, based on communication between three main subsystems of traffic system: driver, infrastructure and vehicle.

The effectiveness of proposed measures depends on the level of implementation of measures and their mutual combination. Four levels of measure implementation are defined [21]:

1. *Limited implementation* – measures are implemented individually on different locations of traffic network, with priority given to critical intersections.
2. *Route level implementation* – measures are implemented on entire route of PT network (all signalized intersections along the route, all PT stops, etc.).
3. *Area wide implementation* – measures are implemented on specific parts of traffic network (urban zones with greater number of PT lines with limited access to private vehicles).
4. *General implementation* – measures are implemented to all PT lines and routes and operative concept of traffic management is changed.

At present, active priority strategies are the most used strategies of adaptive traffic control, but there is a need for detection and identification of PT vehicles which are approaching signalized intersection. The main purpose of giving PT priority on signalized intersection is reduction of operational travel times of PT vehicles. As previously mentioned, present solutions are limited regarding available traffic data and with the implementation of cooperative concept traffic data can be expanded. Cooperation and data exchange between main subsystems of the traffic system is crucial because drivers, vehicles and infrastructure are no longer independent – they cooperate, in order to improve the quality of transport system in general [22,23].

Using the cooperative concept in adaptive traffic control with PT priority, it is possible to upgrade it with an expanded set of traffic information, which is not supported by existing control systems. Besides basic traffic information (PT lane number, delay regarding predefined timetable, present location of PT vehicle, etc.) it is possible to provide additional traffic data [24]:

- frequency of PT vehicles on specific route;
- arrival time of the following vehicle;
- current number of passengers in specific PT vehicle;
- queue lengths on every intersection approach;
- frequency of PT priority activation status, etc.

Figure 14.7 depicts the concept of cooperative PT priority. First, the PT vehicle enters the zone of cooperative traffic control (the traffic road segment with more than two signalized intersections), the traffic management centre (TMC) recognizes and marks 'improved' PT vehicle which sends an advanced set of traffic data to TMC. V2I communication is achieved, and the needed traffic data sets are transferred. Based on received information, TMC calculates if priority is necessary, adaptive PT priority strategy is selected and algorithms for signal plan changes (green extension, red truncation, or other technique) are initiated.

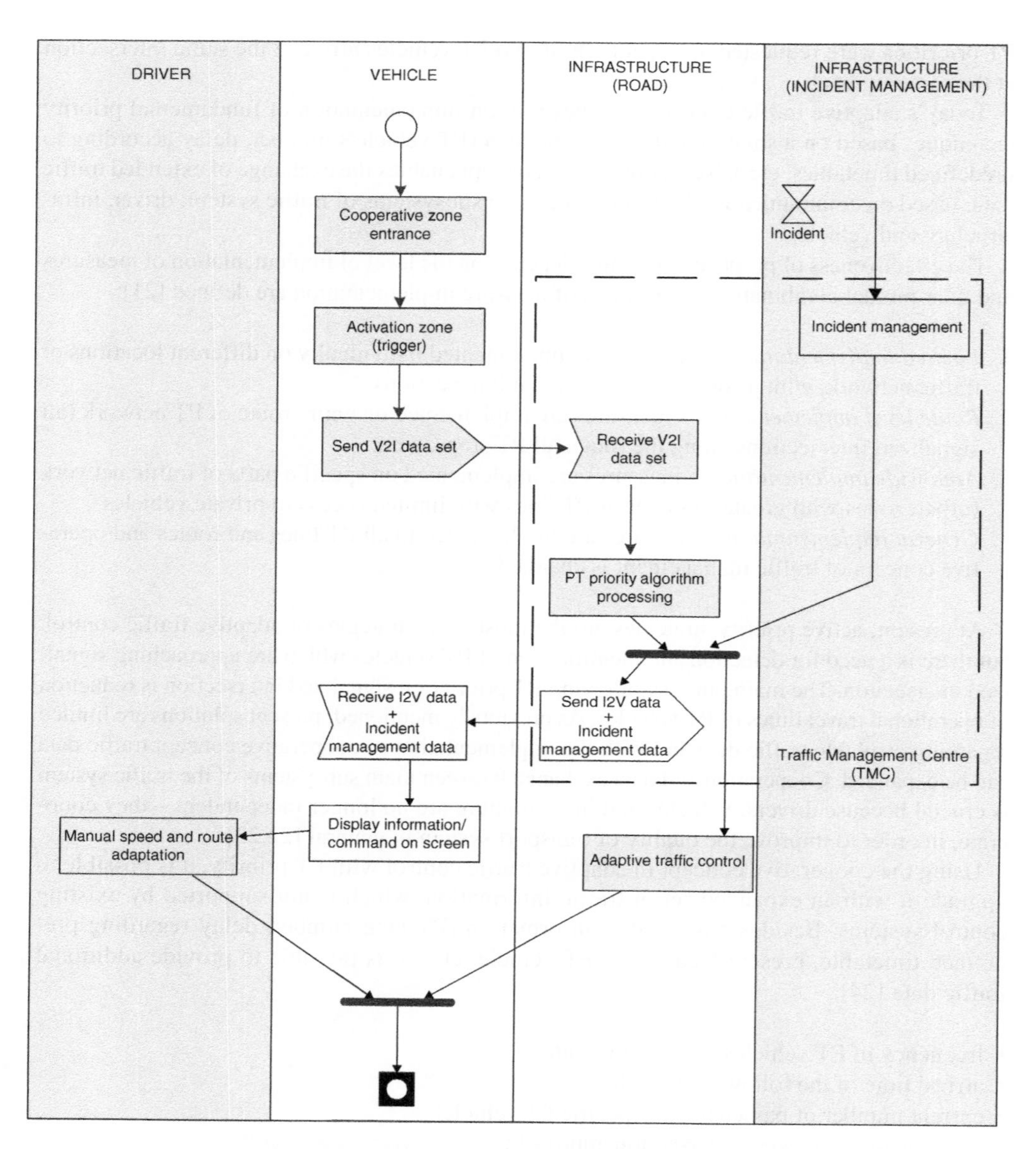

Figure 14.7 Activity diagram of cooperative PT priorities.

There are some specific situations that the present control systems cannot predict and solve such as incidents on PT routes or other unexpected congestion which immediately increases PT operational travel time. In that case, with cooperation between vehicle, driver and infrastructure, alternative routes can be selected and proposed to drivers, together with information about predicted delays to the next PT stop. Drivers who make a decision about an alternative route adjust the PT vehicle's speed (according to the information on screen gathered from TMC) so that the activated PT priority algorithm can be utilized. From this example it is evident that communication and cooperation between all three traffic subsystems may be accomplished, and that exchange of expanded traffic data set is enabled. Cooperation between main traffic subsystems can improve the quality of PT systems and the quality of urban traffic network overall.

A simulation model with the cooperative concept was made on a real demonstration corridor in the city of Zagreb. Zvonimirova Street was selected (2500 m long), with two-way street traffic and separated PT routes. Along the demonstration corridor there are eight signalized intersections with fixed signal control at present. After evaluation of the simulation model, operational PT travel times on corridor were reduced from 696.8 seconds to 650.5 seconds (which represents 31.3 seconds of PT travel time reduction). The depicted travel times were measured for only one simulation hour, and one PT line.

With development and deployment of improved cooperative PT priority strategies (after previously defined levels of measure implementation), direct and indirect benefits can be defined. Direct benefits are reflected mainly through the possibility of using PT priority without ad hoc investments for infrastructure and on-board devices. Also, development of bus lanes application will favour the realization of new bus lanes (with the possibility of giving access to bus lanes to restricted private vehicle categories). The cooperative concept of adaptive traffic control has some indirect advantages. Cooperative systems application will contribute to mobility rationalization and will give benefits in terms of reduction in traffic and congestion [24]. With improved cooperative PT priorities the number of private cars in urban areas would be reduced, and PT system would have an increment in average speed, reduction in operational travel times, delays on signalized intersections, etc.

14.5 Conclusions

Cooperative systems in transport provide great possibilities. Many analyses show that classic methods cannot overcome the challenges in the traffic and transport system set by the European Union. Regarding this subject, during the EU Seventh Framework Programme, investments in cooperative systems were: Call 2 (2008) 57 million euros; Call 8 (2012) 40 million euros. In addition, most standards are currently ready for implementation. One of the basic requirements for these systems is the enabling of the system and services interoperability. The European Union has recently invested significant effort to prevent duplication of activities, not only within the EU, but worldwide (USA, Japan, China). In terms of future implementation, it is important to further encourage innovation and entrepreneurship in the telematics industry sector. Market expansion is particularly important. It will contribute to cost reduction and increase of competition. From the perspective of future users, it is important to create confidence in the products and services in the field of cooperative systems. In any case, the initial results are promising.

Acknowledgment

The research reported in this chapter is supported by the FP7 – Collaborative Project: Intelligent Cooperative Sensing for Improved traffic efficiency – ICSI (FP7-317671) and by the EU COST action TU1102. The authors wish to thank Asier Perallos and Nikola Bakarić for their valuable comments.

References

1. EU Energy, Transport and GHG Emission trends to 2050 – Reference Scenario 2013, European Commission, Directorate-General for Energy, Directorate-General for Climate Action and Directorate-General for Mobility and Transport, 2014.
2. S. Mandžuka, M. Žura, B. Horvat, *et al.* (2013) Directives of the European Union on intelligent transport system and their impact on the republic of Croatia, *Promet – Traffic and Transportation* **25**(3): 273–83.
3. M. Ramage (1998) Evaluation of learning, evaluation as learning, *SIGOIS Bulletin* **17**(3): 77–8.
4. R. Murphey and P. Pardalos (2002) *Cooperative Control and Optimization*, Applied Optimization Series. New York: Springer.
5. M. Hasan, M. Jha and M. Ben-Akiva (2002) Evaluation of ramp control algorithms using microscopic traffic simulation, *Transportation Research Part C 10*, MIT, Cambridge, MA 02142, USA, pp. 229–56.
6. A. Kotsialos and M. Papageorgiou (2010) Coordinated ramp metering for freeway networks – a model-predictive hierarchical control approach, *Transportation Research Part C: Emerging Technologies* **18**(3): 311–31.
7. M. Treiber and A. Kesting (2013) *Traffic Flow Dynamics – Data, Models and Simulation*. New York: Springer.
8. A. Hegyi, B. De Schutter and H. Hellendoorn (2005) Model predictive control for optimal coordination of ramp metering and variable speed limits, *Transportation Research Part C* **13**(3): 185–209.
9. M. Gregurić, E. Ivanjko and S. Mandžuka (2014) Cooperative ramp metering simulation, *International Convention on Information and Communication Technology, Electronics and Microelectronics – MIPRO 2014*, pp. 970–5.
10. M. Papageorgiou, C. Diakaki, V. Dinopoulou, *et al.* (2003) Review of road traffic control strategies, *Proceedings of the IEEE* **91**(12): 2043–67.
11. M. Gregurić, M. Buntić, E. Ivanjko and S. Mandžuka (2013) Improvement of highway level of service using ramp metering, *Proceedings of the 21st International Symposium on Electronics in Transport – ISEP 2013*, Ljubljana, Slovenia, *2013*, pp. 1–7.
12. K. Bogenberger and A.D. May (1999) *Advanced Coordinated Traffic Responsive Ramp Metering Strategies*, Working Papers, California Partners for Advanced Transit and Highways (PATH), Institute of Transportation Studies (UCB), UC Berkeley.
13. A. Kurzhanskiy and P. Varaiya (2008) *CTMSIM Traffic Macro-Simulator for MATLAB, User Guide*, TOPL Group, UC Berkeley.
14. G. Štefančić, D. Marijan and S. Kljajić (2012) Capacity and level of service on the Zagreb Bypass, *Promet – Traffic & Transportation* **24**(3): 261–7.
15. M. Abdel-Aty, K.M. Haleem, R. Cunningham and V. Gayah (2009) Application of variable speed limits and ramp metering to improve safety and efficiency of freeways, *2nd International Symposium on Freeway and Tollway Operations*, Honolulu, Hawaii, *2009*, pp. 1–13.
16. O. Kaan and K. Puskin (2010) *Incident Management in Intelligent Transportation Systems*, Boston: Artec House Inc.
17. P. Škorput, S. Mandžuka and N. Jelušić (2010) Real-time detection of road traffic incidents. *Promet – Traffic & Transportation* **22**(4): 273–83.
18. P. Škorput (2009) Realtime incident management in transport (in Croatian), MSc Thesis, University of Zagreb, 2009.
19. W. Ekeila (2002) Dynamic Transit Signal Priority, Master of Applied Science, University of Sharjah.
20. W. Ekeila, T. Sayed and M. El Esawey (2009) Development and comparison of dynamic transit signal priority strategies. In E. Masad, N.A. Alnuaimi, T. Sayed and I.L. Al-Qadi (eds), *Efficient Transportation and Pavement Systems*. Boca Raton, FL: CRC Press, pp. 1–9.
21. B.A. Nash and R. Sylvia (2001) *Implementation of Zurich's Transit Priority Program*, Minnesota Transportation Institute, San Jose State University.
22. ERTCO (2010) *Cooperative Urban Mobility Handbook*, Project CVIS, ERTICO.
23. M. Vujic (2013) Dynamic Priority Systems for Public Transport in Urban Automatic Traffic Control, PhD Thesis, University of Zagreb, Faculty of Transport and Traffic Sciences, Zagreb.
24. D. Cocchi (2010) How public transport can benefit from cooperative systems – ideas from the Bologna test site (presentation), *CVIS Workshop, Prague 2010*, pp. 1–8.

15

Methodology for an Intelligent in-Car Traffic Information Management System

Nerea Aguiriano, Alfonso Brazalez and Luis Matey
University of Navarra, San Sebastián, Spain

15.1 Introduction

Along the road, drivers come up against unexpected situations; for that reason static and dynamic road traffic signalization, besides RTTI (Real-time Traffic and Travel Information) systems, attempt to warn drivers about all the events happening on the road. However, drivers must consciously take notice about the state of the road, traffic intensity and many other variables that help drivers obtain the most complete *Situation Awareness* (SA); in other words, drivers must achieve the most complete SA of the road environment to perform – efficiently and safely – the driving task. That knowledge of the surrounding situation is necessary for creating the basis of decision-making in the driving task [1]. Therefore, drivers should be able to take the right decisions in such complex and dynamic environments as a road. While driving, drivers perform an automatic cognitive processing that increases the risk of being less responsive to new stimuli. For example, when a new road sign is set on a familiar route, many drivers would proceed without taken notice of it. On the other hand, when the driver pays more attention to other elements such as other vehicles, it could end in a loss of notifications given by road signalization or radio messages. The degraded evaluation of the immediate environment and, consequently, incorrect driving control behaviour, can result in a distraction or even in an accident. There are research studies which have correlated the perception of road traffic signalization with factors such as: vehicle speed [2], and driver's fatigue [3,4], age [5,6] or expectations [7]. In general, it has been noticed that those factors are somehow detrimental variables in the perception of road information.

Intelligent Transport Systems: Technologies and Applications, First Edition. Asier Perallos, Unai Hernandez-Jayo, Enrique Onieva and Ignacio Julio García-Zuazola.

Another drawback in the interaction between drivers and road signalization is related to the memorization of all the information transmitted through signs. Unfortunately, the memorization of road signs is quite poor among drivers [8]. Apart from detecting traffic signs, drivers must remember the road signalization they have just passed and which could refer to a non-immediate event. It is known that drivers' fatigue and the amount of hours they have been awake have the greatest effect on that sign registration [9].

On the other hand, the understanding of posted signs does not increase with experience, nor with the age of drivers [10]. Sometimes drivers find that they do not know if a sign is referring to them or not. For example, 21% of drivers of long trucks do not understand the meaning of 'headroom hazard' [11]. So then, it would be really interesting to inform each driver just with the information concerning the type of vehicle they are driving.

Pingatoro [12] suggested that drivers should be warned with as many traffic signs as necessary. So then, it would be a step forward if traffic information and road signalization are delivered to the driver onboard the vehicle. This way, the driver would have on the display of the instrument panel of his vehicle all the current information, and he would be able to achieve the complete SA, even if he has missed them from the roadside. Motivated by this, nowadays, there is an important research line in traffic sign detection and recognition [13]. However, if all detected information is displayed to the driver on the instrument panel without any filter or prioritization process, the workload of the driver could be increased considerably. It is known that a person's SA is restricted by his limited attention and working memory capacity. Therefore it is necessary to prioritize all the information given to the driver. Moreover, in a given context not all messages are always relevant for the driver or at least may not in the same level. Somehow, the risk level of the information provided by messages depends on the current driving context. Besides, due to the limited cognitive processing of the driver, which is also affected by his fatigue state, there is a limitation in the amount of messages that could be presented at the same time to the driver. Therefore, the prioritization task should include the state of the vehicle, the environment and the driver. Somehow, it could be said that the prioritization should be done in an intelligent way.

In relation to messages' prioritization, the International Organization for Standardization (ISO) and the Society of Automotive Engineers (SAE) have already established some standards. For the prioritization task, SAE J2395 [14] employs a method based on three criterions: safety relevance, operational relevance and time frame. Depending on the combination of those three parameters attached to the information element, the SAE J2395 establishes in a predefined table a Priority Order Index (POI). However, application of the standard SAE J2395 to road signalization and traffic information is not effective enough since it does not take into account the environment conditions and how message relevance is changing according to them; moreover, almost all information is safety relevant and operational relevant in some way. On the other hand, in ISO/TS 16951:2004 [15] it decides the priority index of each message on the basis of values from Criticality and Urgency ratings. The term Criticality is associated with the severity of the impact of the most likely accident or malfunction that can occur when the message is not received or is ignored by the driver; while Urgency is related to the time within driver action has to be taken if the benefit intended by the system is to be derived from the message. However, for the type of messages which encode road signalization and traffic information, it would be difficult to establish their Criticality value.

On the other hand, various scheduling algorithms [16,17] have been developed to prioritize Intelligent Transport System (ITS) messages on board the vehicle. Sohn *et al.* [16] manage the display of an In Vehicle Information System messages (that range from collision warnings and navigation instructions to tire pressure and email alerts) on the basis of their importance, duration and preferred display time. However, even if they consider the dynamic value of the message on the basis of the display time as a variable that changes dynamically, they still consider the relative priority of the messages as a static value; furthermore, their algorithm does not consider environment and vehicle conditions. Zhang and Nwagboso in [17] focus on ITS messages prioritization over CAN bus using a fuzzy neural network. They establish that the priority of the message is related to the incident that occurs on-board or off-board the vehicle, so they include information about vehicle state or conditions in the prioritization process. Unlike in [17], in this methodology the display of incidents are also prioritized on the basis of environment conditions, apart from their relation with vehicle speed.

Being aware of the importance of a smart prioritization of the information shown to the driver, a new methodology is shown for the message prioritization task. From now on, it will be named as the *Intelligent Information Management System* (IIMS). The methodology takes into account the unsettled road environment and vehicle features' influence by applying fuzzy logic. Furthermore, for a given message, this methodology adjusts the type of alert, depending on its own characteristics and driver fatigue. It has to be pointed out that this information scheduling goes beyond the limitations imposed by previously mentioned standards and algorithms. The proposed methodology has been implemented in a traffic information system. Next, the main features of this system are described in order to have a better understanding of the framework on which the proposed methodology has been developed.

15.2 Validation Framework

The proposed methodology has been tested in the framework of a national project CABINTEC, for the development of an intelligent cabin. The project was initialized in 2007 and was funded by the ERDF (European Regional Development Fund) and with assistance of the Ministry of Science and Innovation of the Spanish Government.

The signalling system has had two clearly distinguishable research lines: one line was focused on the HW design for the transmission and reception of roadside beacons, and the other line was based on the information management of the onboard module. Specifically, the onboard module was designed for collecting the information given by the beacons. The beacons transmit information corresponding to static and dynamic messages. Static messages would correspond to static road signs, while dynamic messages are related to Variable Message Signs (VMS) or data related with the changing traffic and road conditions. Static road traffic sings could be classified into two groups; on one hand there are all maximum speed limit traffic signs, and on the other hand there are the rest of signs – prohibition, guidance, regulatory, etc. Furthermore, dynamic messages encode other traffic information such as incidents ('Accident ahead', 'Congestion ahead', 'Roadworks', 'Car ahead driving very slowly', 'Kamikaze car coming') and reminders, apart from traffic density and meteorological condition data (for instance, 'Snowing heavily'), all scheduled by the Transportation Management Centre (TMC).

The onboard module consists on RFID and ZigBee integrated receivers that collect the static and dynamic messages from the environment. So as to have the most complete information of the road onboard the vehicle, there should be a big deployment of RFID and ZigBee beacons along the road, which would all be controlled by the TMC. The vehicle receives the messages when it passes through the action range of the beacons. The main feature of the onboard module, apart from receiving all static and dynamic messages, is to process all the information and decide in an intelligent way which messages are relevant for the driver in each driving situation. And moreover, this module provides the best way to inform the driver. It can be assured that somehow, the onboard module is an IIMS.

However, if all messages received from beacons are presented to the driver, without any filtering or priority assessment, his workload could be overloaded, and consequently it could increase the probability of having an accident. Not to undermine safety, it was considered necessary to add an intelligent information management procedure to the onboard module. The methodology will be explained in detail in the next section.

As regards the information management task, not all messages received on module are supposed to identify the driving situation with the same severity level, and not all of them have information with the same predefined risk level; indeed, not all messages are affected in the same way by the current driving context, which is defined by the state of the vehicle and the environment. Then, the information collected by the onboard module has a specific dependence with the instantaneous state of the *vehicle* and the *environment* entities (V and E, respectively).

On the other hand, the way in which information is presented to the *driver* (from now on referred as D entity) is a key point for the efficient delivery of infrastructure information. For example, driver fatigue should be taken into account. That fatigue data is assessed by an onboard module, which supervises the driver in real time.

The validation of the methodology proposed for the intelligent information management has been carried out in a high performance and full immersive truck driving simulator at CEIT [18] with professional drivers. The main purpose of the simulator is to carry out new designs, and validate new developments in the automotive area. The simulator gives the opportunity to choose the driving environment and vehicle conditions: driving scenario, road conditions, weather conditions, traffic conditions, vehicle type, vehicle load, vehicle faults and errors, etc. Three types of scenarios are available for driving simulation: urban, intercity and mountain roads. The simulator cabin is a genuine Iveco Stralis cabin mounted on top of a six DOF movement platform. Three large screens help the driver to fully immerse in the driving environment with two TFT screens for side mirror replacement to provide rear view; the result is practically 360 degrees of field of view.

15.3 HMI Design Methodology

In this section, the generic methodology of the IIMS is presented. The first step for obtaining a generic solution was to identify IIMS's main tasks. Regarding the functionality of IIMS, it can be said that the principal tasks carried out by the system are the reception of road messages, its information processing, and the later representation of relevant messages

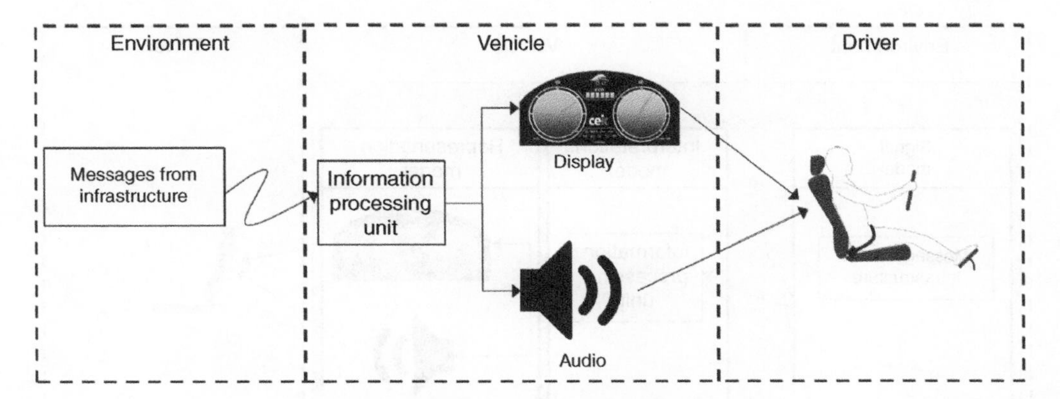

Figure 15.1 Context of the IIMS.

to be delivered to the driver (see Figure 15.1). In that context, road messages are receive from the environment; then all this information processing is carried out onboard the vehicle; and finally, relevant messages are represented to the driver through the communication channels of vehicle's HMI (Human Machine Interface).

The generic solution of the IIMS should not be limited to the predetermined fixed group of input messages, nor to the specific implementation of information representation. For that reason, it was considered essential to give a structure to the format of input messages, and moreover, decide about the representation of messages' information independently from the communication channels available at the vehicle (audio, visual and/or haptic channels). Between the inputs and outputs of the IIMS, the main information processing task of the IIMS is carried out where input messages' classification, relevance assessment and prioritization tasks are executed; all those management tasks could be embraced into a generic interpretation task. That interpretation task should work with a generic input format, and it gives a generic output for being used by any implementation of information presentation system in the vehicle. Then, for its assignments, the IIMS could be considered as a module with three functional parts: input messages' reception, an interpretation task, and a representation of the outputs. Then, having to introduce these functional parts of the IIMS in the context defined by Figure 15.1, it was a requisite to create generic models for each of its functional parts. Specifically, the three functional models of the IIMS were named as follows (see Figure 15.2): the Signal Model, the Interpretation Model, and the Representation Model.

The Signal Model establishes the generic data fields of the input messages required by Interpretation Module to manage the information and moreover contains other useful data about how messages must be given to the driver. In the corresponding tasks, the information of input messages is managed in a smart way since tasks are executed taking into account the driving context – given by DVE context – as can be seen in Figure 15.3.

The main model of the IIMS is the Interpretation Model; it is responsible for classifying input messages, deciding whether input message information is relevant to the driver, and furthermore, assesses the risk value of relevant message information depending on the driving context. It has to be pointed out that each input message (m_i) is associated with a predefined

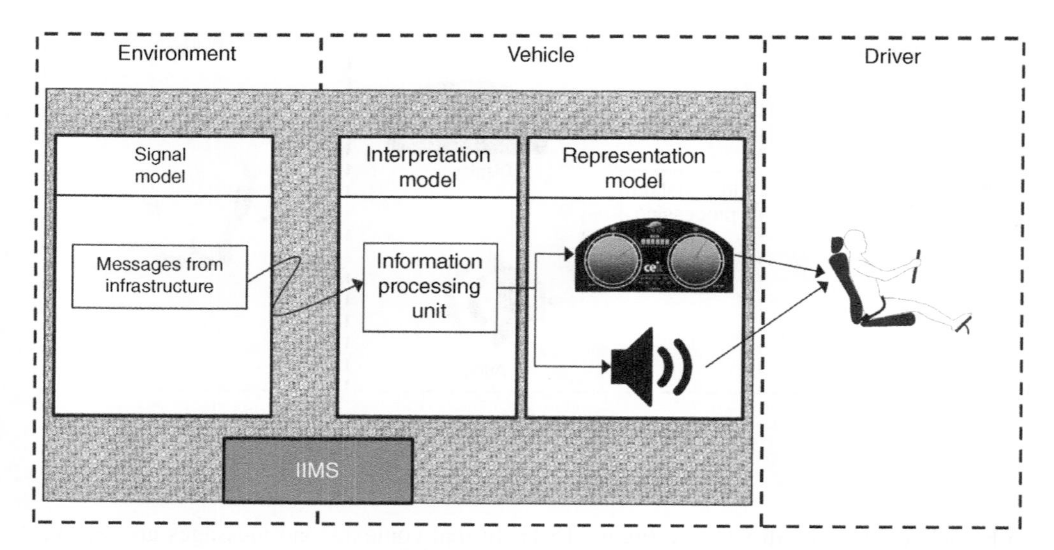

Figure 15.2 Models of the IIMS.

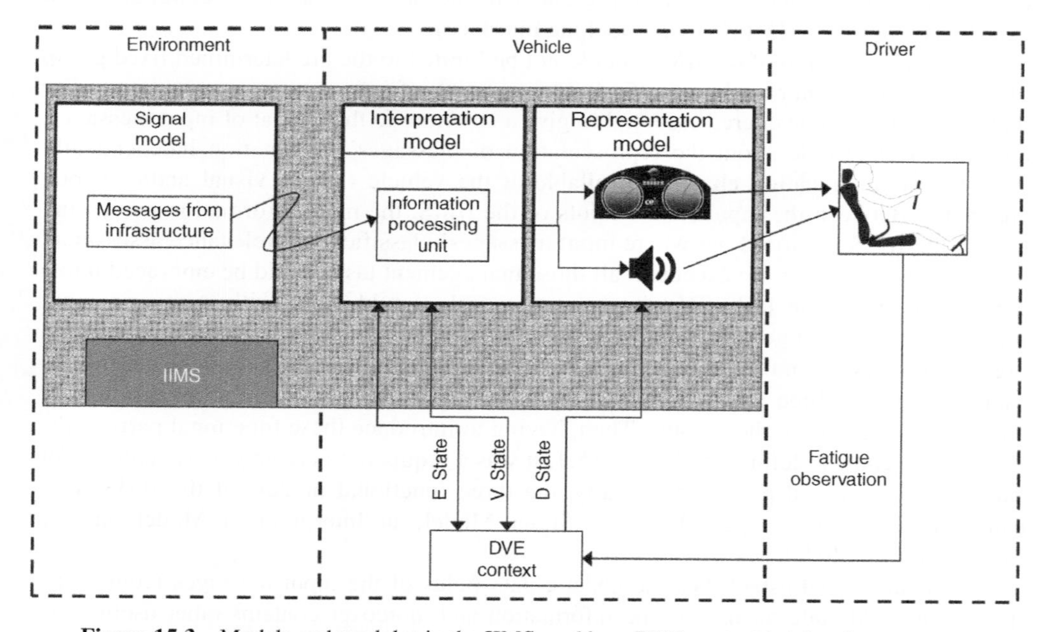

Figure 15.3 Models and modules in the IIMS, and how DVE context is taken into account.

Initial Risk value (R_i), obtained from the Signal Model. However, the IIMS, being a dynamic system which considers the driving context, updates each R_i value of relevant messages into the corresponding Final Risk value (R_i'), on the basis of VE state (see Figure 15.3). Specifically, E state is characterized by *Traffic Density* (TD) and *Meteorological Conditions* (MC) values; while, V state is characterized by vehicle *Speed* (SP) value. That assessment of the current risk level of each input message is done by modifying their predefined R_i value into R_i', by adding the influence of the current TD, SP and MC values. In order to apply the influence of those VE state defining variables over R_i, fuzzy logic has been used.

The output of the Interpretation module will be a ranking of relevant messages on the basis of their R_i' values. The Representation Model then receives that ranking of relevant messages. Sometimes, however, the queue of relevant messages to be displayed on the HMI is too long. If all of them were to be displayed at the same time, driver's mental workload could be incremented drastically, and that could lead to driver distractions. However, the maximum amount of information to be given to the driver is beyond the scope of the present publication and should be defined by human factors analysts. On the other hand, the Representation Model defines the warning level which best fits for the presentation of message information to the driver. When deciding that warning level, the corresponding icon and/or text of the message are required, plus its R_i' and K_c, besides the current driver's fatigue data stored in the DVE module. The fatigue of the driver is an instantaneous characteristic of D entity which defines its state and is periodically assessed by the Driver's Supervision Module.

The proposed methodology for the IIMS will be validated with the onboard module's prototype which has been described previously. In this case, the inputs are the static and dynamic messages sent by beacons located along the infrastructure of the road, and then, received on the vehicle by the onboard module.

In the following subsections, the three models employed in the IIMS are described in detail.

15.3.1 Signal Model

The Signal Model is a description of generic fields that facilitate the classification and integration of any input message. Those fields contain information for the intelligent module and are listed below:

- *Identifier of the message (or just ID):* identifies the message sent.
- *Type of message:* indicates if the message is a maximum speed limit message, one related to environment data (such as fog, rain, snow, ice, wind or traffic density), or other incident and road traffic signalization.
- *Type of road:* the type of road the beacon is at. This data is directly sent to the DVE module.
- *Driving lanes affected:* the lanes of the roads to which message information is addressed.
- *Distance:* distance at which the vehicle will come up with the hazard or message information becomes valid.
- *Duration:* duration of the validity of message information.
- *Description:* description of the situation. The description of the situation is extracted to categorize to which situation the message corresponds to. These predetermined situations have been considered: ('Accident ahead', 'Congestion ahead', 'Roadworks', 'Car ahead driving very slowly', 'Kamikaze car coming' or 'Normal situation').
- *Severity:* severity level of the situation that the message corresponds to. The predetermined field of severity level of the situation is part of the Signal Model which is necessary in the Representation Model.
- *Activation:* array of activation factors the message information depends on (such as wind, fog, ice, snow, rain and traffic). The Activation field has the information about whether a

message is active or not according to some factor such as rain, snow, fog, ice, wind or traffic; that data field is necessary because some message information should not be taken into account if the value of a factor is equal to zero. For example, that is exactly what happens with the road sign of 'Slippery road due to snow or ice'; if the factor of ice and snow is zero, there is no point showing the corresponding icon to the driver regardless other factors.

- *Relation with MC, TD and SP:* array of values which define the membership function of the relation between the message and Meteorological Conditions (MC), Traffic Density (TD). When saying the relation, the term is referring to how it is aggravated the predefined R_i associated with the message on the basis of those values. The aggravation is enough to be quantified at three levels: 'few', 'medium' or 'high'.
- R_i: initial risk value of the information.
- K_c: knowledge rate of message information. This field about knowledge of the signal (K_c) shows the knowledge rate that people have about the message. A study carried out by FESVIAL (The Spanish Foundation for Road Safety) [19] interviewed 1.723 drivers in Spain, and concluded the knowledge rate of the meaning of some road signs included in driving manuals.
- *Icon:* picture embedded in the message.
- *Value:* numerical value embedded in the message.
- *Text:* text embedded in the message.

It has to be pointed out that in the membership functions of the relation with MC, TD and SP, the transition between the three different levels of risk alteration due to the value of the corresponding variable is soft. For that reason, each variable – MC, TD and SP – has been decided to be considered as a linguistic variable with its corresponding universe of discourse. Furthermore, the way in which the combination of factors affect R_i will be carried out in the Interpretation Model on the basis of some base rules.

Next, Table 15.1 depicts an example of all the data extracted in the Signal Model –just listed above – from a road message which is received on the onboard module; specifically, the message corresponds to a static road traffic signalization about 'Roadworks' zone.

15.3.2 Interpretation Model

The Interpretation Model is the second model of the proposed methodology for the IIMS. The model consists of three functional blocks (see Figure 15.4): Filter module, Message Relevance assessment module and Final Relevance calculation module based on Fuzzy Logic.

The Filter module assesses the type of message received from the infrastructure. It decides whether the received message is a road traffic signalization (given by static or dynamic messages), or other traffic information such as reminders, occasional incidents or hazards (encoded in dynamic messages). Furthermore, the Interpretation Model can receive dynamic messages which encode information about meteorological conditions (such as rain, fog, ice, snow or wind levels) or the traffic density level at that moment; the information of these type of messages are sent directly to the DVE module which is the module in charge of storing E state. Moreover, it has to be pointed out that among road traffic signs, the Interpretation Model differentiates if messages concern a maximum speed limit sign or the rest of road traffic signs. It is assumed that the maximum speed limit information must

Table 15.1 Example of the Signal Model for a road message.

ID	Type of message	Type of road	Driving lanes affected	Dist. (m)	Dur. (m)	Description	Severity	Act.
P_18	Road traffic sign	Conv. 2 lanes/ direc.	All	0	2000	Road work	6	[0,0,0,0,0,0]
Relation MC	**Relation SP**	**Relation TD**	**R_i**	**K_c**	**Value**	**Text**		
[0,30,50,75,85,100]	[0,40,45,70,75,120]	[0,30,40,75,80,100]	7.2	98	null	Road work zone		

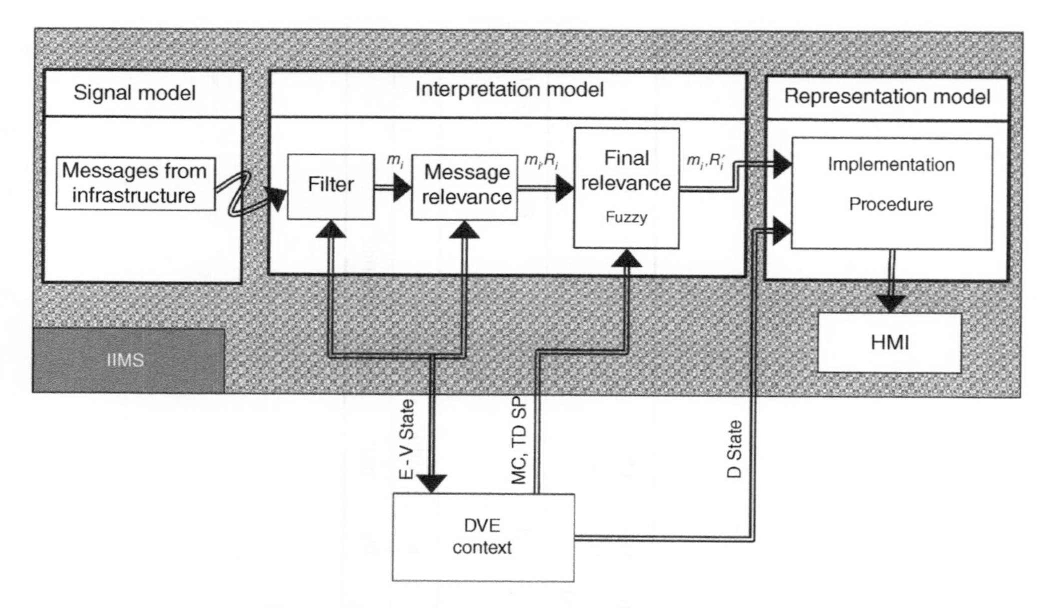

Figure 15.4 IIMS methodology for onboard module.

always be shown to the driver whereas not the rest of traffic signalization. For that reason, all messages with maximum speed limit information should be directly represented on the speedometer display.

However, before displaying the maximum speed limit data, the Interpretation Model maps the received speed limit value into the corresponding specific speed limit for the vehicle in question (task carried out by Filter module). In other words, the maximum speed limit value needs to be mapped on the basis of the type of vehicle is being driven and the type of road it is driving through. For instance, when driving through a Spanish highway with a truck which transports dangerous goods, a maximum speed limit of 120 kph is adapted into a value of 80 kph.

Other dynamic and static messages continue to the next step of the Interpretation Model, the Message Relevance assessment block, where messages' relevance is assessed depending on some environmental characteristics (such as fog, rain, wind, etc.), the driving lane, and vehicle's static characteristics (such as height, weight, weight per axis, etc.).

The next task of the Interpretation Model is to make a prioritization of the relevant input messages (m_t). That scheduling is done depending on the risk associated with the information of those messages, given by R_i. However, it can be assumed that somehow, the status of V and E entities aggravate the initial relevance of the messages, so then it could not be considered R_i as the actual risk value of the input message. The real risk value would be defined as R_i', which includes VE state influence. For that reason, it has been considered necessary to define a factor which would integrate that VE state. From now, that factor will be called W_{VE}.

When trying to determine the variables that would be needed to define W_{VE}, each of the entities were considered separately. On one hand, MC and TD could be considered as the variables which define in detail the current context of E entity; specifically, on the basis of rain, snow and fog conditions, a predefined value of MC is determined. On the other hand, since vehicle speed increases drivers' risk feeling about the driving task [20], SP value

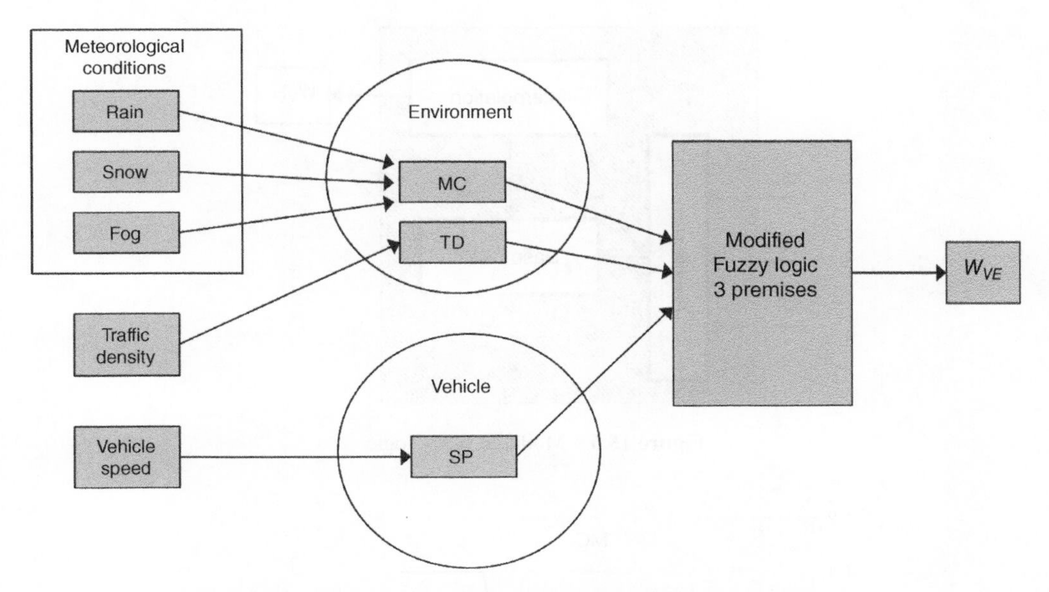

Figure 15.5 Integration of V and E entity information into a W_{EV} factor.

would be included to characterize the state of V entity. The proposed Interpretation Model defines the final relevance as the result of multiplying the initial relevance with its W_{VE} factor (see Equation (15.1)).

$$R_i' = R_i * W_{VE} \tag{15.1}$$

In order to estimate W_{VE}, modified fuzzy logic proposed by Castro [21] was applied. In his research work, Castro applied the modified fuzzy logic just for two premises. Nevertheless, the proposed methodology for the IIMS requires the use of three premises (see Figure 15.5), each one for the fuzzyfied values of MC, TD and SP. Subsequently, a more detailed description of this modified fuzzy logic for three premises will be given and its implementation.

15.3.2.1 Modified Fuzzy Logic for Three Premises

The modified fuzzy logic explained by Castro employs just three modules (see Figure 15.6) instead of four as the traditional fuzzy logic does. The three modules used in the modified fuzzy logic are: (1) Fuzzification, which modifies MC, TD and SP input variables in order to apply the base rules, (2) Base Rules, which have the knowledge about the aggravating relation between V-E entities and R_i, and (3) Interpolation, which integrates in a unique module the function of inference and defuzzification of the traditional fuzzy logic. The output of the Interpolation phase is the multiplicative W_{VE} factor, necessary to the calculation of R_i'. With this modified fuzzy logic, the processing time is considerably reduced, and facilitates the system setting up.

In the first step, MC, TD and SP input values are quantified in linguistic numerical values of the aggravating relation of R_i of the message with the current context of VE. This fuzzification is done by using trapezoidal member functions of Figure 15.7. Depending on the input value of each

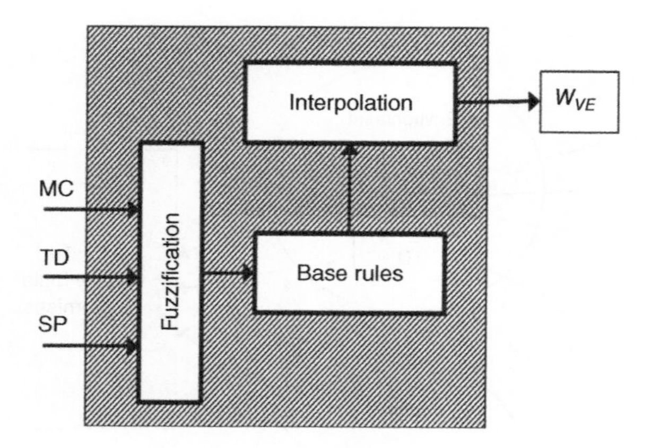

Figure 15.6 Modified fuzzy logic.

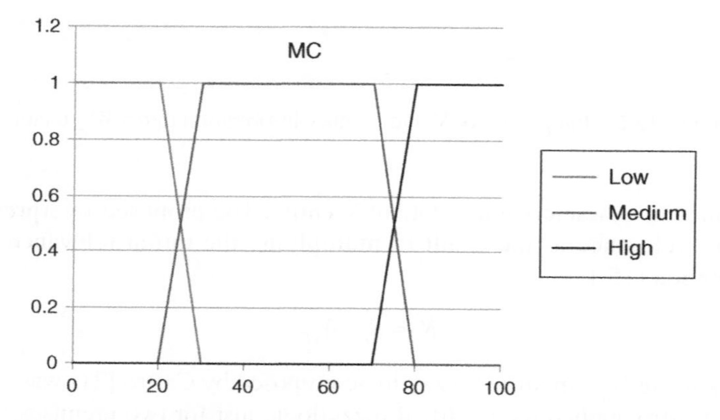

Figure 15.7 Member functions for Meteorological Conditions (MC).

variable, the aggravating strength relation could be determined as 'little', 'medium', or 'high'; each linguistic value is quantified as 1, 2 or 3 linguistic numerical values, respectively. It has to be pointed out that Figure 15.7 also represents how the arrays of values which define the membership functions of the relation between the input message and MC, TD and SP, are translated into their corresponding graphic of membership functions. In this case, Figure 15.7 shows how the specific values of $m_i[MC]=[0,20,30,70,80,100]$, $m_i[TD]=[0,15,25,40,45,100]$ and $m_i[SP]=[0,50,60,80,95,100]$ are translated into the corresponding membership functions.
The next step is carried out when assessing the base rules which express the knowledge of experts. Those rules are numerically tabulated in a 3D hypermatrix, where its values correspond to linguistic numerical output values of W_{VE}.

The interpolation module includes inference mechanisms and defuzzification tasks. Through interpolation surfaces, the module interpolates directly the values obtained from member functions (c_i values) after fuzzifing MC, TD and SP input values. This way, VE context is translated into a single output coefficient (W_{VE}) that multiplies the R_i of the received

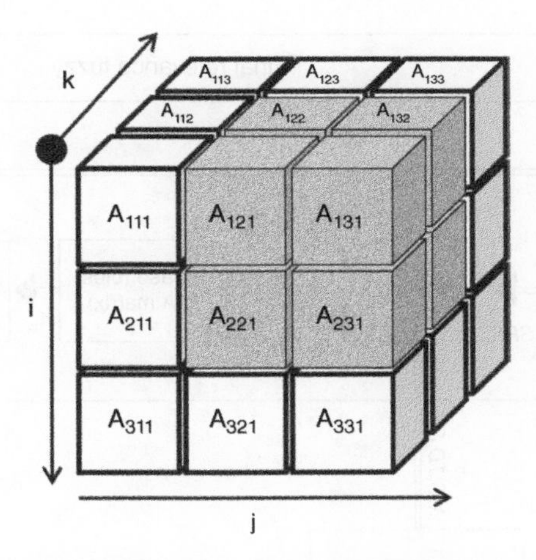

Figure 15.8 Fuzzification for three dimensions and two rules.

message; it is necessary to point out that W_{VE} is calculated for each single message (m_i), since each message is affected independently by MC, TD and SP.

Sometimes, more than one base rule is obtained for each dimension because the input value is associated with more than one membership function value. Four possible cases have been identified for the combination of MC, TD and SP. A general case, where three dimensions are involved, is shown in Figure 15.8. The corresponding W_{VE} value for this case is given by Equation (15.2).

$$\frac{c_1A_{121}+c_2A_{131}+c_3A_{221}+c_4A_{231}+c_5A_{122}+c_6A_{132}+c_7A_{222}+c_8A_{232}}{c_1+c_2+c_3+c_4+c_5+c_6+c_7+c_8} \tag{15.2}$$

15.3.2.2 Implementation of the Modified Fuzzy Logic in the Interpretation Model

To summarize how the Modified Fuzzy Logic is integrated in the Interpretation Model, so as to calculate R_i' of input relevant messages, Figure 15.9 has been included, which shows data required by the main functional parts of the Interpretation Model.

Each message R_i value is obtained from the Signal Model on which basis R_i' value will be assessed; for that purpose, W_{VE} factor is required. So as to assess that W_{VE} factor, MC, TD and SP values, which are VE context defining variables, are fuzzified. The shape of the membership functions with which MC, TD and SP values will be fuzzified, are given by the vectors which define the relation (alteration levels) of the message with each of the variables (m_i[*relation with MC*], m_i[*relation with TD*] and m_i[*relation with SP*]). Those arrays are also included in the Signal Model.

Once having MC, TD and SP values fuzzified on the basis of three alteration levels, which are quantified with the linguistic numerical values, the next step would be to assess which base rules should be considered. The base rules are expressed on matrix A of 3 × 3 × 3 dimensions, where its vertex are named as A_{ijk}; i subscript corresponds to MC dimension, while j subscript is for the dimension of SP, and k subscript corresponds to TD dimension. Each element A_{ijk}

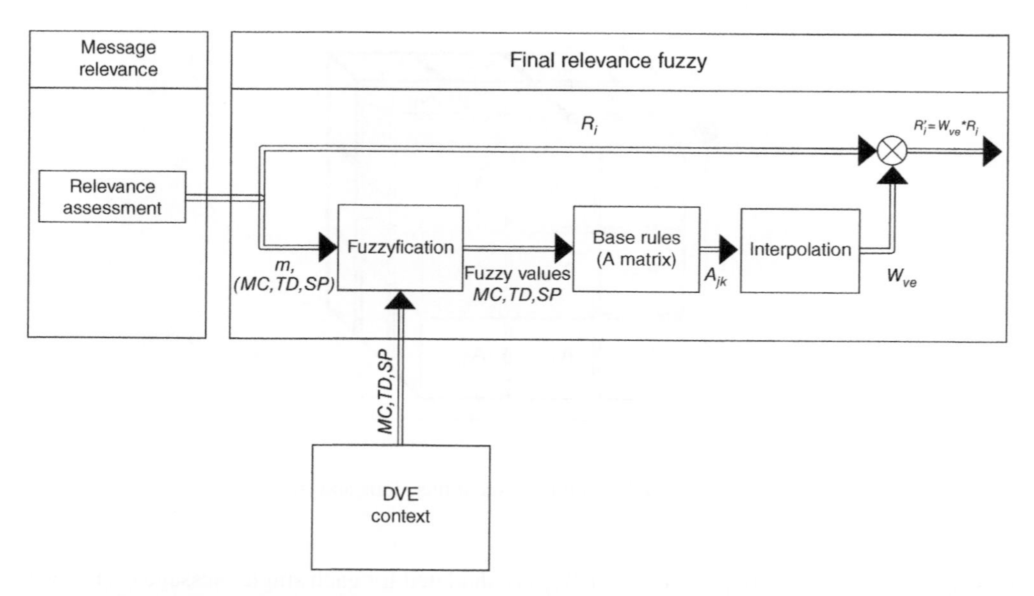

Figure 15.9 Implementation of the modified fuzzy logic in the Interpretation Model.

takes a value ranging between 1 and 4. As more than one base rule could be implied since the input values of MC, TD and SP values can be associated with more than one membership function, the A_{ijk} values involved are interpolated in the Interpolation phase. Once the base rules are interpolated, W_{VE} value would be obtained.

Finally, W_{VE} factor (obtained from the Modified Fuzzy Logic module), and R_i (extracted from the Signal Model) are multiplied so as to have the real risk level of the message that results in R_i'.

15.3.3 Representation Model

From the Interpretation Model, the queue of relevant messages arranged by R_i' is sent to the Representation Model, whose aim is to display those messages. For the corresponding display task, the Representation Model decides the representation mode of relevant messages on the basis of some warning levels.

Each proposed warning level has been defined with regard to the warning intensity required by the risk involved in the message's information and the driver's state to perform the driving task. It has been assumed that for the IIMS system, four warning levels are required. It has to be pointed out that the definition of the warning levels must be independent from the specific implementation of the HMI available on the vehicle. This way, the proposed methodology for the IIMS would be applicable in any type of vehicle or any type of advising system: display, audio device, haptic device, etc.

It is proven that driver's fatigue deteriorates the driving task; therefore, when having a high level of fatigue, a higher warning level should be used to represent relevant messages. On the other hand, the risk involved by relevant messages can be assumed not just to be defined by their R_i' value assessed taking into account VE state. Moreover, the driver's knowledge about the information given by the message should be considered. Since poor knowledge of the message would increment the risk associated with the driving task, a higher warning level should be used to deliver the message through the HMI. It has to be pointed out that the worst case would be when the message to be displayed has a high risk value with a low knowledge rate.

With regard to the issues that have been explained above, it is assumed that they should be considered when choosing the appropriate warning level. Therefore, on the basis of those, the next four warning levels have been defined: (1) Informative Warning Level, (2) Cautionary Warning Level, (3) Alerting Warning Level, and finally, (4) High Danger Warning Level.

15.3.3.1 Implementation of the Representation Model

For the implementation of the Representation Model for the proposed IIMS in HMI module, there were two different communication channels available to implement the four warning levels: the visual channel and the audio channel. The visual channel gives the opportunity to represent both message's icon and text on the virtual instrument cluster display (or VIC Display), while vehicle interior ambient lighting permits to modify its colour with a soft changing profile. On the other hand, through the audio channel, beeps and recorded explaining voices could be transmitted.

In the proposed methodology, R_i' indicates the risk associated with the message, while K_c determines driver's knowledge about the information embedded on the message. The value of K_c is included in the Signal Model as was mentioned previously in the Signal Model description, and the value of R_i' is determined in the Interpretation Module. On the other hand, in order to know driver's fatigue, a request will be made to DVE module which stores the data assessed by the onboard module responsible for driver supervision.

In the case of fatigued drivers, so as to get the driver's low attention, a beep audio reinforcement could precede the display of icon and text. Furthermore, it has been proposed to add an explaining voice message to those icons whose K_c is lower than 70%.

The design of the warning levels is summarized in Table 15.2. However, all of them have in common that they display icons (with the text associated when having) and they adapt the vehicle interior ambient lighting colour, which indicates the risk level of the context given by

Table 15.2 Warning levels for the Representation Model of the IIMS.

Warning level	Icon (and text)	Colour reinforcement	Beep reinforcement	Voice reinforcement
Informative Warning Level	Yes	Yes	No	No
Cautionary Warning Level	Yes	Yes	No	Yes
Alerting Warning Level	Yes	Yes	Yes	No
High Danger Warning Level	Yes	Yes	Yes	Yes

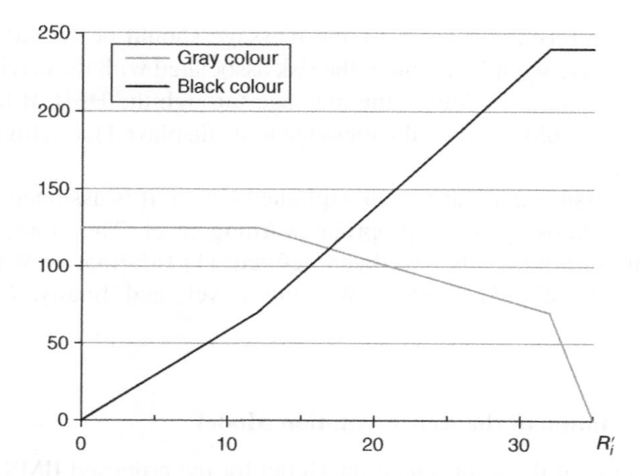

Figure 15.10 Soft colour changing profile on the basis of R_i' value.

the highest R_i' value at the specific road point. The difference between warning levels is given by the audio reinforcement. For example, the Informative Warning Level, the most basic warning level, will be used when the driver is detected as fully awake and he knows the meaning of the message with an acceptable level. The Informative Warning Level does not have beep and voice reinforcement. The Cautionary Warning Level introduces voice reinforcement in case of K_c low value. Moreover, the Alerting Warning Level would correspond to the framework where the driver is fatigued, but it is supposed that he should know the significance of the icon. Finally, the High Danger Warning Level would correspond to the risky driving context where the driver is fatigued and the message's K_c value is low (< 70%).

For the establishment of the ambient lighting colour, the Representation Model makes use of the list of R_i' values given by the Interpretation Model. The ambient lighting colour is set by the highest R_i' value of the messages to be displayed at the same time, which is mapped into a soft colour code profile. That colour changing profile goes from a blue colour which corresponds to a low R_i' value to a red colour which belongs to a high value of R_i' (see Figure 15.10).

As a first step for the proposed IIMS module, it has been decided to establish a maximum of two icons that will be presented to the driver. Actually, the maximum number of icons that could be represented at the same time should be determined by psychologist; so then it is an issue that goes far beyond the present scope of this chapter. But for the present proposal, messages belonging to the same situation ('Accident ahead', 'Congestion ahead', 'Roadworks', 'Car ahead driving very slowly', 'Kamikaze car coming' or 'Normal situation'), where the severity level remains constant, the first two messages from the ranking with the highest R_i' values will be delivered to the driver. In the case of having messages belonging to different severity levels, messages with the highest severity level will always be considered with higher priority when having to be displayed.

The model permits us to show a message more than once on the VIC Display, but only if the message continues being relevant after the first time it has been displayed and its R_i' value is still assessed as having to be displayed.

15.4 Case Study

In real life, drivers can find themselves driving in a wide range of situations. Roadwork zones concentrate many hazards on road environments, not only for vehicles and drivers, but also for road employees. Those driving situations can be degraded with bad weather conditions and high traffic density, poor visibility, etc. Each modification on roadwork zones can produce serious safety problems by generating unexpected situations for drivers. Some of the road-works can cause lanes number reduction or lanes change, while others could also modify the profile of the asphalt.

To see how the proposed IIMS methodology works, the case study of Figure 15.11 has been selected. It is a specific conventional road with two lanes for each direction, and the fast lane is closed due to roadworks. Through the trajectory of the vehicle, the onboard module receives messages, consecutively, sent by beacons with the road traffic signs. Each road signalization is a symbolic representation of specific hazards, regulations or guidance indications for the driver.

Next, the proposed methodology for the IIMS is highlighted by the message management carried out in the three models that compose the IIMS – the Signal Model, the Interpretation Model and the Representation Model – for the case of Figure 15.11. The performance of the IIMS has been highlighted for different driving situations defined by different traffic density levels, meteorological conditions and vehicle speeds.

15.4.1 Signal Model for Received Messages

The key performance of the intelligent module is based on how the signs are affected by the changing environment. It is summarized in the Signal Model by the relationship between the sign and the Meteorological Conditions, Traffic Density or Vehicle Speed. And it is given by a table of numbers which represent the Influence shown in Figure 15.7. For this case study, in Table 15.3 those parameters are shown for some signs.

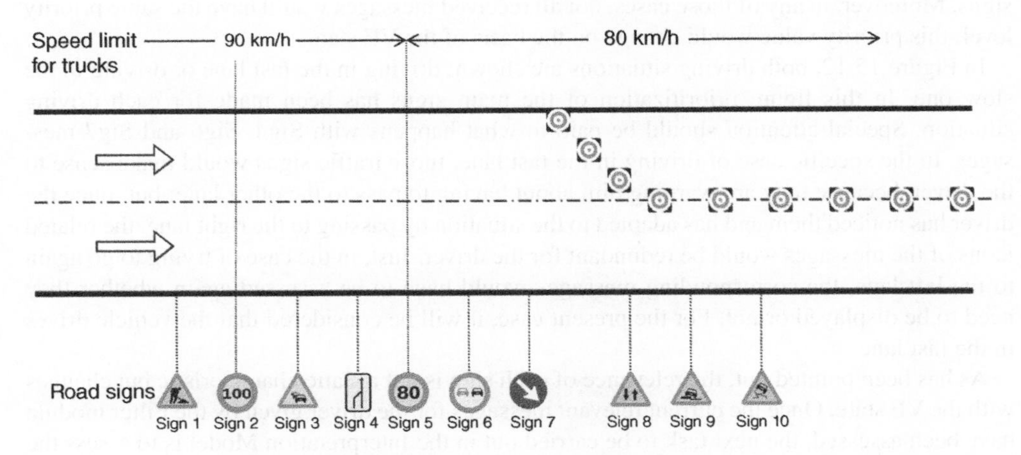

Figure 15.11 Case study of a road work.

Table 15.3 Signal model for the traffic signs of the case study.

SIGN	Relation with MC	Relation with SP	Relation with TD
Sig1	[0,30,50,75,85,100]	[0,40,45,70,75,120]	[0,30,40,75,80,100]
Sig3	[0,30,40,60,65,100]	[0,53,55,65,75,120]	[0,20,40,60,70,100]
Sig10	[0,15,25,35,40,100]	[0,50,55,70,75,120]	[0,80,85,90,95,100]

It is interesting to point out that Sig10 is more sensitive to adverse meteorological conditions than Sig1 and Sig3, as it is reflected on the Relation with MC field. For instance, Sig1's and Sig3's risk level would be altered absolutely highly with a value of MC equal to 85 and 65, respectively; whereas, Sig10 undergoes that change with a lower value of MC (equal to 40).

On the other hand, in relation to the traffic density and vehicle factors, it is obvious that Sig3 is more correlated with the TD factor rather than Sig1 and Sig10; whereas, vehicle speed factor-wise, all the messages are affected in a quite similar way.

15.4.2 Interpretation Model

When the driver reaches the roadwork zone in the case study, it is a different traffic situation if he or she is driving in the fast lane or in the slow one. Some of the signs have more relevance while driving in the fast lane and vice versa. For instance, in the case of Spanish roads, while driving in the slow lane, some road traffic sign information would turn out to be redundant for the driver; while driving through the fast lane, the driver may have to be aware of more road signs. Moreover, in any of those cases, not all received messages would have the same priority level; this priority value would change on the basis of the VE state.

In Figure 15.12, both driving situations are shown: driving in the fast lane or driving in the slow one. In this figure prioritization of the main signs has been made for each driving situation. Special attention should be paid to what happens with Sig4, Sig6 and Sig7 messages. In the specific case of driving in the fast lane, those traffic signs would make sense to the driver, because they are warning him about having to pass to the other lane; but, once the driver has noticed them and has adapted to the situation by passing to the right lane, the related icons of the messages would be redundant for the driver. Just, in the case of trying to go again to the left lane, the corresponding messages would have to be assessed again whether they need to be displayed or not. For the present case, it will be considered that the vehicle drives in the fast lane.

As has been pointed out, the relevance of each sign is not a static characteristic but changes with the VE state. Once the current relevant messages for the driver given by the Filter module have been assessed, the next task to be carried out in the Interpretation Model is to assess the relation of those messages with the VE state.

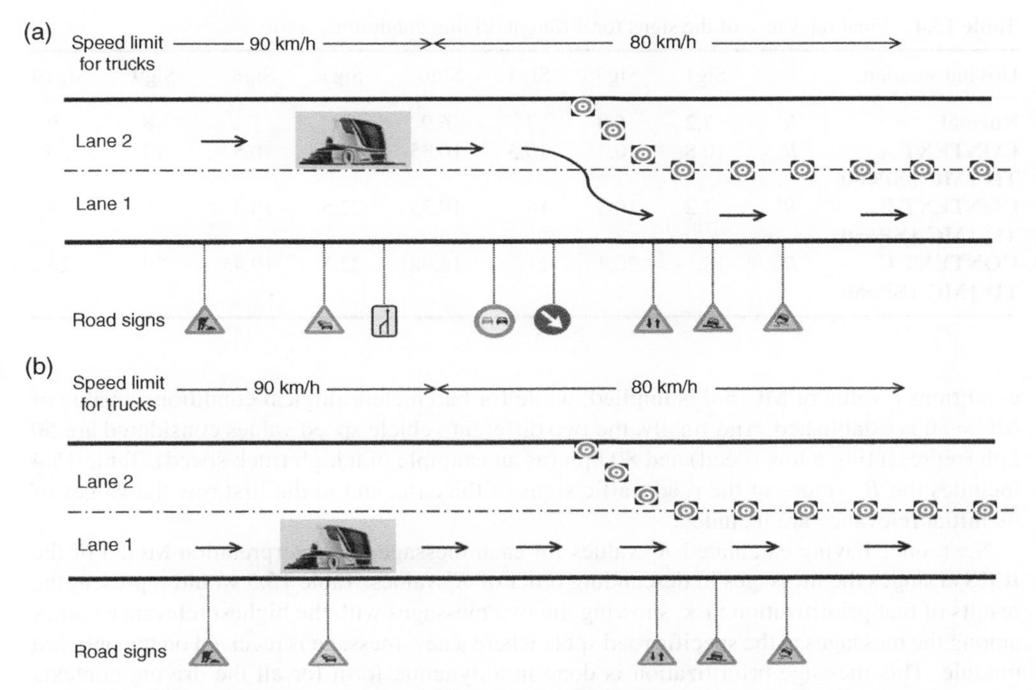

Figure 15.12 Relevant traffic signs for the truck depending on the lane it goes.

15.4.2.1 Influence of Vehicle-environment Conditions

When the truck is getting to the roadwork zone, the onboard module receives the message containing the roadwork traffic sign Sig1, which identifier will help to define a situation category of 'Roadworks'. However, the risk level of this driving situation can change under bad weather conditions or overspeeding. This means that not all road traffic signs of the study have the same relevance under different VE contexts (defined in the relationship with MC, SP and TD). Moreover, the initial relevance of every traffic sign, given by R_i, changes to R_i' depending on the influence of those factors.

On the other hand, it must be pointed out that information is continuously required by the driver; that is the case for the maximum speed limit information (messages Sig2 and Sig5). The IIMS does not calculate the final relevance of those messages; instead, their values are mapped depending on the vehicle type and shown directly to driver. The final relevance of the rest traffic signs must be assessed. Those values try to solve the priority conflicts that occur when multiple traffic sign messages are received in the vehicle. Furthermore, the current highest relevance value helps to determine the colour of the vehicle interior ambient lighting which indicates the risk level of the driving situation.

In order to notice how meteorological conditions, traffic density and vehicle speed affect the relevance of the traffic signs, some driving situations are considered. These driving contexts are a combination of different values of TD, MC and SP values in the range 0–100. For example, low traffic density is given by a value of TD = 10, and a high traffic density is considered for a value of TD = 75. On the other hand, when referring to good meteorological

Table 15.4 Final relevance of the signs for different driving conditions.

Driving stuation:		Sig1	Sig3	Sig4	Sig6	Sig7	Sig8	Sig9	Sig10
Normal	R_i	7.2	6.8	7	6.9	9	7	8	9
CONTEXT A TD ↓MC ↓SP=80	R'_i	10.8	10.3	10.5	10.35	13.5	10.5	12	20
CONTEXT B TD ↑MC ↓SP=50	R'_i	7.2	10.2	14	10.35	22.5	10.3	8	15
CONTEXT C TD ↑MC ↓SP=80	R'_i	18	20.4	21	18.98	22.5	19.95	20	26

conditions a value of MC = 0 is implied, while for bad meteorological conditions a value of MC = 80 is established. And finally, the two different vehicle speed values considered are 50 kph (representing a low speed) and 80 kph (as an example of a high truck speed). Table 15.4 includes the R'_i values of the road traffic signs of the case, and in the first row the values of its initial relevance are included.

Next, once having calculated R'_i values for each message, the Interpretation Model of the IIMS arranges the messages in descending order of R'_i values. Table 15.5 would represent the results of that prioritization task, showing the two messages with the highest relevance values among the messages at the specific road spots where a new message is received on the onboard module. This message prioritization is done in a dynamic form for all the driving contexts defined in Table 15.4, while for the baseline case, where no VE specific context is taken into account, a static prioritization is carried out on the basis of R_i.

In order to regard the influence of traffic density, let's imagine the next contexts: driving in a sunny day with the asphalt totally dry (which translates in low MC) and at high speed (SP = 80), for both cases, when driving in the rush hour (high TD) and with low traffic density (low TD). In the specific case of driving in with high traffic density in a sunny day and at 80 kph, once receiving Sig3 message ('traffic delays in the road') warning about the risk of encountering vehicles driving at low speed or stopped on the road, Sig3 should be continuously shown to the driver as most important sign. However, in the other case of driving with low traffic density in a sunny day and at 80 kph, Sig3's R'_i value is seen as not high enough compared to other messages to finally determine that it should be shown to the driver.

On the other hand, considering the influence of the vehicle speed, it has to be pointed out that in general a higher vehicle speed increments the risk level of any driving situation. To notice how this vehicle speed affects the prioritization task, let's focus on the contexts where the truck is driving in the rush hour and in very good weather conditions for both low speed (SP = 50) and high speed (SP = 80). In the specific case of driving at 50 kph, the signal about 'traffic delays in the road' would not be shown to the driver while driving through the zone with two lanes when Sig6 and Sig7 messages are received, since it is supposed that the vehicle's low speed reflects the fact that the driver might have perceived the possible traffic congestion in the road, but not the ban on overtaking. Therefore Sig4, Sig6 and Sig7 are considered more relevant for the driver than Sig3. However, in the zone of one line for each direction and driving at 50 kph, the high traffic density is considered more relevant than other context-relevant messages, such as Sig9.

Table 15.5 Prioritization of incoming messages for different driving situations.

Message reception:	Sig1	Sig3	Sig4	Sig6	Sig7	Sig8	Sig9	Sig10
Baseline Arrangement:								
Arrangement in Context A TD ↓ MC ↓ SP = 80								
Arrangement in Context B TD ↑ MC ↓ SP = 50								
Arrangement inContext C TD ↑ MC ↓ SP = 80	Sig1							

On the contrary, when driving at 80 kph in that sunny day in the rush hour, a Sig3 message should be delivered to the driver continuously; moreover, since the vehicle's high speed could destabilize the truck due to irregularities on the asphalt, the context provokes a Sig9 message (which means 'lateral step on the left') to gain position in the list of prioritization, even with a higher risk level than Sig8 (meaning 'traffic in both directions'), contrary to what happened when driving at 50 kph.

So as to consider the influence of meteorological conditions, firstly, let's focus on the road point of Figure 15.12 where Sig10 message ('slippery road') is received. It is obvious that when there are good weather conditions, that is, without rain, snow and ice, it is not worth showing Sig10 to the driver since it is not relevant for him or her regarding the driving context. However, when there are bad weather conditions, Sig10 passes to warn the driver about a very risky situation as it involves the loss of adherence; therefore, its warning message should be delivered to the driver. However, in the case of driving in the rush hour and very bad weather conditions at 50 kph, the Sig3 message should be considered more relevant for the driver rather than Sig10, while in the case of driving in the rush hour and very bad weather conditions at 80 kph, the higher priority level Sig10 message should be considered rather than Sig3.

15.4.3 Representation Model

The representation model uses the changing colour of the interior ambient lighting. The colour coding is shown in Figure 15.10, where in the abscissa axis is represented the final relevance value R_i'. For example, if we see the results of the final relevance value in the Context C (Table 15.6), as the signs are received in the onboard module the colour of the ambient light would change as the values of the table change.

On the other hand, as the previous beep sound is determined by driver's fatigue level, here the warning level of each message is not specified. However, it can be pointed out that in the case of displaying Sig1 and Sig4, no voice record would be transmitted because they have a low K_c value; while, when representing Sig3, Sig8 and Sig10 a recorded explanatory message would be given to the driver in order to give a definition of the icon displayed on the instrument panel.

When driving at 50 kph, just in Context C the red colour would be set as the vehicle interior ambient lighting, whereas in Context B a purple colour would be set, and for Context A a blue colour is set. On the other hand, as happens when driving at the maximum speed limit, when displaying Sig1 and Sig4, no voice record would be transmitted while, when representing Sig3, Sig8 and Sig10 a recorded explanatory message would be given to the driver.

Table 15.6 Final relevance of the signs in Context C.

Driving situation:		Sig1	Sig3	Sig4	Sig6	Sig7	Sig8	Sig9	Sig10
CONTEXT C TD ↑MC ↓SP=80	R_i'	18	20.4	21	18.98	22.5	19.95	20	26

15.5 Conclusions

In order to assure that drivers achieve the most complete situation awareness of the road context, messages from beacons about road signs, and additional traffic and meteorological information, are sent to the onboard module. However, it is well known that there are limitations on drivers' attention resources and working memory capacity; moreover, there is a space limitation in the instrument panels of vehicles that make it impossible to represent all the information at the same time. Therefore, for these reasons, it is necessary to prioritize all the information and deliberate on which are the most important active messages to be given to the driver.

Unlike ISO and SAE standards, which consider message priority as a static value, the proposed methodology considers the priority of messages over time. The prioritization task is carried out by an Intelligent Information Management System. The system takes advantage of three models: the Signal Model, the Interpretation Model and the Representation Model. Basically, the Signal Model establishes the main structure of the information included in the messages. The Interpretation Model assesses which are the active messages (m_i) and updates their relevance values from R_i to R_i' values depending on VE context. In order to assess how the priority of a message is changed by the real VE context, a modification factor is calculated using fuzzy logic. Finally, the Representation Model decides the warning level for the representation of messages on the basis of their relevance and the driver's fatigue status.

References

1. M.R. Endsley (1995) Toward a theory of Situation Awareness in Dynamic Systems, *Human Factors and Ergonomics Society* **37**(1): 32–64.
2. J. Rogé, T. Pébayle, E. Lambilliotte, *et al.* (2004) Influence of age, speed and duration of monotonous driving task in traffic on the driver's useful visual field, *Vision Research* **44**: 2737–44.
3. Y. Liu and T. Wu (2009) Fatigued driver's driving behavior and cognitive task performance: Effects of road environments and road environment changes. *Safety Science* **47**: 1083–9.
4. J. Rogé, T. Pébayle, S. El Hannachi and A. Muzet (2003) Effects of sleep deprivation and driving duration on the useful visual field in younger and older subjects during simulator driving. *Vision Research* **43**: 1465–72.
5. L. Schmidt (1982) Observance and transgression of local speed limits. *Arbeiten-aus-dam-verkehrspsychologischen-Institute* **19**(6): 107–16.
6. H. Otani, S.D. Leonard, V.L. Ashford and M. Bushore (1992) Age differences in perception of risk. *Perceptual and Motor Skills* **74**(2): 587–94.
7. A. Borowsky, D. Shinar and Y. Parmet (2008) Sign location, sign recognition, and driver expectancies. *Transportation Research Part F* **11**: 459–65.
8. J. Fisher (1992) Testing the effects of road traffic signs' informational valued on driver behavior. *Human Factors* **34**(2): 231–7.
9. A. Drory and D. Shinar (1982) The effects of roadway environment and fatigue on sign perception. *Journal of Safety Research* **13**: 25–32.
10. H. Al-Madani (2000) Influence of drivers' comprehension of posted signs on their safety related characteristics, *Accident Analysis and Prevention* **32**: 575–81.
11. M. Galer (1980) An ergonomics approach to the problem of high vehicles striking low bridges. *Applied Ergonomics* **11**(1): 43–6.
12. L. Pingatoro (1973) *Traffic Engineering: Theory and Practice*. New Jersey: Prentice-Hall, Inc.
13. A. Ruta, F. Porikli, S. Watanabe and Y. Li (2011) In-vehicle camera traffic sign detection and recognition. *Machine Vision and Applications* **22**: 359–75.
14. SAE J2395, ITS In-Vehicle Message Priority, The Engineering Society for Advancing Mobility Land Sea Air and Space, SAE International, 2002.

15. ISO/TS 16951, Road vehicle – Ergonomic aspects of transport information and control systems (TICS) – Procedures for determining priority of on-board messages presented to drivers, International Organization for Standardization, 2004.
16. H. Sohn, J.D. Lee, D.L. Bricker and J.D. Hoffman (2008) A dynamic programming algorithm for scheduling in-vehicle messages. *IEEE Transactions on Intelligent Transportation Systems* **9**(2): 226–34.
17. A. Zhang and C. Nwagboso (2001) *Dynamic Message Prioritisation for ITS Using Fuzzy Neural Network Technique, Transportation and Automotive Systems Research Center*, Society of Automotive Engineers, Tech. Paper Series 2001-01-0068.
18. CEIT, Centro de Estudios e Investigaciones Técnicas de Guipúzcoa, www.ceit.es (last accessed 3 May 2015).
19. Complete inform, 2010. Las señales a examen: la opinión de los conductores españoles sobre las señales de tráfico. FESVIAL (Fundación Española para la Seguridad Vial).
20. B. Lewis-Evans and T. Rothengatter (2009) Task difficulty, risk, effort and comfort in a simulated driving task – Implications for Risk Allostasis Theory, *Accident Analyisis and Prevention* **41**: 1053–63.
21. M. Castro (2007) Nueva Metodología de Control No Lineal Basada en Lógica Difusa. PhD Thesis, San Sebastián.

16

New Approaches in User Services Development for Multimodal Trip Planning

Asier Moreno, Itziar Salaberria and Diego López-de-Ipiña
University of Deusto, Bilbao, Spain

16.1 Introduction

Transportation systems' efficiency is essential for economic development. Intelligent Transportation Systems (ITS) are usually associated with technological systems for infrastructure operations, vehicles and goods in urban transportation and roads. Their use is widespread and there are a wide range of research works and applications. So, ITS can be defined as a set of applications within computer science, electronics and communications aimed at improving mobility, security and transport productivity, optimizing the use of infrastructures and energy consumption and improving the capacity of the transport systems [1].

Therefore, technological progress within the scope of software applied to transport services is enabling the development of new applications types. The aim of these applications is usually to improve the management of urban mobility and the quality of life in the cities.

At institutional level, promoting the use of public transport is a priority, so that it becomes a preferential transport option, favouring economic development [2]. Thus, according to data from the International Public Transport Association [3], it is expected that public transport will duplicate its market share in 2025 in regard to 2009, completing the transition to a more sustainable transport model.

Consequently, during the last years there has been a remarkable change in transport habits and urban mobility. People are adopting gradually as a viable alternative the use of public transport versus private transport (due to environmental sensitization and fuel costs), and as another alternative the use of multimodal transport [4].

Intelligent Transport Systems: Technologies and Applications, First Edition. Asier Perallos, Unai Hernandez-Jayo, Enrique Onieva and Ignacio Julio García-Zuazola.

16.1.1 Multimodal Transport

In order to transport people and goods, several modes of transport exist. A *mode of transport* is a particular typology for transporting goods or people. Three modes of transport can be generically distinguished: air, sea and land. On the other hand, the sequence of modes of transport applied for the movement of the load from the origin to the destination, including one or more transhipments, is known as *transport chain* [5].

However, the transport network lacks connection of times and timetables between modes. This leads to inconsistencies with the needs of the freight transport chain and involves operational requirements. The solution to these issues can be reached by applying a multiple transport system, combining different transportation models, contributing at the same time to the improvement of the internal market and strengthening economic and social cohesion.

At this point the *intermodality* concept arises. *Intermodal transport* means the movement of goods in a single unit or vehicle using successively two or more modes of transport without handling the goods when exchanging modes. Hence, the intermodality term has been used to describe a transportation system where two or more modes of transport are involved in freight transport in an integrated manner, without loading and unloading processes, in a transport chain from door to door (Figure 16.1).

On the other hand, *multimodal transport* means the movement of goods using two or more modes of transport covered by a contract of multimodal transport between locations. So, intermodal transport is a type of multimodal transport. In addition, *multimodality* designates the transportation organization by the simultaneity of different modes for the same itinerary or a specific geographical area (Figure 16.2).

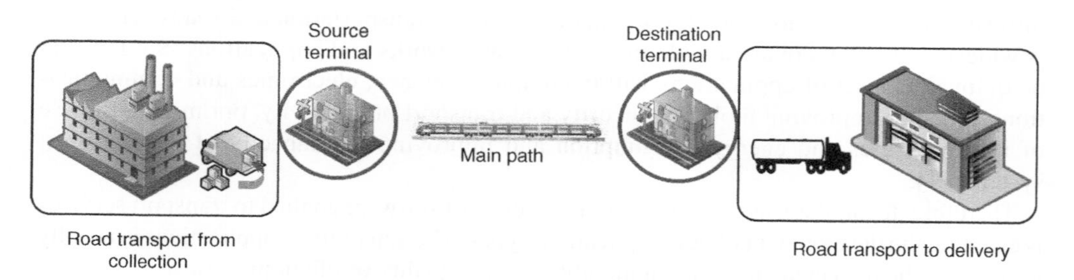

Figure 16.1 Intermodal freight transportation.

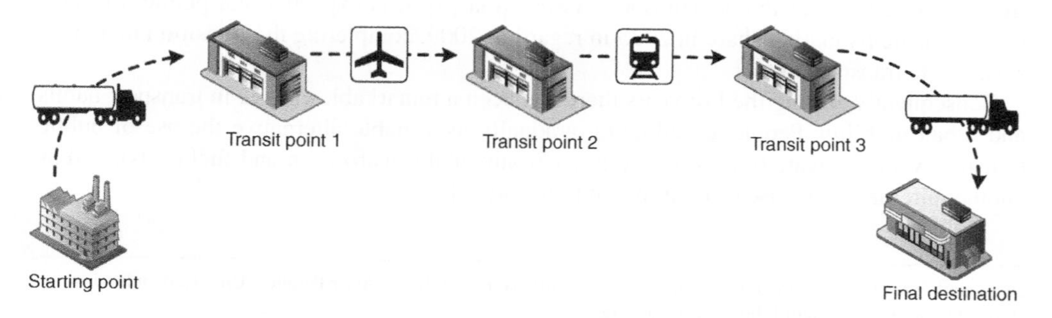

Figure 16.2 Multimodal transport chain.

This kind of transportation system is beneficial to countries for several reasons. It provokes backlog seaports, lower costs in goods control, increased security of taxes collection, reduction in costs for customs tax collection, increased competitiveness of products in international markets and lower prices of imported goods.

Besides, from the user point of view multimodality presents the following benefits: lower transportation total costs, minor travel times, better travel times scheduling, certainly in operation compliance, having a single contact with full responsibility, cargo handling technical care, reduced risk of loss for sacking or theft, and negotiation capacity.

However, related to these concepts, a lack of integration of different modes of transport can be identified, in addition to inefficiencies existing in each of the specific modes. These issues provoke an increase in costs and prices, and the growth of time which is required to reach from the origin to the final destination. So, this inefficiency results in the freight being longer in transit, assuming more risk of damage, and making the administrative procedures more complex.

So, being the final objective to improve the mobility of people and freight, an intelligent support system capable of improving the multimodal transportation should ensure a continuous and rational information flow. Furthermore, technological progress within the scope of software services applied to transport enables the development of new application types to improve citizens' mobility management, as well as the life quality of the cities.

In this sense, the European Commission (EC) white paper on transport marks a strategy route in order to achieve a competitive and sustainable transportation [2]. Focusing on multimodality topics, EC aims to optimize the efficiency of multimodal logistic chains. The objective is to get a basic efficient network for intercity and multimodal transportation.

Integration improvement of modal networks leads to better modal choices. There will be more connections between airports, ports, rail, metro and bus stations, and they will be transformed in multimodal transport platforms for passengers. Online information and electronic reservation and payment systems that involve all modes of transport must facilitate multimodal travelling.

So, the EC contemplates the definition of necessary steps to further integrate different modes of passenger transport to offer travel continuity from the origin to the destination. From the technological point of view, it is considered advantageous to create the conditions to promote the development and use of interoperable and multimodal intelligent systems in order to define timetables, travel information, online booking systems, and smart tickets expedition. These actions could include a legislative proposal to ensure access of private service providers and travel information for real-time traffic.

16.1.2 Travel User Services

Currently, a traveller has the option of planning a long-distance trip through the combination of different types and transport operators. But making such combinations involves several hours of browsing and even some phone calls, needing to link the different routes, schedules and fares, without ensuring the traveller selects the best combination for the desired trip. This is the scenario where a multimodal planner can help select the best choice for travellers.

These issues would be focused on the category of vehicle and infrastructure optimization within ITS solutions. This optimization is oriented by the integration of different information sources and the generation of new knowledge. The most relevant research works in this area are focused on ATIS (Advanced Traveller Information Systems), designed to assist travellers in planning their trips and route optimization.

So, these systems use information and communication technologies in order to collect, process and distribute the latest traffic information, road conditions, trip duration, expected delays, alternative routes and/or weather conditions, giving travellers the opportunity to make informed decisions about when to travel, what mode of transport use and what alternative route to select.

Therefore, multimodal trip planning solutions exist in order to offer and provide added value software transport services to the users.

16.2 Travel Planning Information Systems

16.2.1 Standard Travel Planning Services

There are solutions that have made important progress in facilitating trip management and planning to users. In this section some of these solutions will be described, as well as their main functionalities.

16.2.1.1 Google Transit

Transit in Google Maps [6,7] is a transport planning tool that combines transport companies' updated information with the power of Google Maps. This tool integrates information about stops, routes, scheduling and tickets in order to make route planning.

Regarding transport information, Google Transit works with the GTFS [8] specification. Thus, a carrier can use this specification to publish services information offered in a given geographic area.

Google Maps is a cost-effective solution focused equally on transport entities and experimented travellers' requirements. Google Transit, as a Google Maps feature, is available on several mobile devices through Google Maps for mobile application. In addition, public transportation information is also included in Google Earth.

Figure 16.3 shows the Google Transit user interface for a multimodal transport search. Map data: ©2014 Google, based on BCN IGN Spain.

16.2.1.2 Moveuskadi

The transportation sector is responsible for almost 25% of greenhouse gases emitted into the atmosphere in the Basque Country (on the north of Spain), provoking the impairment of the environment.

Therefore, the Basque Government is making a clear commitment to promote public transport through investments in new infrastructures or policies, and also seeking the complicity and changing habits of all citizens.

Focused in this topic, Moveuskadi [9] was developed. It is a website which is accessible from web browsers and mobile devices, and provides information about public transport in the Basque Autonomous Community. Thus, this application offers information for route

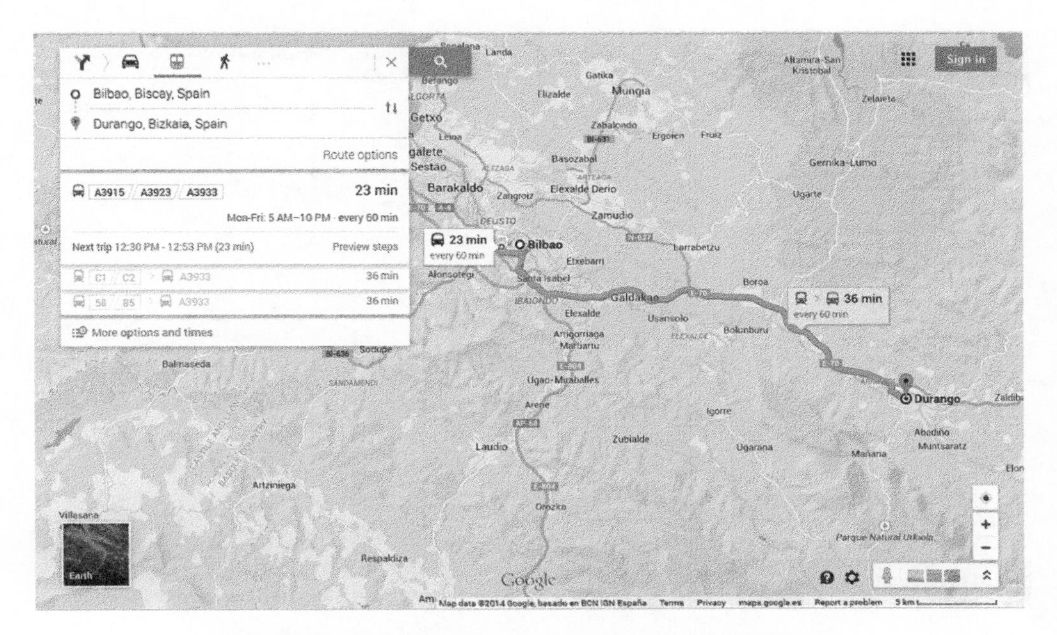

Figure 16.3 Google Maps web application © Google Maps.

planning (schedules, stops, time, cost, etc.) in an intermodal way, for example involving different types of public transport (metro, tram, city and intercity buses, commuter trains, long distance boats, ferries, funicular, suspension bridges, elevators, and so on) to get from the origin to the destination. The routes offered also provide links to transport companies.

Thus, this planner allows knowing which alternatives are more ecological to go from one location to other. It informs about the CO_2 emissions quantity differences depending on the route choices, as well as the graphical representation of the CO_2 emitted in each planning depending on the selected mode of transport, comparing it always with respect to the same situation but using a car.

So, the aim of the Basque Government is to promote sustainable mobility, showing to the citizens which are the greener alternatives to travel, and making them aware of the environmental impact provoked by their movements, and the savings they could get using public transport.

16.2.1.3 Open Trip Planner

OpenTripPlanner (OTP) [10] is a multimodal open-source trip planner based on geographical data from OpenStreetMap [11]. It runs on Linux, Windows, and basically any other platform running Java Virtual Machine. Since spring 2012, the code has been under continuous development, and many performance demonstrations are being made throughout the world.

In addition to planning trips, OTP provides detailed instructions for journeys made by bike or on foot so that users can plan their trip completely (door-to-door).

Related to transport carriers, OTP can be customized to match companies' look and brand, so that their drivers feel the same familiarity with online maps as well as paper

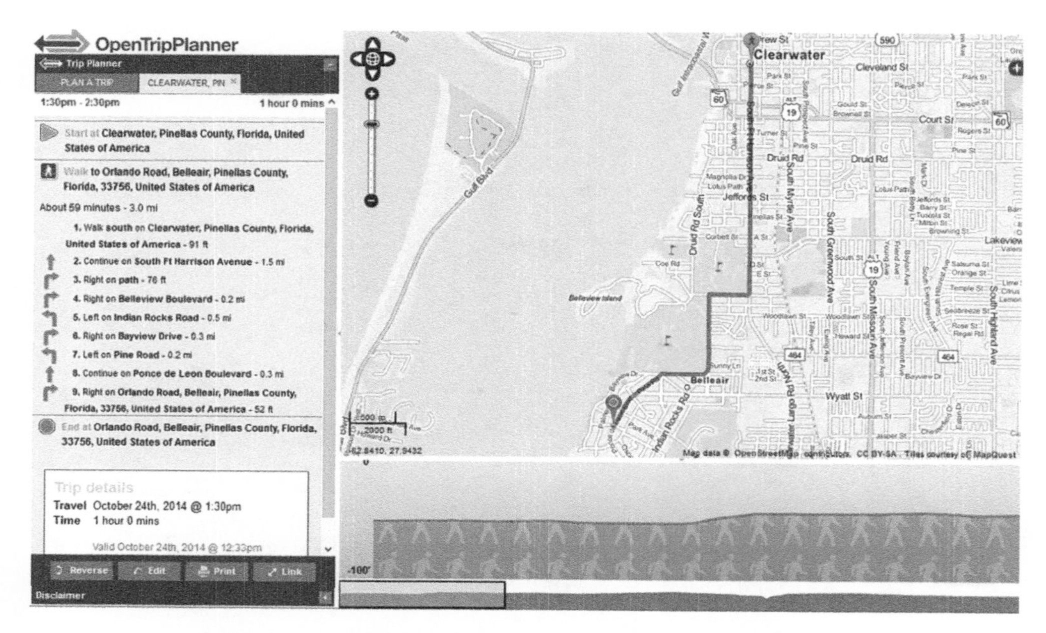

Figure 16.4 OpenTripPlanner web application © OpenTripPlanner.

maps. OTP also applies conventional navigation aspects that are common to other online mapping tools and trip planners, facilitating the adoption of this type of tools by the members of the company.

Further work is in progress to enable mobile access, and real-time planning, as well as advanced tools for internal management. Figure 16.4 shows the Open Trip Planner user interface for a multimodal transport search. Map data: © OpenStreetMap contributors, available under the Open Database License, cartography is licensed as CC BY-SA

16.2.1.4 SITI Project

The SITI [12] (Information System Intermodal Transport) is a project funded by the Ministry of Public Works and Transport for planning trips through intermodal transport.

This project develops a prototype that responds to the following requirements: geocoding, door-to-door route calculation, textual information and route map representation. Moreover, the idea was to design a scalable system that operates in both urban and interurban areas, developing a low-cost system based entirely on open source mapping.

The developed prototype shows that development is possible of an itinerary planner based on open source and standards with low cost and low risk. As a result, a transport operator can offer, applying the same data source, a simple but world-renowned travel planning service through Google, and a local service from its own site. Thus, they can adapt this service to their own needs in many ways, for example the rules of route optimization, desired modes of transport selection in the route calculation, fare information, ticket sales, etc.

16.2.2 Transit Information Formats and Standards

Transportation companies have their own information about service planning related to routes and schedules made by its fleet of vehicles. But as J.L. Campbell *et al.* [13] indicate in their work, transport systems' effectiveness (as well as the planners described before) depends largely on the ability to integrate information from various sources and on the adequacy of the information provided to specific users.

Currently there are initiatives for the publication and exchange of carriers' transit information which is allowing developers to consume this information and to integrate it in their applications in an interoperable way. The most used transit information publishing formats are described below.

16.2.2.1 General Transit Feed Specification (GTFS)

The General Transit Feed Specification defines a common format for public transportation schedules and associated geographic information. GTFS has established itself as the de facto standard for the representation of transit data, thanks in part to the support received from Google.

Regarding the format, GTFS information publishing is performed compressing ZIP files composed of a set of CSV (Comma Separated Content) format text. Each file models a particular aspect of transit information: stops, routes, tracks and other planning data.

16.2.2.2 Web Feature Service (WFS)

The Open Geospatial Consortium Web Feature Service Interface Standard (WFS) provides an interface allowing requests for geographical features across the web using platform independent calls. The XML-based GML furnishes the default payload encoding for transporting the geographic features.

The WFS specification defines interfaces for describing data manipulation operations of geographic features. Data manipulation operations include the ability to get or query features based on spatial and nonspatial constraints and create/update/delete new feature instances.

16.2.2.3 Ad-hoc Solutions

There are also other systems to represent and/or provide transit information, mostly defined by agencies or institutions that manage their own data. One of the most prominent is TransXChange [14] used in the United Kingdom and Australia to interchange bus service planning information.

Therefore, various systems coexist in order to represent transit information. WFS specification is focused to make queries on geospatial data stored in GIS and its use is complex. Ad-hoc solutions are optimal in their domain but they are not interoperable. GTFS is the de facto standard, although the treatment and consultation of its data is not trivial, because they are not structured and do not allow the inclusion of qualitative attributes of the route (such as ecology or tourist interest), which are increasingly relevant when making schedules.

16.2.3 *New Trends in Transit Information*

In the last years, the efforts on user services and transportation information have been focused on data interoperability in order to obtain value added information. Figure 16.5 shows the evolution of the transport information and its transformation through different processes making it more relevant for the user.

In order to support these processes and provide solutions to new information (quantitative and qualitative) integration, sharing and aggregation limitations, there are studies about the possibilities offered by relatively recent emerging paradigms in the computing field, such as ontologies and the Web of Data.

16.2.3.1 Ontologies

Work performed since the 1970s by the Artificial Intelligence community showed that formal ontologies could be used as a mechanism to specify and reuse knowledge between different software entities. In computing, the ontology term refers to the formulation of a comprehensive and rigorous conceptual schema given within one or more domains [15]. Being independent of language and understandable by computers, ontologies are useful because they help to achieve a common and integrated understanding of descriptive information.

16.2.3.2 Semantic Web

Ontologies are the heart of the Semantic Web, an extension of the World Wide Web in which the meanings (semantics) of information and services are defined [16] allowing the requests of people and machines using web content to be satisfied. On the Web there are millions of resource accesses, which have brought a lot of success, but also provoke problems of information overload and sources' heterogeneity. The Semantic Web helps to solve these problems by reducing the cognitive effort of users, delegating them to agents that can reason, and make inferences to offer the right information to the people.

In order to obtain the proper definition of the data, the Semantic Web essentially uses RDF, SPARQL, OWL and mechanisms that help turn the web into a global infrastructure where data and documents can be shared and reused between different user types.

16.2.3.3 Linked Data

The concept of Linked Open Data arises as the mechanism to make interconnected RDF datasets available on the Internet using the HTTP protocol, as with HTML documents.

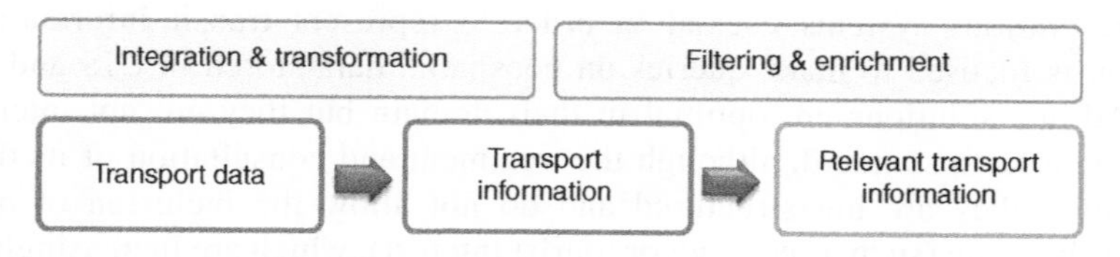

Figure 16.5 Evolution of transport information.

Table 16.1 *5-star* Linked Data rating system.

Stars	Description	Type	Example
★ OL	Available on the web	On-Line	PDF
★★ OL RE	Available as machine-readable structured data	Readable	XLS
★★★ OL RE OF	Nonproprietary format	Open Format	CSV
★★★★ OL RE OF URI	Using open standards from W3C	URI	RDF
★★★★★ OL RE OF URI LD	Link data to related datasets	Linked Data	RDF

In computing, data describes a method to publish structured information, so that it can be interconnected and thus be more useful. It is based on standard web technologies such as HTTP, RDF and URIs, but instead of using them to serve web pages to people, it extends them to share information in a way that can be processed automatically by computers. This allows data from various sources to be connected and retrieved [17].

In order to standardize and measure the quality of the published datasets which are beginning to be huge, in 2010 a metric known as '5 star linked data' was introduced. Basically, it consists of a set of incremental characteristics that should meet the published data to be considered as linked data under community criteria.

The more functionality the data set meets, the better the quality of this. Table 16.1 shows that classification.

This concept gained momentum after the publication of large semantic resources such as DBpedia [18] or Bio2RDF [19] and the announcement by some governments about its decision to make public their data on a set of Open Government initiatives.

In October 2007 datasets of more than two billion triples were counted, interlinked by means of more than two million RDF links [20]. In September 2011 this information had grown to 31 billion RDF triples and 500 million links.

Such efforts, both institutional and personal, are resulting in the publication of data having the characteristics and infrastructure of the Semantic Web. Figure 16.6 shows the state of the linked open data map [21] in August 2014.

16.3 Integrating Linked Open Data for Multimodal Transportation

Capabilities to promote knowledge sharing, structuring such knowledge and interoperability between systems, have favoured the use of ontologies and LOD in many areas of study and different application domains (e.g. medicine [22,23], education [24,25], logistics [26] or tourism [27,28]).

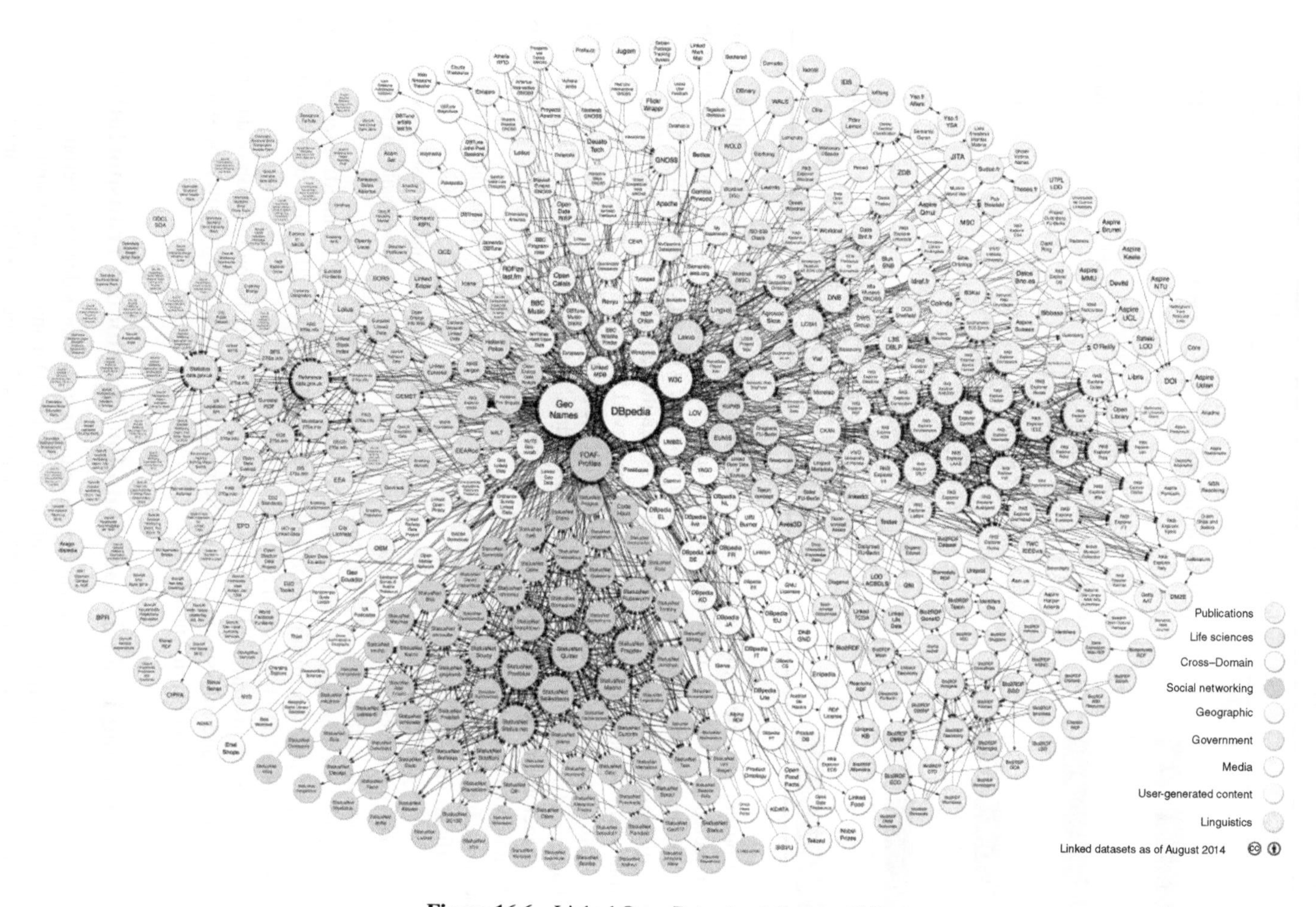

Figure 16.6 Linked Open Data cloud diagram [21].

Therefore, it should not be surprising to find similar work in the field of transportation information, motivated by various objectives, methods and expected results, but all under the premise of semantic data modelling. This chapter will describe the most relevant research conducted in this area, after which the solution proposed by the authors will be presented.

16.3.1 Related Work

16.3.1.1 Ontological Geospatial Data Management

Interoperability is becoming essential for geographic information systems whose information is usually stored in geospatial databases, accessible only via GIS (Geographic Information System). However, these sources are increasingly heterogeneous.

Therefore it is necessary to consider this heterogeneity and favour methods that enable interoperability between geographic tools in order to satisfy the growing demand in the use and sharing of geospatial data.

Traditional GIS systems perform spatial queries using a method based on keywords. This approach is unable to fully express the user's needs, because of the lack of geographic concepts (semantics) in the data set.

In this context, the most promising approach to end this ambiguity is the implementation of geospatial semantics using ontologies for geographic datasets.

B. Lorenz has performed an exhaustive analysis in his study 'Ontology of transportation networks' [29], which points out the efforts made by international and European institutions in order to standardize geographical information. These include the definition of GDF (Geographic Data File), developed through the research and development of European Digital Road Map (EDRM) project, and being the leader in its scope until this, with providers like NAVTEQ, Bosch, Philips or Volvo, using these maps for their navigation systems for automobiles.

The thesis work presented by R. Lemmens [30] delves into the need to seek interoperability between different data sets and web services based on geographic information. It exposes how the paradigm change in terms of software architecture (from a centralized mode to a distributed and interconnected one) has influenced substantially in the available geographical information tools (interactive maps, route planners, statistical information, etc.). Thus, the need to integrate and reuse available information (both internal and external) arises in the organizations, appearing to be the problems associated with data heterogeneity.

On the other hand, T. Zhao [31] presents a paper which attends to the need to geographical data interoperability from a distinct perspective. While the authors cited above focus their efforts on geographic information semantic modelling, designing one or more ontologies to try to get a direct translation to a richer semantic model for geospatial formats, Zhao seeks to reuse existing standard formats and geospatial protocols and enhance them using semantic in the query layer.

He argues that the direct conversion of all present geospatial data (stored mainly in geographical databases) to an ontological model is not a viable alternative because the process would be prolonged in time, being subject to the occurrence of errors and inefficient. Furthermore, existing tools for ontology management (such as Protégé) would not be able to bear such a heavy burden of instances, mainly due to required memory consumption.

With this in mind, he proposes the inclusion of semantics in an upper layer, using an adapter or interface implemented as RDF ontology and located between web queries made by users and the underlying geospatial formats.

16.3.1.2 Semantic Modelling of Transport Information

Although several authors have tried to manage GIS systems' geographical information by ontologies, it has been proven that the complexity of this type of information, as well as its capability requirements make it impractical to compute large-scale application of such solutions. Therefore, the current approach is to direct the research towards domains or more specific areas within the transport information.

Thus, K. M. Oliveira [32] has performed work focused on the use of ontologies in information modelling for multimodal transport with the ultimate goal of generating personalized user interfaces based on user characteristics (characteristics that are also represented according to a semantic context model). This work is a good example of how ontologies applied to transport information can be used to give more intelligence to complex software systems.

Another example is the research by Houda [33], which focuses on the information required by the passenger for preparing a journey, choosing the best way to move from one point to another using multimodal transportation. This work also takes into account services such as restaurants, libraries, etc., available on the route and useful for the passenger. In order to support the passenger's journey planning by providing such services, several kinds of journey patterns are modelled on a taxonomy based on the transportation network.

Work by Gunay [34] presents a more general solution, which achieves semantic modelling public transport information based on PDTO (Public Transport Domain Ontology), generating at the same time a geoportal based on the INSPIRE (Infrastructure for Spatial Information in the European Community) data theme. The aim of this work is to investigate the use of semantics to empower the traditional GIS approach.

The author relates a semiautomatic loading data into existing transportation ontology to generate individuals (specific instances). It also defines mechanisms for managing geospatial query requests and selection of the optimal route. Furthermore, as points for improvement, it indicates the high computational cost of the solution and transport limitation incorporated into this, recommending the use of distributed techniques as a possible solution to these problems.

16.3.2 Management and Provision of Multimodal Transport Semantic Information

The objective of the work presented below is to provide the user with an efficient alternative based on the use of public transport in order to plan both urban and intercity travel. The developed platform will permit information to be provided about the optimal route for the user, taking into account parameters such as time, distance, energy consumption, or the cultural and/or tourist interest, using public and efficient modes of transport. For this, in addition to currently available information which will consume the system (routes, distances, etc.), it will be necessary to have an architecture that enriches this information with collaborative information provided by possible users, and permits this data to be exploited, filtered and displayed according to its interest or search context.

Our final goal is to construct a distributed software architecture that allows, through the formalization of transit data acquired from heterogeneous sources together with the integration of relevant information, enabling software services related to multimodal mobility.

16.3.2.1 Multimodal Transport Ontology (MTO)

Several limitations related to the definition of a language that allows the formal modelling of transit data have been found. This characteristic is very satisfactorily resolved using ontologies as demonstrated in other research fields, so creating an ontology for transport information provision and management should be conducted.

A survey of existing transportation vocabularies that could be reused for the ontology has been conducted. Thus, vocabularies widely supported by the community, such as Geonames [35], Geosparql [36], Time [37] or WGS84 [38], have been used.

One of the most relevant characteristic of the designed architecture is that it defines the extensibility (allowing the ontology to be extended to give support to broader and/or more complex domains) and the reusing and re-engineering of nonontological resources (NORs). Through the links made to these domains the proposed ontology is able to service advanced geographic requests regarding both geospatial (e.g. POIs within 5 km of a given route) and geopolitical (e.g. POIs located in Biscay) data.

Given that GTFS is currently the de facto standard for transit data representation, it has been chosen to undertake a reformulation of CSV files to entities within the ontology. The adapter is a portable and multiplatform desktop application developed in Java. Its functionality is to generate semantic information from GTFS files. It has an ontological base in which the transport domain is defined and a number of automated rules to perform a direct conversion between formats and resolve the corresponding relationships.

As indicated above, the basis for the transport data model will be taken from GTFS specification. Figure 16.7 shows the main classes composing the developed transportation ontology.

One of the key characteristics for a system which aims to model transportation information is the design used to store the geographical data. In this aspect, the decisions that have been taken are based on the standardization and future extensibility and/or reusability of the designed ontology.

Geo-referenced points have been defined by using widely supported properties like latitude and longitude, following the WGS84 standard. Since it is intended to add functionality to these points, relationships with other relevant ontologies and services have been established enriching the information provided to the end user.

16.3.2.2 Distributed Architecture

The representation of transport-related information as an ontology facilitates the use of advanced features, like the ability to link data with other data sources, which can be relevant to the specific domain. This supposes an improvement with respect to the quality of the information by integrating its nonquantitative aspects.

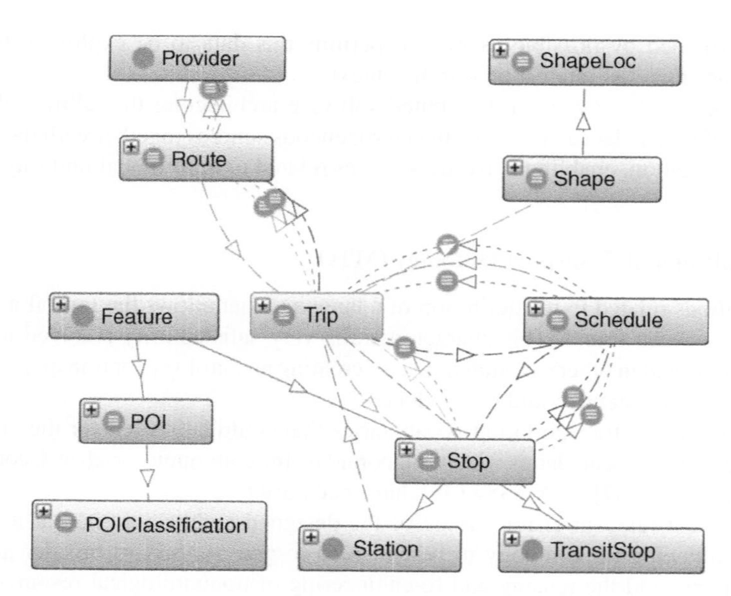

Figure 16.7 MTO main concepts and relationships.

In the particular case of the proposed ontology, it was decided to extend it by linking and classifying the collaborative points of interest (POIs) provided by the Linkedgeodata [39] ontology, as well as the ones provided by Geonames. Linkedgeodata uses the information collected by the OpenStreetMap project and makes it available as an RDF knowledge base according to the Linked Data principles.

An outline of the proposed architecture is shown in Figure 16.8.

The described data model has to be complemented with an architecture that supports it. This is conducted by several distributed SPARQL servers. SPARQL is a query language able to retrieve and manipulate data stored in RDF format. Each one of the servers maintains its own transit information and is managed in a local way, but facilitates the interoperability by means of its connection with the remaining servers by URIs, thus allowing distributed queries to be performed in a transparent way.

That is, when a user makes a request about a particular intermodal route to one of the servers, this provides the same query to the servers that have registered and so on until the query can be met because the destination server knows the data requested or until all servers are queried.

16.3.2.3 Semantic Trip Planner

As pointed out, in the state of the art one of the software tools that have advanced most in recent years in the field of ITS is the multimodal trip planner. It is therefore considered appropriate to validate the developed solution to the implementation of a multimodal and semantic trip planner running on the proposed architecture (Figure 16.9).

The proposed multimodal trip planner makes use of the distributed architecture and the ontology-based data model to give users relevant information related to multimodal

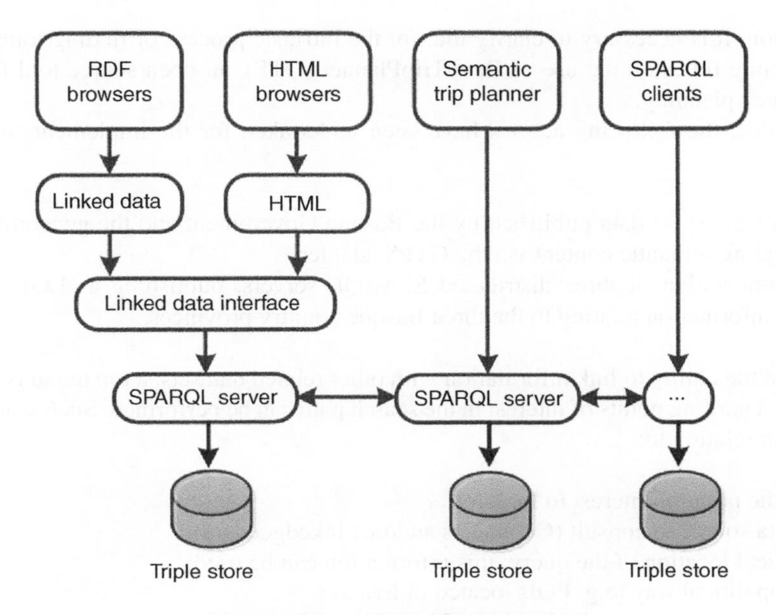

Figure 16.8 Distributed architecture.

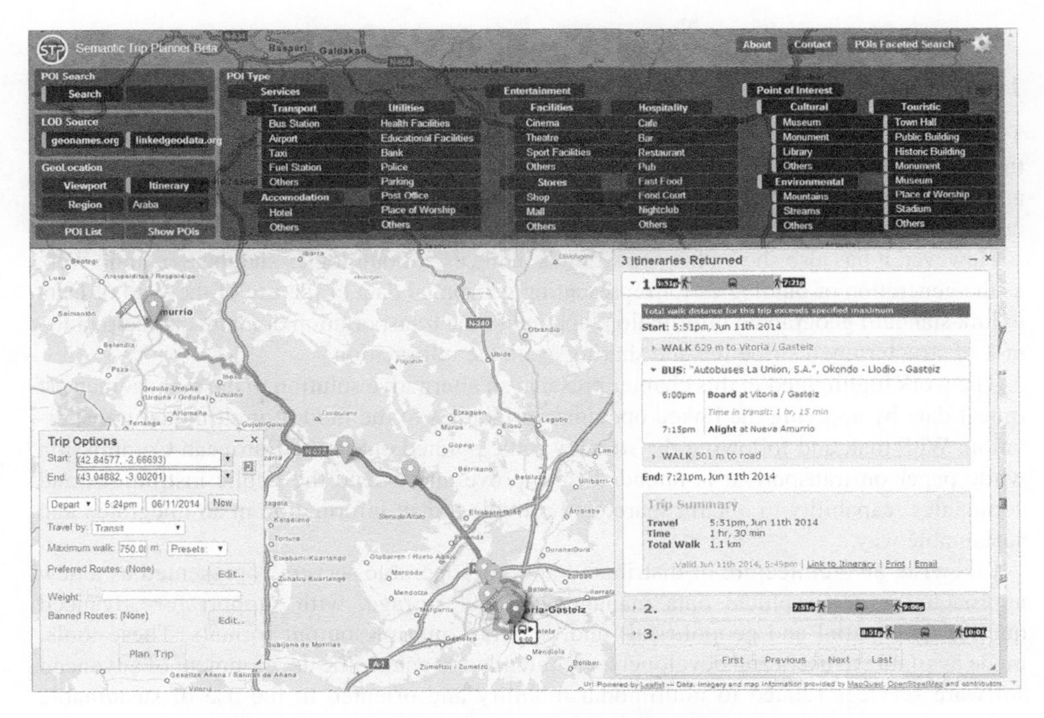

Figure 16.9 Semantic trip planner © Map Quest OpenStreetMap.

transportation. It is necessary to clarify that for the intrinsic process of finding transit routes the architecture relies on the use of OpenTripPlanner (OTP), an open source tool for multi-modal journey planning.

Besides this, the following actions have been undertaken for the implementation of the planner:

- the load of transport data published by the Basque Government and the generation of the corresponding semantic content via the GTFS adapter;
- the implementation of three distributed SPARQL servers, publishing as LOD semantic transport information relating to the three Basque country provinces.

Thanks to the ability to link information with other related datasets, a top menu is provided so searches regarding points of interest in the search path can be performed. Such searches can be filtered in relation to:

- name of the point of interest to look for;
- linked data source to consult (Geonames and/or Linkedgeodata);
- geographical location of the query; this information can be used:
  - in a geopolitical way (e.g. POIs located in Biscay);
  - in a geometric way (e.g. POIs within 5 km of a given route);
- hierarchical classification of the POIs:
  - Services, Entertainment and Points of Interest.

16.4 Conclusions

This chapter has conducted a review of software solutions and services designed to promote the use of multimodal transport. This study led us to address the challenge of the management and representation of transport information, often based on the use of heterogeneous databases or closed noninteroperable formats.

However, it has also shown how new trends in the use of ontologies and the Semantic Web, as the generation of ontologies for representing geospatial data or the use of the Web of Data to link standard geographic information, applied to the transportation domain, can help in the task of structuring, distributing and sharing available information.

The work highlighted in this chapter presents an alternative solution to the management of transit data by applying the linked open data principles to the field of multimodal transportation. With that, and following the strategy route pointed out in the European Commission white paper on transport, it is intended to improve and extend the public institutions and companies' capability to obtain, share, and provide transit information in an integrated and sustainable way.

Likewise, an architecture of distributed and interoperable servers is presented as a new approach for geographical data management and storage, with support for advanced queries (geospatial and geopolitical) and enabling multiple output formats. These tools, made available to users and developers, are intended to enable the development of advanced software services related to multimodal mobility and oriented to the use of sustainable transport.

References

1. General Secretariat for Transport, Spanish Ministry of Public Works and Transport (2010) *Los Sistemas Inteligentes de Transporte*.
2. Comisión Europea (2011) *Libro blanco del transporte: Hoja de ruta hacia un espacio único europeo de transporte: por una política de transportes competitiva y sostenible*. Oficina de publicaciones de la unión europea.
3. International Association of Public Transport (UITP) (2009) Doubling the market share of public transport worldwide by 2025, *58th UITP World Congress, 2009*.
4. International Association of Public Transport (UITP) (2014) Local public transport trends in the European Union, *60th UITP World Congress*.
5. Ministerio de Fomento de España. *El lenguaje del transporte intermodal, vocabulario ilustrado*. http://www.fomento.gob.es/NR/rdonlyres/17FBCF00-91E0-4761-A11C-88A16277D8A4/1550/01_lenguaje_transporte_intermodal.pdf (last accessed 3 May 2015).
6. Google Transit, https://developers.google.com/transit/google-transit (last accessed 3 May 2015).
7. Google Transit Partners Program website, https://developers.google.com/transit/gtfs (last accessed 3 May 2015).
8. GTFS Data Exchange, http://www.gtfs-data-exchange.com (last accessed 3 May 2015).
9. Moveuskadi, http://moveuskadi.com (last accessed 3 May 2015).
10. OpenTripPlanner, http://opentripplanner.com (last accessed 3 May 2015).
11. OpenStreetMap, http://www.openstreetmap.org (last accessed 3 May 2015).
12. Sistema multimodal para la planificación de viajes en transportes públicos de código abierto y basado en estándares 'de facto', http://dugi-doc.udg.edu/bitstream/10256/1418/1/C37.pdf (last accessed 3 May 2015).
13. J.L. Campbell, C. Carney and B.H. Kantowitz (1998) Human factors design guidelines for advanced traveler information systems (ATIS) and commercial vehicle operations (CVO). National Technical Information Service.
14. TransXChange, http://www.dft.gov.uk/transxchange (last accessed 3 May 2015).
15. T.R. Gruber (1993) A translation approach to portable ontology specifications, *Knowledge Acquisition* **5**: 199–220.
16. T. Berners-Lee, M. Fischetti and M.L. Dertouzos (2000) *Weaving the Web: The Original Design and Ultimate Destiny of the World Wide Web by Its Inventor*. HarperInformation.
17. T. Berners-Lee, C. Bizer and T. Heath (2009) Linked data: the story so far. *International Journal on Semantic Web and Information Systems* **5**(3): 1–22.
18. DBpedia, http://wiki.dbpedia.org (last accessed 3 May 2015).
19. Bio2RDF, http://bio2rdf.org (last accessed 3 May 2015).
20. D. Fensel, F.M. Facca, E. Simperl and I. Toma (2011) *Semantic Web Services*. New York: Springer.
21. M. Schmachtenberg, C. Bizer, A. Jentzsch and R. Cyganiak (2014) Linking Open Data cloud diagram, http://lod-cloud.net (last accessed 3 May 2015).
22. M. Ashburner, C.A. Ball, J.A. Blake, *et al.* (2000) Gene ontology: tool for the unification of biology. *Nature Genetics* **25**(1): 25–29.
23. A. Mol (2002) *The Body Multiple: Ontology in Medical Practice*. Durham, NC: Duke University Press.
24. C. Guangzuo, C. Fei, C. Hu and L. Shufang (2004) OntoEdu: a case study of ontology-based education grid system for elearning, *GCCCE2004 International conference, Hong Kong*. Citeseer, pp. 1–9.
25. H. Jia, M. Wang, W. Ran, *et al.* (2011) Design of a performance oriented workplace e-learning system using ontology. *Expert Systems with Applications* **38**(4): 3372–82.
26. H. Dong, F.K. Hussain and E. Chang (2008) Transport service ontology and its application in the field of semantic search, *IEEE International Conference on Service Operations and Logistics, and Informatics*, vol. **1**, pp. 820–4.
27. G.D. Abowd, C.G. Atkeson, J. Hong, *et al.* (1997) Cyberguide: A mobile context-aware tour guide. *Wireless Networks* **3**(5): 421–33.
28. D. Buján, D. Martín, O. Torices, *et al.* (2013) Context management platform for tourism applications. *Sensors* **13**(7): 8060–78.
29. B. Lorenz, H.J. Ohlbach and L. Yang (2005). Ontology of transportation networks. REWERSE Deliverable A1-D4. University of Munich, Institute for Informatics.
30. R.L.G. Lemmens (2006) Semantic interoperability of distributed geoservices. PhD thesis.
31. T. Zhao, C. Zhang, M. Wei and Z.-R. Peng (2008) Ontology-based geospatial data query and integration, *Geographic Information Science, Lecture Notes in Computer Science* **5266**: 370–92.
32. K.M. De Oliveira, F. Bacha, H. Mnasser and M. Abed (2013) Transportation ontology definition and application for the content personalization of user interfaces. *Expert Systems with Applications* **40**(8): 3145–59.

33. M. Houda, M. Khemaja, K. Oliveira and M. Abed (2010) A public transportation ontology to support user travel planning, *Fourth International Conference on Research Challenges in Information Science (RCIS), May 2010*, pp. 127–36.
34. A. Gunay, O. Akcay and M.O. Altan (2014) Building a semantic based public transportation geoportal compliant with the INSPIRE transport network data theme. *Earth Science Informatics* **7**: 25–37.
35. GeoNames Ontology, http://www.geonames.org/ontology (last accessed 3 May 2015).
36. GeoSPARQL, http://www.opengeospatial.org/standards/geosparql (last accessed 3 May 2015).
37. Time Ontology in OWL, http://www.w3.org/TR/owl-time (last accessed 3 May 2015).
38. Basic Geo (WGS84 lat/long), http://www.w3.org/2003/01/geo (last accessed 3 May 2015).
39. LinkedGeoData, http://aksw.org/Projects/LinkedGeoData.html (last accessed 10 May 2015).

Index

Figures and tables are indicated by *italic page numbers*, boxes by **bold numbers**, and text in the Foreword shown as roman numbers (e.g. "xxiv")

Intelligent Transport Systems: Technologies and Applications, First Edition. Asier Perallos, Unai Hernandez-Jayo, Enrique Onieva and Ignacio Julio García-Zuazola.

Printed and bound by CPI Group (UK) Ltd, Croydon, CR0 4YY
16/04/2025
14658473-0002